BREED YOUR OWN VEGETABLE VARIETIES

This book is dedicated to Henry M. Wallbrunn,
beloved teacher, mentor,
and friend.

BREED
YOUR OWN
VEGETABLE
VARIETIES

*The Gardener's and
Farmer's Guide to
Plant Breeding and
Seed Saving*

CAROL DEPPE

*Chelsea Green Publishing
White River Junction, Vermont*

Designed by Barbara Werden and Ann Aspell.

Printed in the United States.

First printing, November 2000.

22 21 20 19   10 11 12 13

Printed on acid-free, recycled paper.

Library of Congress Cataloging-in-Publication Data

Deppe, Carol.
  Breed your own vegetable varieties : the gardener's and farmer's
  guide to plant breeding and seed saving / Carol Deppe.
      p. cm.
  ISBN 1-890132-72-1 (alk. paper)
  Orginally published : Boston : Little, Brown, and Co., c1993.
  Includes bibliographical references (p.)
  Vegetables—Breeding. 2. Vegetables—Varieties. 3. Vegetables—Seeds.
  SB 324.7 D47 2000   00-060261
  635'.0423—dc21

Chelsea Green Publishing
85 North Main Street, Suite 120
White River Junction, VT 05001
(802) 295-6300
www.chelseagreen.com

Contents

Part I: An Introduction to Plant Breeding and Seed Saving

Part II: Seed Saving Practice

Part III: Developing Crops for a Sustainable Future

Acknowledgments

I welcome the opportunity to thank the following people for information, assistance, and encouragement: Ken Allan, Jan Blüm, Dan and Ruth Borman, Rebecca Brown, Helen Deppe, Glenn Drowns, Miriam Ebert, Ewald Eliason, Ianto Evans, John Gale, Rich Hannan, Howard Haynes, Karl Huber, Alan and Linda Kapuler, Damon Knight, Cy Dratzer, Rob Mabe, Rose Marie Nichols McGee, Ralph McNees, Tim Peters, Forest Shomer, Anita Sullivan, Jeff and Joy Taylor, Lee Wallingford, Kent Whealy, Kate Wilhelm, and Jimmy Williams.

I am indebted to Roger Swain, science editor at *Horticulture,* who shepherded this book in its initial stages, when it was only a magazine article with ambitions. I am especially grateful to Catherine Crawford and Christina Ward, my editors at Little, Brown and Company, for their enthusiasm, support, and guidance. I also thank Barbara Jatkola for her excellent copyediting, Jan Blüm for use of three of her photos, and Abigail Rorer for her illustrations in Appendix B.

Finally, I acknowledge and thank Richard Balkin, my agent.

Chapter 1 of this book originally appeared, in somewhat different form, in *Horticulture.*

Acknowledgments to the Second Edition

I acknowledge and thank Christina Ward, who started as this book's editor in the first edition and continued as its agent in the second. I am grateful to everyone at Chelsea Green Publishing Company for their enthusiasm, skill, and support for making a second, expanded edition of this book possible. Most especially, I thank Ben Watson for his kind, gentle and skillful editing.

A few paragraphs of Section II (here and there) were first published by *National Gardening Magazine* (Parts I and II of 'Custom-Crafted Vegetables,' the June/July and August issues, 1998).

Introduction to
the Original Edition

ALL OUR major food crops were originally developed by amateurs. Until recently, all gardeners and farmers saved their own seed; all gardeners and farmers were automatically amateur plant breeders — and amateur plant breeding was the only kind of plant breeding there was.

This book has two major purposes. First, I want to encourage all of you gardeners and farmers to rediscover the excitement, the pleasure, and the rewards of developing your own vegetable varieties. Second, I want to show amateurs how to breed plants more efficiently. Our ancestors often needed generations to accomplish what a modern professional plant breeder does in just a few years. But the major techniques modern professionals use are simple, and they require no special equipment. Any gardener can do them.

This book extracts out of modern genetics exactly those methods needed for practical small-scale plant breeding and presents them with a minimum of theoretical trappings. I hope that this book will enable the modern amateur to breed plants with very nearly the efficiency of the modern professional.

I feel I am ideally situated to write this book because I am both a professional plant breeder and an amateur. I have one foot firmly planted in the soil of the professionals because I am a geneticist. I have a Ph.D. in biology from Harvard University with a specialty in genetics and more than twenty years of experience as a working geneticist.

I have my other foot even more firmly planted in the soil of the amateur plant breeder. No one pays me to breed vegetables. I do it just for fun. I learned plant breeding on my own only after leaving academia and only after finding that I could not get many vegetable varieties that I wanted for my garden. They didn't exist. As a geneticist, I could see that much of what I wanted would be easy to develop. But no one was doing it.

The primary focus of this book is on vegetables and their wild relatives and on wild plants that have the potential for being bred into vegetables. The techniques presented herein can easily be applied to breeding other plants, however. Those interested primarily in plants other than vegetables should note that Appendix A, "801 Interesting Plants," gives basic breeding information for most common food plants of the world as well as for many uncommon ones.

This book is for all gardeners everywhere. It's for every gardener who has ever dreamed of bigger, better, tastier, earlier, more productive, or more beautiful vegetables. It's for every gardener who has ever been told, "You can't grow that here," but who wants to anyway. It's for people who may not have a lot of space or time, but who have some ideas — ideas about what they wish they could grow that isn't available now.

This book is for every gardener who likes to grow unusual vegetables, unusual varieties of vegetables — rarities. It's a lot of fun to discover a plant that does well in your area and to introduce it to your friends and neighbors. And what could be rarer than the plant that nobody in the world has ever grown before, because you've just developed it?

This book is for every seed saver. Saving seed is the basic method of plant breeding. If you save seed, you are already a plant breeder, already practicing selection, already choosing what genetic material to perpetuate. This book will help you do it better.

In addition, it's a short step from saving seed to deliberately breeding plants. Why limit yourself to preserving the achievements of the amateurs of the past? They didn't. You can both preserve the fine old heirloom varieties and use them to create new varieties of your own.

Growing and preserving heirloom and open-pollinated varieties combines well with plant breeding. Heirloom varieties often serve as sources of breeding material, ideas, and ideals. Breeding projects give you more excuses to grow and evaluate heirlooms. After all, you don't know what you want to breed for until you've tried a number of varieties and have some favorites.

You can grow and maintain heirlooms and open-pollinated varieties and use them to develop your own hybrids and produce your own hybrid seed. You can have hybrids whose parents you know, hybrids you yourself control. In addition, you can use simple plant breeding methods to take any desirable commercial hybrid and convert it into a stable open-pollinated variety.

This book is for every gardener or farmer who grows organically, who wants plants that yield well without pesticides, herbicides, and chemical fertilizers. It's for everyone who is interested in the sustainability of agriculture, who wants varieties designed for use in sustainable agricultural systems. It's for everyone who cares about strengthening the regional basis of gardening and farming, who wants plants that are fine-tuned to produce well with minimal cost or labor in specific localities and under specific conditions. The surest way to get such varieties is to breed them yourself.

But above all this book is for every adventuresome gardener — every gardener who likes to try things, who likes to experiment — who just likes to play around.

Introduction to the Second Edition

Welcome to the new, expanded second edition of *Breed Your Own Vegetable Varieties*.

This book is in many ways two books written by two different people.

Part I, the core of the first edition, is full of the joy of first discovery. My formal academic background was in genetics and molecular biology, not in plant breeding. I used the process of writing the book to learn about plant breeding myself.

Part I is set in the years of about 1990 to 1993, when the first edition was written. The techniques of standard plant breeding were fully developed by then, and haven't changed since. However, some of the seed companies I refer to in Part I no longer exist. Others have merged or transmogrified. And seed people have changed too — migrating from one company to another, starting their own companies, and so on.

In reviewing Part I for the second edition, I found it impractical to try to change the specifics about people and seed companies to reflect the realities of the year 2000. So I left Part I largely intact in the second edition, allowing it to represent people and institutions as they existed in the early nineties. I have updated the appendixes, however, to reflect current information about seed companies and seed people.

Parts II and III, written in spring of 2000, are the work of greater experience. They reflect my current interests and situation — an increasing concern with the overall sustainability of agriculture and the security of our food supply as well as my own transition from amateur plant breeder to professional, and from backyard gardener to farmer and food company.

In between the writing of the first and second editions of this book, the entire world of plant breeding has changed almost beyond recognition. In 1993 there was a serious dearth of university-based vegetable breeders working on projects aimed at the public interest. Now there are almost none. Changes in patenting laws have allowed publicly funded institutions to patent plant varieties created with public money.

The result of these legal changes is a new generation of university-based breeders who create mostly patented varieties and F_1 hybrids — who choose to work on developing varieties resistant to herbicides, for example — instead of varieties that taste better or grow better, that are more agroecologically adapted, or that enhance the viability and independence of the farms in their regions instead of destroying them.

In short, most university-based plant breeders now behave in exactly the same way as agribusiness breeders. Profit, not the real needs of gardeners or farmers, is their primary motivation. But what is most profitable for a corporation is not necessarily what is most profitable, let alone most important, for farmers or gardeners.

In addition, genetically engineered crops and foods were only on the horizon in 1993. Now they are a reality. In the universities, the last generation of plant breeders skilled in standard field methods has retired or is retiring. They are being replaced with genetic engineers. Greater perspective would have added genetic engineering to a full repertoire of standard plant breeding. Instead, we now have universities full of "plant breeders" who only have one tool — genetic engineering — and who are prepared to conduct only those projects that use the genetic engineering approach.

There are many limitations as to what can be approached with genetic engineering, however. Much of what is most important to home gardeners and farmers is not very amenable to the genetic engineering approach. And almost all of what matters in breeding plants for organic or sustainable agriculture is best and most easily approached using standard plant breeding methods. In their blind leap onto the genetic-engineering bandwagon, the universities and multinational seed companies have virtually abandoned nearly all of the most important problems.

Crop varieties incorporate the values of their creators. When you grow varieties bred by others, you propagate their values along with their varieties. Today's professional plant breeders — university and corporate — are breeding plants to facilitate and serve the modern megafarm agribusiness pattern. These varieties produce well in huge monocultures grown with massive doses of herbicides and chemical fertilizers. Bred into the varieties are the values of their creators — that more is always better, that monocultures are best, and that pollution, biodiversity, and sustainability don't matter.

It is time for new patterns — new patterns for agriculture, and new patterns for plant breeding. It's time for the rising up of a new generation of plant breeders out of the very soil of our farms and gardens. It is time for farmers and gardeners everywhere to take back our seeds, to rediscover seed saving, and to practice our own plant breeding. It is time to breed plants based upon an entirely different set of values.

How to Use This Book

The information, concepts, and stories in Part I build consecutively. Thus, it's best to read the chapters in Part I in the order presented.

Parts II and III assume familiarity with Part I, but don't depend upon it entirely. I believe they stand alone well enough that many people will be able to read the three parts independently. Chapters within each part build somewhat consecutively, however.

Chapter 5 in Part I comprises a thorough but nonstatistical introduction to doing variety trials, and to garden- and farm-based plant and crop research.

Table I and the appendixes constitute a reference book to plant breeding that should be useful to both amateur and professional plant breeders. The appendixes also comprise an advanced course in standard plant-breeding methods. The appendixes fill the gap between the amateur and professional level.

Table I gives basic breeding information on more than eight hundred plant species. It is organized by scientific name. If you know only the common name of a plant, look it up in the index. The index cross-references common and scientific names.

PART I

An Introduction to Plant Breeding and Seed Saving

ONE

Amateur Vegetable Breeding

EVERY GARDENER should be a plant breeder. Developing new vegetables doesn't require a specialized education, a lot of land, or even a lot of time. It can be done on any scale. It's enjoyable. It's deeply rewarding. You can get useful new varieties much faster than you might suppose. And you can eat your mistakes.

There has never been a better time to get involved in amateur vegetable breeding. The seed saver exchanges that have emerged during the past decade provide a rich source of raw materials for plant breeding. The smaller seed companies, many also founded recently, are eager to help perpetuate and distribute the creations of amateurs. And the professional plant breeders are all busy elsewhere. They are engaged, almost exclusively, in developing *commercial* varieties of vegeta-

bles — vegetables bred for uniformity and once-over picking so they can be harvested by machines, for tough skin and hard flesh so they aren't ruined by those machines, and for good storage and shipping characteristics so they can be transported long distances. These are not usually the qualities home gardeners need. Yet most of the new varieties that are released annually with such fanfare are commercial cultivars.

Gardeners buy only small amounts of seed compared to commercial growers, so seed of varieties that are best suited for gardeners is sold in only small amounts. Large seed companies often can't afford to carry it. No one can make a profit developing it. So no one is. If we gardeners want good new *garden* varieties, we'll have to breed them ourselves. But this is as it should be. Gardeners have been developing their own varieties for centuries. Besides, why should we let the professionals have all the fun?

This chapter is an introduction to how to breed your own vegetable varieties and focuses on three amateurs who have done it. None of them had any special education in genetics or plant breeding. Yet each has produced good new garden varieties — varieties that have been formally tested by a seed company and have been introduced or scheduled for introduction; varieties that have been found worthy of being listed and sold side by side with the best developed by anybody anywhere.

Glenn Drowns was only sixteen when he started breeding plants, but he was already an experienced and enthusiastic gardener. It all started when he was two and a half, and his family planted some flower seeds. He was fascinated. By the age of four he was crawling through the fence to help his neighbor, who had a bigger garden. At eight he had a 500-square-foot garden of his own and was ordering his own seed. About then he developed a passion for vine crops. Each year he ordered every squash variety he could find. By age eleven he was selling all the produce his family didn't need and using the money to buy seed. By age sixteen he was growing fifty different varieties of squash and ten of cucumbers. He also tried about half a dozen of the shortest-season watermelons he could find.

More than anything else, Glenn wanted a ripe watermelon. His family lived in the extreme northern part of Idaho, where the growing season is short, cold, and unpredictable. His melons wouldn't ripen. "I tried everything," he recalls. "I even grew them inside little plastic tents all summer, just really baked them. But I never got a ripe melon. The only fully ripe watermelons I had ever tasted came from the stores — which didn't count."

Then Glenn took high school biology, and his class discussed crop improvement and hybridization. "Wow, maybe I could get a ripe melon that way," he thought. So he tried a cross. One of the parents was probably 'Sugar Baby'. The other was from a package of seed some friends had given him. The package was labeled "watermelon." Glenn didn't keep much in the way of records. He knew about record keeping, but he thought he was only playing around.

Within four years, growing no more than a dozen plants per year, Glenn Drowns produced a new stable variety — a variety that reliably produces similar plants from seed to seed and year to year. 'Blacktail Mountain' is a round, deep green melon with very faint stripes. Vines grow out to form a plant about 10 feet in diameter. The average melon is 8 inches across, weighs 8 to 10 pounds, and has a rind about half an inch thick. The flesh is orangish red, crisp, and very sweet.

What is essential about 'Blacktail Mountain' is that it's early — quite possibly the earliest watermelon ever grown. It's about five days earlier than 'New Hampshire Midget', for example, one of the earliest watermelons. 'Blacktail Mountain' has been scheduled for introduction by Seeds Blüm, of Boise, Idaho, in 1993 or 1994.

Glenn, now twenty-seven and a high school science teacher with a degree in biology, laughs when he describes the breeding program that led to 'Blacktail Mountain'. Glenn hadn't been able to find out anything at all about how to do crosses. "So I just sort of guessed," he says.

Watermelons, like most cucurbits, have separate male and female flowers. Glenn noticed that some flowers had what looked like little fruits underneath and others didn't. He figured the flower buds with fat bases must be females and the ones with skinny bases must be males. He made that first cross exactly the way he does watermelon and squash crosses today, eleven years and thousands of hand-pollinations later.

Glenn starts by taping the male and female flower buds shut with masking tape in the evening before the buds open for the first time. The timing is part convenience and part necessity. Late afternoon of the day before opening also is OK, but any earlier and the buds might still be growing fast enough to damage themselves on the tape. The following day Glenn untapes the buds. If they are ready to open that day, they will slowly expand after untaping. He plucks the male flower and uses it to sprinkle pollen onto the stigma of the female flower. (The stigma is at the top of the pistil.) Then he retapes the female flower with fresh tape and labels it. He never untapes the female flower; it just shrivels at the end of the developing fruit.

Glenn crossed the putative 'Sugar Baby' to the unknown watermelon. Then he saved the seed and planted it. The plants of that first generation, as he recalls, were similar to the 'Blacktail Mountain' of today. He self-pollinated each. "That was 1978," Glenn remembers. "Nineteen seventy-eight was a gruesome

year." It was unusually cool, and his garden suffered a lot of deer damage. Only a few plants survived. They gave him exactly one ripe watermelon. "Selection was easy," Glenn says. "There was just one ripe melon. So I selected it."

He planted the seed from that melon in 1979. The plants that resulted, his second generation, looked pretty similar to those of 1978. He self-pollinated each and got two or three ripe melons. He planted the seed from those in 1980, but the vines were all frozen out on August 4. Fortunately, Glenn had known better than to plant all his seed in one year. He replanted in 1981 and continued to inbreed — that is, to self-pollinate each plant. By 1982 he was offering 'Blacktail Mountain' through the Seed Savers Exchange. It was already stable and was basically the same variety it is today. But he grew it for a number of years before he felt confident enough in its stability to offer it to a seed company.

Part of the fun of breeding your own varieties is the surprises. Glenn had deliberately selected for a very early watermelon. Unexpectedly, 'Blacktail Mountain' also proved to have unusual keeping qualities. Glenn found that out completely by accident. "There are long-storage melons, the Christmas types," Glenn says. "I'm surprised more people don't grow them. I always do, and I eat my last watermelons in February. But the storage types all seem to have that white rind. I've always thought of the green-skinned melons as an immediate eating thing." Storage melons are harvested at just under ripe and finish ripening during storage. Glenn usually had no reason to harvest 'Blacktail Mountain' that way. But Iowa, where Glenn lives and gardens now, intervened.

In 1988 an early heat wave and drought killed all Glenn's melon vines, so he had to replant in late June. Then, the last week in August, came the torrential rains. The melon field was in a low place. It flooded. Only the earliest half dozen varieties had melons that were mature enough to harvest. Glenn grabbed those and put them in his garage. Many were not quite ripe.

Several weeks later, Glenn opened one of his 'Blacktail Mountain' melons to look at the seed. To his surprise, the melon popped open. And it was just as crisp and tasty as if it had been harvested in its prime and eaten immediately. The melons from the other five varieties had long since turned to mush. Intrigued, Glenn used the rest of his melons in a storage trial. He found they kept nicely for up to two months. That's not as long as a winter-storage type, but it's unexpected for a standard green-skinned type, and it means that Glenn can eat his last 'Blacktail Mountain' watermelons on Thanksgiving.

The development of 'Blacktail Mountain' was unusually easy. The plants of the first generation after Glenn's cross already had the characteristics he was looking for. It's much more common for the desired type to show up only in a later generation. In addition, there was little variation among the plants in subsequent generations. That also was unexpected.

A more typical example is Glenn's development of 'Mexigold' squash. He started by crossing 'Mexican Banana', a blue-green banana squash, to 'Golden Delicious', which is orange and heart shaped.

Glenn usually grows just a few plants of that first generation after a cross, because the plants are usually all pretty similar. He plants as many of the second generation as he has room for, though, often fifty or more. Each second-generation plant is different from the others, each showing a different combination of the characteristics of the original parents.

Glenn doesn't remember what the first generation of plants after the cross were like. He self-pollinated them to obtain a second generation. That seed grew into plants of all sorts, each with its own kind of squash. There were bananas and hearts of all hues, pinkish and bluish and orangish. He selected for pink hearts. Each year he self-pollinated plants that produced pink hearts and eliminated the rest. After a number of years, all the plants produced pink hearts. Glenn had "stabilized" the strain — had developed a new pure breeding variety.

'Mexigold' is a pink heart-shaped squash that weighs 10 to 12 pounds. It's very productive. But like 'Blacktail Mountain', 'Mexigold' produced a surprise: it's extremely early. Glenn had selected for squash shape and color, not earliness, and neither of the two original parents were particularly early. Yet 'Mexigold' is consistently earlier than 'Buttercup', one of the earliest squashes. In fact, the only variety Glenn knows of that is earlier than 'Mexigold' is 'Arikara', which doesn't have good table quality. 'Mexigold' will be introduced by Seeds Blüm when enough seed becomes available.

It's not always necessary to start a plant breeding project by doing a cross. Glenn sometimes simply notices and uses genetic accidents. One vine on one of his bush acorn squash plants, for example, had white fruit. It was a mutant. So Glenn saved the seed from that plant and is inbreeding to develop a white bush acorn line. He's selected over a number of generations and has mostly white-fruited plants now, but occasionally the strain will still throw a plant that produces green fruit. Glenn is still inbreeding and stabilizing the line.

You don't have to start out as a youngster to become an amateur plant breeder. Nor must you necessarily take pains to do carefully controlled crosses. You might not even have to do any inbreeding to stabilize new varieties. A lot depends on the vegetable you're working with.

Ewald Eliason, age seventy-three, is a retired Minnesota dairy farmer. He started breeding his own potatoes seventeen years ago. "I've always enjoyed gardening," he says, "and potatoes are my favorite."

Potatoes are much easier to work with than squash or watermelons. For either of those, Glenn Drowns has to control pollination to develop or maintain a cultivar. Otherwise, bees would cross all his cultivars at random, and the seed would not breed true. Maintaining a dozen melon or squash varieties entails hand-pollinating each variety each year. But since potatoes are propagated from tubers, not seed, all Ewald has to do to maintain a variety is plant its tubers. The resulting plants are clones of and genetically identical to the mother plant, whether insects are cross-pollinating the flowers or not.

Ewald started breeding potatoes just by saving true seed from the plants in his gar-den. The tuber pieces that are used as starts are called seed, so potato breeders say "true seed" when they mean the product of sexual reproduction. Since potato cultivars are maintained by vegetative propagation, they are usually not genetically pure. When true seed is planted, the seedlings are individuals, each genetically different from its parents and all others. When such a plant forms tubers, it is an instant new variety.

Fertilized potato flowers form berries that resemble small green tomatoes. Each contains about two hundred seeds. Ewald picks the ripe berries, puts them in a food blender with water, and blends very briefly. The seeds sink, and the pulp floats. He plants the seeds in pots in a cold frame and transplants to his garden.

As Ewald's potato hobby grew, he accumulated more and more varieties. When the Seed Savers Exchange started, he joined and obtained varieties — including many heirlooms — from fellow members. (An heirloom variety is one that has been handed down from generation to generation.) He received potatoes from friends. He visited Minnesota's North Shore Experiment Station one summer and obtained experimental material. His sister went to Finland one summer and brought him back some true seed from European types. And he started doing his own crosses.

Potatoes, unlike watermelons, have perfect flowers — flowers containing both male and female parts. Potato flowers are usually pollinated by pollen from other potato flowers, but they do self-pollinate to a significant extent. Professional potato breeders emasculate the flowers when they do crosses — that is, they remove the stamens before the pollen is shed.

Ewald, however, doesn't bother emasculating. He just takes a flower from the variety he wants as the male parent and dusts its pollen onto the stigma of a flower of the other variety. Sometimes the offspring will be from his cross, and sometimes they won't. "I just dust the pollen on there and take what I get," Ewald says. "Most of life is like that, I've found." Whatever the seedlings are, Ewald knows they're sure to be interesting.

Just sprinkling pollen on there and taking what you get is a perfectly valid breeding method. When professionals do it, they call it a "fertile × fertile cross." Dignified with that fancy name, it sounds much less cavalier. All it means, though, is that the breeder felt the task of emasculating was not worth the effort. In some vegetables, for example, the flowers are so tiny that emasculating is very difficult or even impossible. For practical breeding work, it's usually not necessary that all the offspring of an attempted cross really be hybrid, as long as enough of them are.

Ewald plants fifty to one hundred seeds from each cross he does. Virtually all the plants from his various crosses give him hills of usable potatoes. Some plants don't yield well, however, and others are scabby. Most plants are pretty good, he finds, but not all are worth increasing and maintaining. During the growing season Ewald evaluates the potato plants — how vigorous they are, and how disease free. At the end of the season he digs the potatoes and looks at them. He likes potatoes that "grow clean" — that are scab free and blight resistant. He also makes notes about the yield. Then, during the winter, he evaluates the potatoes for keeping and eating quality.

Ewald grows twelve 100-foot rows of potatoes per year, but since he has about three hundred varieties, he doesn't usually have very much of any of them. His soil is sandy loam, and he has subsurface water. He doesn't have to irrigate, even during droughts. He never sprays, either. He gardens organically. Presumably, he is selecting for varieties that perform well under organic gardening methods.

Ewald's favorite variety — "so far," he says emphatically — is 'Mesabi Gold', which Seeds Blüm plans to introduce when enough tubers are available. It's a result of a cross between 'Swedish Mandell', a yellow fingerling type, and 'Minnesota #30'. 'Mesabi Gold' has yellow skin, yellow flesh, and good keeping and eating quality, Ewald says. Yield is good, too — about seven or eight full-size tubers per hill. The growing plant is distinctive. Unlike most potatoes, it has upright stems about 30 inches high. (Both parents had the normal bushlike form.) "This is like a throwback to a tomato," Ewald jokes. "It's more a starch-type tomato."

'Blossom', another of Ewald's varieties, was introduced by Seeds Blüm in 1989. It's one of Jan Blüm's favorites. "It's a beautiful potato," she says. "It has red skin and pink flesh and is a very nice producer. And it has lots of big flowers and looks just wonderful in the garden."

'Blossom' came from a random berry off a plant of 'Poorlander', an heirloom variety. 'Blossom' has good keeping and eating quality, says Ewald, and yields potatoes "of more than ordinary size." But unlike most potato plants, which flower only on the ends of their stems, 'Blossom' also tends to flower at the nodes. Ewald thinks it could be used to breed an ornamental potato. "I'm thinking it has the potential of being developed into a plant you could grow in your flower garden," he says, "and at the same time, you could enjoy the potatoes."

Ewald Eliason is still breeding more and more potatoes. "When you grow all the various plants and dig them up, every one is a surprise," he says. "They're like children in a family. You just don't know what you're going to get."

Like Glenn Drowns, Tim Peters became an amateur plant breeder while still in high school. Glenn had the good fortune to start with a vegetable that was easy to work with. Tim began with broccoli, a plant breeder's nightmare. And although Glenn seems to have been born a gardener, Tim had it thrust upon him.

At the age of fifteen, Tim was not particularly interested in gardening or in plants. He liked animals and was learning how to breed fish. His father, though, was an avid gardener. Then the family moved from California to Myrtle Creek, a small town in the foothills of the Cascade Mountains in southwestern Oregon. The soil was poor. The insects were terrible. There was no rain. After a couple of years, Tim's dad took him to the garden and said, "It's your job."

Tim was not pleased. He wanted to spend his time with his fish. Instead he hauled manure, weeded, and watered the garden by hand. "Boy, I hated that," he recalls. "But after a while I decided that since I was going to be stuck there in the garden all summer, I'd better start learning to like it. And after a while I did start liking it some."

At the end of that summer Tim saved the seeds from a tomato plant some friends had given him. It was those seeds that changed him. When he held them in his hands, they seemed magical, mysterious, profound . . . and he could hardly wait for the next year to come.

The following year Tim made his first cross, a self-pollination of 'Green Comet Hybrid' broccoli. It was a disaster. He saved the seed and planted it, but the plants hardly grew at all.

Broccoli, like many of the brassicas, has a genetic incompatibility mechanism — a physiological mechanism that prevents self-pollination. The plants have perfect flowers and are anatomically capable of pollinating themselves, but the pollen doesn't usually fertilize well. In some cases it won't germinate at all on the stigmas of flowers on the plant from which it was derived. In other cases the pollen germinates but grows so slowly that it is usually beaten out by foreign pollen. Even where self-pollen grows and achieves fertilization, seed may not develop properly or, if it develops, may produce only weak, sickly seedlings.

Tim didn't give up. He had read in a magazine that the Japanese practice bud pollination of brassicas. He didn't know what bud pollination was, but it sounded like a good idea. So he started pollinating buds, and it worked. Only much later did he find out that the incompatibility property of the stigma of brassicas develops fairly late during flower formation, so that by pollinating in the bud stage, he was bypassing the incompatibility system entirely.

Later in his breeding project Tim needed to isolate broccoli plants to prevent cross-pollination. A professional would have had isolation farms where each variety could be grown far from any others. Tim developed his own approach. He put the plants in pots and loaded them onto his bicycle. Then he rode around looking for isolated homes. When he found one, he knocked on the door with a potted broccoli in his hands, explained what he was trying to do, and asked the owners if they could help him out. Usually they did. Tim's early work with broccoli didn't lead to any new introductions, but he learned a lot about breeding plants.

After he graduated from high school, he received a plea for help from a friend and fellow Seventh-Day Adventist in Africa, and off he went. He spent several years setting up a farm for a mission school and breeding tropical corn. He returned in 1980, during a "minidepression" in Oregon. There were no jobs. There was also no money for college. Tim got a part-time job doing landscaping and found a position as caretaker on a farm. The caretaking job included a trailer for him to live in and a 4-acre field where he could grow and breed vegetables. Oregon State University, just a few hours north, had a fine agricultural library. "I'm going to learn everything I can from this field and that library," Tim resolved. And he did.

Tim works with broccoli, tomatoes, corn, melons, mustard, cauliflower, and many other plants. In 1988, Territorial Seed intro-

duced the first Peters-bred vegetable variety to be released commercially. It is, appropriately, a broccoli.

'Umpqua', named after the river in southwestern Oregon, is a very early, deep green broccoli with especially large central heads. "In our trials," the Territorial catalog declares proudly, "'Umpqua' has outperformed all other open-pollinated varieties that we know of." And it's early — earlier than all other open-pollinated varieties and as early as the earliest hybrids. (An open-pollinated variety will come true from seed; a hybrid — the offspring of a cross between two true breeding varieties — will not.)

It took more than just inbreeding from a single cross, however, to produce 'Umpqua'. Like most plants that have incompatibility mechanisms, broccoli exhibits strong inbreeding depression. This means that the more genetically uniform the plants are, the smaller and weaker they tend to be and the lower their yield. Thus inbreeding, which is done to increase genetic uniformity, can be counterproductive. Instead of working toward genetic uniformity, the breeder tries to obtain uniformity for the important agricultural characteristics while simultaneously maintaining enough genetic heterogeneity to produce vigorous plants.

Tim started by crossing several hybrids with each other. Then he inbred the progeny from each cross to develop lines and selected the type he wanted — very early broccoli with medium height, big central heads, and good yield — within each line. Next, he pooled all the lines and mass-selected — that is, he let all the lines he had developed separately grow together and cross-pollinate. For several generations thereafter, he selected for the type he wanted until his new open-pollinated variety had become stable. It was a mess but a practical approach to dealing with the plant that had become his priority.

Tomatoes are one of Tim's other favorites. Tomatoes are much easier to work with than broccoli. Because tomatoes are almost totally self-pollinating, you don't have to hand-pollinate to maintain a number of different varieties or to inbreed from selected individuals. The plants inbreed themselves. They don't usually show inbreeding depression, either. It's fairly easy to produce good inbred varieties that perform as well as or better than the commercial hybrids.

With tomatoes Tim does a lot of what could be called "dehybridizing the hybrids." "The professional breeders have put lots of good things into those modern hybrids," Tim says. "I like those good things." So he tries to get the good things out of the hybrids and into open-pollinated varieties. He might cross a very disease-resistant hybrid, for example, with an heirloom variety that has spectacular flavor, then inbreed and try to select a line that has the best characteristics of both parents.

Tim often starts by saving seed from a good hybrid. The seed company has already done the original cross between two different cultivars. Seed companies usually keep the identities of the parents secret so that no one else can produce the hybrid seeds. Customers often believe that they can't save hybrid seed, that it's not viable. But you can save the seed from most hybrid vegetables.

The seed from Tim's hybrid broccoli failed not only because the parent was a hybrid but also because of the incompatibility system. The hybrid broccoli plants were genetically identical to each other, like clones of one plant. The incompatibility mechanism prevented the plants from pollinating themselves or from being pollinated by genetically identical others. Tomatoes, corn, squash, melons, cucumbers, and many other vegetables are self-compatible. Seed from their hybrids is good; it just doesn't breed true. For a commercial grower, a row or field of plants that are all individuals could be a disaster. But for the gardener, it might represent opportunity.

For example, Tim has done a fair amount of work with 'Early Girl Hybrid' tomato, which does well in the Pacific Northwest. It's very early, it is resistant to cracking and verticillium wilt, and it has good flavor. But the fruit isn't very big. Tim saved and planted the seed from an 'Early Girl' fruit. All the resulting plants were pretty early, and all the tomatoes tasted good. There was a tremendous range in fruit size, however, from almost cherry types to great big ones. He saved the seed from plants with big tomatoes and has developed several lines that have the virtues of 'Early Girl' but are open-pollinated and bear big beefsteak-type tomatoes.

Tim Peters had been an amateur vegetable breeder for more than a decade when he read that Territorial Seed Company was looking for a vegetable breeder. He could not imagine anything more wonderful than being paid to breed vegetables. Although he had only a high school diploma, he screwed up his courage and went to talk with the people at Territorial. He now handles the seed company's trial and breeding program.

*   *   *

Glenn Drowns, Ewald Eliason, and Tim Peters all developed useful new vegetable cultivars — ones that met their requirements better than anything else available. You can do it, too. But where should you start? What kinds of things are worth doing? Are you likely to make any money at it? Should you patent your plant? And how do you get a new variety out where others can use it?

Many amateur breeders start by accident. They simply notice something unusual and save the seed. If you want to start deliberately, choose a vegetable you like to eat; you could be eating a lot of it before you're done. The amateurs profiled in this chapter all suggest that you limit yourself to one or two vegetables or projects, but none of them did. You probably won't either, and it is perhaps just as well. Not all projects work out.

John Gale, president of Stokes Seeds, has two suggestions based on his experience with amateurs. First, he says, an unstable cross straight out of the garden isn't useful, no matter how interesting the individual plant is. You have to grow a new variety for several years and cull out the off types each generation. Only after the variety is yielding all desired-type plants is it ready to offer to a seed company.

Second, find out what is already available. There are tomatoes of almost every color, for example — pink, orange, yellow, green, brown, purple, bronze, white. They aren't popular, but they're certainly not new.

Rose Marie Nichols McGee, of Nichols Garden Nursery, concurs, and mentions that she is especially unthrilled by the idea of a "new" white carrot. "Carrots started white,"

she points out. "Breeders worked for a long time to get the beautiful orange color. And wild carrots are white, and any accidental cross with them gives you a white, so they turn up all the time. And anyway, we already carry white carrots."

There are a lot of things that Rose Marie does want. There has never been much breeding work with herbs, she says. Almost anything that's different is interesting. Foliage differences; variegation; color differences in plants, flowers, or stems; differences in size, growing habit, or form; differences in flowering characteristics or patterns; and differences in odor are all worth paying attention to. Disease resistance also matters with herbs.

With vegetables, Rose Marie is interested in early maturity, compactness, and disease resistance. If, for example, you have an eggplant that is doing nicely when all the rest are dying from verticillium wilt, you might have something there. "The sophisticated amateur is just real well positioned to notice these things," Rose Marie says. "You may not know what the disease is, but you can see one plant is behaving differently." (An Agricultural Extension Service agent can probably help you identify the disease.)

"With vegetables," Rose Marie says, "we look for quality and also beauty. Many people are gardening on small plots, and the attractiveness of their vegetables as plants is an increasingly important part of the pleasure.

"Genetic accidents occur all the time," Rose Marie says. "They are not uncommon. But what is important isn't just the plant. It's whether there is someone around to notice — a person with a keen eye, with the ability to observe and recognize what's special."

Seeds Blüm carries many unusual vegetables for which no professional breeding work is done. "Unless amateurs do it," Jan Blüm notes, "it's not going to be done." Is anyone interested in breeding sunroots (aka Jerusalem artichokes, *Helianthus tuberosa*), for example, American groundnut (*Apios americana*), or camases (*Camassia* species)? Camas is a native plant the American Indians used. "There are lots of varieties," Jan says, "and they're good, and the plants are beautiful. But very few people have played around with them."

Professionals usually spend most of their time working on crops that are already important. Amateurs can afford to gamble. They need not justify their efforts to superiors or grant agencies. Sunroots, groundnuts, and camases aren't major vegetables now, but perhaps they could be if someone did the breeding work.

With the common vegetables, Jan cares about flavor above all else. She also loves vegetables that are used for particular dishes. "If you cherish a tomato because it makes the very best lasagna, or whatever, that matters," she says.

Jan also has two suggestions for amateur plant breeders. First, like Rose Marie Nichols McGee, she emphasizes that good things frequently come from accidents — the spontaneous mutation, the accidental cross. "So many people, when you talk about this sort of thing, will tell you about some wonderful thing that showed up in the garden one year," Jan says. "But they didn't save the seed

from it. Sometimes I really ache when I hear about these things. You just know you're not going to see that again." Make records of anything unusual in your garden, Jan advises. Save the seed and grow some of it out next year.

Second, Jan emphasizes that selection, by itself without making crosses, is a valid breeding method. People do crosses to generate the genetic heterogeneity from which to select new types. But heirloom varieties often already contain a lot of genetic heterogeneity. When they do, you can guide the variety toward your own breeding goal just by culling out the plants you don't like each generation and saving seed from the best. "Everyone who saves seed is a plant breeder," Jan says. "They choose which genetic material to perpetuate."

What about selling seed of your new variety? And what about plant patenting? Do you have to patent your new variety to be able to sell its seed? Can you legally breed new varieties using a patented variety or a hybrid as one of the parents? The answers to these questions depend partly on what country you are in.

Breeding new garden vegetables is usually a matter of love, not money. In the United States, plant variety protection for seed-propagated plants is provided for by the U.S. Plant Variety Protection Act of 1970. This act isn't of much relevance to those who develop home garden varieties, though. Licensing a cultivar is too expensive — $2,000 for eighteen years. In most cases garden varieties don't sell well enough to cover that, so even those developed by seed companies are usually unprotected. Breeding them is not profitable, which is one of the reasons amateur efforts are so important.

Some new vegetable varieties, especially those intended for commercial production, may have enough profit potential to warrant patenting — if you believe in it. Many people don't; they feel the patenting of plant varieties is immoral. Germplasm, they feel, is the property of all humankind. I find the issue more complicated than I might wish. An excellent coverage of all points of view on plant patenting can be found in Jack Kloppenburg's *Seeds and Sovereignty: The Use and Control of Plant Genetic Resources* (see bibliography).

Many amateur breeders do make money by growing and selling seed of their new varieties without patenting them. Usually, the breeder simply serves as the grower of the variety for the retail company that sells it. Breeders actually are being paid for the growing, not the breeding, however, and would be paid just as much for growing other varieties they hadn't developed themselves. The alternative seed companies in the United States are often in need of additional reliable growers. The work isn't usually highly lucrative, but it can provide a modest return for an enjoyable activity.

Although the Plant Variety Protection Act is usually of little benefit to amateur breeders, it's also of little hindrance. You can grow your own seed from protected varieties as long as you don't sell it. You can sell seed of your own varieties without protecting them. You can use protected varieties in developing your own. You are not allowed to use a protected variety to make hybrids to sell com-

mercially, however, nor may you merely select from a protected variety without doing any crosses. But you can cross a protected variety to something else and then work with the progeny to develop your own variety. And you can use hybrids in any way you want. They are protected by secrecy, not by law.

In the third world, plant patenting and variety protection isn't recognized. There are usually no restrictions on using or selling seed.

In Europe, official or national lists profoundly restrict the activities of amateur breeders and seed savers. To protect patented varieties, there are laws that prevent the sale of everything that isn't on a special list. These laws are far more restrictive than those that apply to all other intellectual property rights in Europe or elsewhere.

It costs time and money to do the evaluations necessary to get a new variety on the official list. You can afford to do it if you are a seed company with a new patented commercial variety that is broadly adapted and is expected to sell in large amounts. But the official-list policy is a barrier to amateurs in getting their varieties recognized and made available. It doesn't prevent them from offering their creations through seed saver exchanges, however. (The Seed Savers Exchange in the United States welcomes foreign members, by the way.)

Once you've developed a new variety, you'll want to get it out where others can use it. The easiest way is probably to offer your new creation through a regional or national seed savers exchange.

The Abundant Life Seed Foundation is both a seed company and a nonprofit foundation dedicated to preserving germplasm of the Pacific Northwest. It handles native, naturalized, and open-pollinated domesticated varieties. Forest Shomer, the director, says that the foundation would be happy to carry new amateur-bred varieties that do well in the Northwest.

The Seed Savers Exchange preserves and distributes open-pollinated varieties that are not available commercially. That includes the one you've just developed, director Kent Whealy says.

The Seed Savers Exchange is a marvelous source of breeding material, contacts, and information. With the exchange's *Garden Seed Inventory* in one hand and its most recent annual *Winter Yearbook* in the other, amateur breeders have descriptions of all varieties available either commercially or through members. They can see what has been accomplished in their vegetables of choice and can figure out whether anyone has ever developed anything like the variety they're thinking about. In addition, members often list their projects and interests, exchange information, and trial each other's varieties.

You don't have to live in the United States to join the Seed Savers Exchange. Since small amounts of seed can be mailed across national borders, you can participate no matter where in the world you live.

You may want to offer your new variety directly to a seed company. Most of the smaller, family-owned, or alternative seed companies are happy to introduce varieties bred by amateurs. In most cases you should approach just one seed company. Trialing and introducing a new variety is work, and the introducing company has no legal protection for a public-domain cultivar. If a new introduction is successful, many companies will soon be offering it. The introducing company receives no reward for its work — other than the pride of being the first to offer something special.

By the time you've developed a good new variety, you'll often know which seed company to offer it to. Look first to the companies from which you buy seed. If you like what they're doing, they might like what you've done, too. Territorial Seed Company was the logical choice for Tim Peters's new broccoli. It is a regional seed company that sells only varieties that do well in the maritime Northwest, and the company has trialed, introduced, and popularized many new cultivars of the brassicas. Glenn Drowns and Ewald Eliason both met Jan Blüm through the Seed Savers Exchange. Her company specializes in open-pollinated and heirloom varieties. Many of the newer or smaller seed companies were founded specifically to serve home gardeners. Among them, any good new garden variety is likely to find a home.

A century ago, seed catalogs were full of new introductions that had been bred by amateurs. Today such contributions are rare. Yet there are not very many professional vegetable breeders, and their work is focused largely on just a few of the economically most important vegetables. In addition, professionals usually must concentrate on varieties intended for commercial producers. Commercial vegetable production is what feeds most people. We gardeners are lucky the professionals develop anything for us at all. Fortunately, some commercial varieties are excellent home garden varieties as well. Also, many professionals deliberately spin off garden varieties, whenever possible, from their commercial breeding programs. And sometimes they spend part of their time working on varieties just for us, whether it is economically justifiable or not. Most are gardeners themselves, or at least started out that way.

The heyday of the professional breeding of garden vegetables is over. It lasted from about 1900 to about 1950. Before then, there was no genetics, so there was no professional plant breeding, in the sense in which we use the phrase today. During the first half of the century, as the basics of genetics came to be understood, commercial vegetable production was still largely just a scaled-up version of garden production. Most vegetables were sold near the farms where they were grown. Commercial varieties often were similar or identical to home garden ones. This was a golden age for the development of garden varieties. Professional breeders, with their newfound knowledge of genetics, created much of the best we gardeners have today. By 1950, however, most commercial vegetables were mass-produced and such production was machine oriented. The needs of commercial growers and home gardeners diverged. Since then the contributions of professional breeders to gardening have dwindled.

Unfortunately, amateur vegetable breeding also has dwindled. It became almost a lost art during the decades when the professionals were so active. Now the professionals have largely left the field, and there is a vacuum. It's time for amateurs to move back in and fill that gap.

Breeding plants is something every gardener did once. It's something every gardener should do again. Developing your own vegetable varieties can add a new dimension to your experience as a gardener and to your satisfaction.

TWO

How Much Space Do You Need?
How Much Time?

GLENN DROWNS did his first watermelon cross in 1977. It took him only a few minutes to guess how to cross watermelons and to do it, and that's all he did that year. In 1978 all he did was to plant a few seeds from his cross instead of a few more seeds of the standard varieties he was trying. That year Glenn got his first ripe melon — from one of his own F₁ hybrid plants. *It took Glenn only a few minutes of work, no extra space, and only one year to produce breeding material that performed better for him than anything that was commercially available.*

In subsequent years, Glenn inbred his variety. Hand-pollinating the flowers required several hours per year. The breeding project still didn't require any extra space, though, since he wanted to grow melons anyway and could both eat the melons and keep the seed. It was a number of years before 'Blacktail Mountain' was ready for commercial release, but almost from the beginning it was doing what Glenn wanted most — giving him tasty, fully ripe melons in a climate where one wasn't supposed to be able to grow watermelons.

It took Ewald Eliason just a couple of years to develop his 'Blossom' potato variety. He made the cross one year and raised plants from the seed the next. When fall came, there was 'Blossom'. All he had to do was keep tubers from it. Since tubers are actually part of the original plant, they are genetically identical to it. For a vegetatively propagated plant, as soon as you find an individual you like, you have an instant new variety.

The space cost of Ewald's breeding could

be considered zero, since he grows and eats a lot of potatoes anyway. He just grows some plants from his own crosses instead of more plants from an established variety. The experiments, even when they don't lead to something new and special, are still edible.

When people say it took them ten years to develop a new variety, they are usually including several years after the actual breeding was complete, during which time the new variety was evaluated and increased for commercial release. Ewald's 'Blossom' took just two years to breed, but it was several years more before it was ready for release. Ewald grew the potato for a number of years before he felt confident that he had something new and special. Then Seeds Blüm needed to grow and evaluate it, as well as to increase the amount of seed material for a few years, before the company had enough seed to be able to list and offer it. Ewald was eating 'Blossom' potatoes right from the beginning, though.

It took ten years for professional plant breeder Calvin Lamborn, of Rogers Brothers Seed Company, to develop the 'Sugar Snap' pea. He actually had snap peas, though, in the F_2 generation — in other words, two years after he did the original cross.

Calvin made that first cross — a cross between a mutant with thick pods and the snow pea 'Mammoth Melting Sugar' — in 1969. He grew the seed from the cross, the F_1 seed, in 1970. In 1971 he planted the F_2 seed and examined and evaluated the plants. A number of plants had thick flesh and round pods; the thick flesh always went together with round pods. He saved the seed from individual round-podded plants. In 1972 he grew out the seed and found some plants that had what have come to be known as snap peas.

Calvin increased the seed material he had in the next few years and began sending samples to seed companies. They weren't interested. No one knew what to do with a round, edible-podded pea. In 1977 'Sugar Snap' won a gold medal in the All-American Selection competition. Suddenly everybody was interested. There was not yet enough seed to introduce 'Sugar Snap', so 1978 was spent increasing the seed — that is, planting most of it and saving all the resulting seed without additional genetic work.

A seed company will not usually list a new variety until there is enough seed so that it can fill the orders that are likely to result. This may require 10 pounds of a new pea or bean for a small company, or hundreds or thousands of pounds for a larger one. All-American Selection winners are not released for introduction until there is enough seed to satisfy every company that wants it. Sometimes a number of generations of seed increase is required before enough seed is available.

'Sugar Snap' was introduced in 1979. The rest is history. Ten years. Ten years of Calvin Lamborn's life? No, not at all. Developing 'Sugar Snap' was only one of many breeding projects he had, and, from the viewpoint of an amateur, the project had largely succeeded by year 3 or 4, because an amateur would be eating the new pea that year while continuing to refine it and increase the seed. The amateur would probably then provide a modest seed sample to a seed company or offer seed through a seed savers exchange.

Others would take care of increasing the seed for commercial release.

Developing your own varieties may not require nearly as many years as is generally stated, when looked at from the perspective of the amateur. It does, however, require time. Ewald Eliason, for example, needs time for making his potato crosses. Making the crosses is easy, but he grows about fifty plants from the seed of each cross, and it takes more time to grow potato plants from true seed and transplant than to plant tuber pieces directly in the field. In addition, while each true seed leads to an instant new variety, most varieties aren't anything special. It takes time to compile records on each plant, evaluate all the potatoes, and decide which ones are worth increasing and maintaining. Ewald doesn't do just one cross either; he does a lot. But, then, he can afford to. He's retired. He has the time. And he loves his potato hobby.

You don't have to have a whole field of potatoes to breed them, however. You can choose the scale that you want to operate at. You don't have to make dozens of crosses. You can make just one. You don't have to grow fifty plants from each cross. You can grow five or ten — or one. With smaller numbers, you might not be as likely to find something worth keeping, but you never can tell. Glenn Drowns made only one watermelon cross and planted just a few seeds from it — and look what happened!

Some plants are much smaller than potatoes, and you can raise and evaluate hundreds in a modest garden bed. Others are big and space greedy. Even if your garden space is modest, you need not limit your

breeding projects to just the smaller plants. There are tricks you can use when you need to grow large numbers of big plants.

Here is a procedure that Tim Peters recommends for tomatoes: Plant about fifty seeds of your first generation in a half-gallon pot of soil. When they come up, select for early emergence and seedling vigor, and eliminate all but about ten. As the plants grow, remove all suckers and branches, and limit each plant to just one vine. In addition, you may need to prune the leaves — that is, remove all the older, bigger leaves to keep the plants from shading each other out. When the plants start flowering and have set a few fruits, remove the tops. Then continue to remove all suckers and branches. Pay careful attention to soil fertility. Use good soil initially and fertilize as necessary.

The plants will be stunted compared to field-grown ones, flowering will be late, and the fruit will be small. But they ripen. You can determine fruit colors and shapes. You can tell whether the plants are determinate (shorter bush types) or indeterminate (continual-bearing vine types). You can tell earliness relative to other plants grown similarly. And you can select for disease resistance. For the latter, just shred some infected plant material on top of the soil in the pot.

During the second year of your breeding project, plant fifty seeds from each of your original ten plants in separate half-gallon pots. Thin, grow, prune, and evaluate as before. This time, though, save the seed from just one plant per pot, because you're going to use only ten pots per year. In subsequent years, choose just one plant per pot for inbreeding. In this way you can inbreed ten different lines out of a cross with just 5 gallons of soil.

You can use this pot method with peppers also, Tim says, and with some of the brassicas. It doesn't work with corn, though, or with the vine crops. They need more space to ripen a fruit at all.

Tim has other tricks for melons, squash, and cucumbers. For example, plant the seeds in a bed, leaving just 6 inches between them in all directions. Then proceed as for tomatoes. Limit the plants to one vine, remove all older, bigger leaves to prevent overshading, and eliminate the growing tip after the first fruits start to form. You'll be able to select for fruit colors and shapes, earliness, disease resistance, and many other characteristics — but not, of course, for fruit size or yield.

How many years a project takes and how much space and time it requires depends on you and how much you want to do, but it also depends on the plant you're working with and on the project. Some projects will be impossible no matter how much space and time you have. Some will be possible but might require more space (or time) than you're willing to put into them. Others might be relatively fast and easy.

If you are interested in producing your own F_1 hybrids, for example, you can make crosses between different heirloom or open-pollinated varieties your first year and trial the hybrids the second. If any of the plants is better than anything currently available, your project is a success as of year 2. From then on you just maintain a few plants of the parent variety so you can continue to produce

the hybrid. Then most of your planting will be of your own hybrid.

In most (but not all) cases, when you cross two different edible cultivars of a vegetable, all the offspring will produce edible vegetables. The fact that you can eat your experiments cuts down on the amount of extra space required. But not all experimental material is edible. If you cross an edible cultivar with a wild relative, for example, the first few generations of plants may not be edible. In such cases space costs can mount rapidly. Even when both parents are edible, most of the offspring may be so inferior that you need to plant a larger space to get the same amount of food. Developing a useful new variety may be easy, but there's no guarantee that it will be.

Even when you are working with thoroughly domesticated varieties, you can't always eat and breed with the same plants. Some of my own breeding work involves peas, and you cannot, alas, eat a pea and save it too. In addition, peas must be picked to be kept in production. So when you leave pods on the plant to let seed develop, the entire plant goes out of production.

Worse yet, I have found that if I try to eat and breed with the same plants, I always end up eating many of my crosses by accident. The little tags that are best for marking pea crosses blow around in the wind and get lost in the foliage. They're obvious enough when you're collecting all the dry pods, but on the live plant they are easy to miss.

So with peas, at least, breeding definitely requires some additional space above and beyond what you would plant just for the kitchen. Fortunately, peas are small plants, and they like to be crowded. I did the first three years of three different pea breeding projects with about 2 feet of row the first year, about 6 feet the second year, and about 20 feet the third. That was actually for five or six projects initially, which got trimmed to three by the third year.

Even major projects involving big, space-greedy plants may not require much space or time for the first year or so. After the second or third year, you often have an idea of whether the project is going to work. If it looks good but is going to take more space or time than you have, you can often borrow land or enlist help for the critical year or two.

Tim Peters didn't have separate, isolated fields to grow his broccoli plants. But he could get the same effect by hauling his potted plants around on his bicycle and recruiting the assistance of his neighbors. Glenn Drowns has often used fields of abandoned farms for isolation fields; he receives permission to use the land gratis. Alan Kapuler, whose Peace Seeds is a private plant introduction station and germplasm collection, evaluates and introduces rare plants, grows and sells hundreds of heirloom varieties, and conducts dozens of his own breeding projects. Until 1990 he did it all on borrowed land.

Until recently I had only a few hundred square feet to grow and breed vegetables. Now I have access to more land than I can use — both garden plots and entire farm fields — made available to me at no charge by gardeners and farmers in Corvallis (Ore-gon, where I live) and outlying areas. Developing new varieties of garden vegetables fascinates most people, and it's a public service. Many people will help if you explain what you're doing and ask for their help.

Some projects do require huge amounts of space and time. A professional breeder may need to consider a dozen characteristics of a plant at once and to grow thousands or even hundreds of thousands of plants to have much chance of finding even one that meets all the requirements. Often the professional is working on some major, important commercial crop plant that has been worked on extensively by hundreds of other professionals who were breeding toward similar goals. In such cases, to make even a slight improvement in an established variety might be a major undertaking.

We amateurs will often be working with plants that have received little or no attention before, so we will often find we can make major improvements quickly and with just a little work. Even when we work with the common vegetables, we usually have goals that are different from those of the professionals. Potatoes are one of the most worked on vegetables. But Ewald Eliason was interested in potatoes with colored flesh that tasted especially good and did well under organic gardening conditions.

Commercial cultivars of major crops are difficult to breed in part because it is usually not profitable to produce the seed unless it can be grown and distributed in huge amounts. This isn't possible unless the variety has very broad regional adaptability. That is, a good commercial variety needs to perform well under a wide range of climates and growing conditions. Amateurs, though, need not be concerned about broad regional adaptability. They care most about truly excellent performance in their own specific regions. If a variety also has broad regional adaptability, that's fine, but it's not essential.

An amateur creation does not need to be of agribusiness significance to be valuable. If an excellent variety is suitable for only a limited region or for specialized production methods or uses, it will probably never be carried by the largest seed companies. But amateurs don't care about that. They can maintain the seed themselves. They can proudly pass it around to all their friends. They can offer it through seed saver exchanges. In addition, many alternative, family-owned, or regional seed companies are eager to introduce unique, regionally adapted varieties.

A final reason why we amateurs can produce useful new varieties so much more easily than one might expect, given the heroic efforts sometimes required of professionals, is that we usually aren't as fussy about what we're after. We have the luxury of being versatile, of being, even, fickle. We don't *have* to produce exactly what we intended to produce. If need be, we can change tracks.

In many cases all we really want is something different, something new and interesting. And it's easy to get something interesting.

THREE

Roles and Goals for Amateurs; Wish Lists and Wild Ideas

WHEN YOU grow vegetables that have been bred by others, you are limited by the priorities and the vision of others. When you do the breeding, it is *your* vision that matters, and *your* priorities.

The quest for flavor is a big part of what home gardening is all about. Flavor is often low on professional vegetable breeders' list of priorities, but for the amateur, flavor is usually somewhere near the top of the list. You can do more than just breed for good flavor; you can breed for exactly the flavor that *you* prefer, flavor that develops under *your* growing conditions and climate, flavor relevant to the exact way in which *you* use the vegetable.

Flavor is a complicated characteristic that is made all the more complicated because evaluating it is so subjective. My idea of perfect tomato flavor, for example, is a vine-ripened 'Early Girl Hybrid' tomato. I like tomatoes that taste both sweet and tart and that have a lot of flavor and aroma in every bite. Many people prefer mild-flavored tomatoes such as the beefsteak types. When two different people talk about good flavor, they are often talking about different things.

Many professional breeders work on tomatoes, but they seem to be concentrating on producing what are called low-acid types. Low-acid tomatoes may or may not have less acid than normal tomatoes, but they taste as if they do. They're what I consider mild flavored. When professional breeders do breed for flavor, they breed for what tastes good to them or to some average consumer. I am not that average consumer, and what I want is tomatoes that taste good to me.

Flavor is profoundly affected by growing conditions and climate. Part of what makes 'Early Girl Hybrid' so popular in maritime Oregon is that it is able to develop its full sweet-tart flavor even with cool spring and summer days and downright cold spring and summer nights. Many tomatoes that are considered especially flavorful elsewhere either don't ripen at all or ripen but are tasteless. There are many professional tomato breeders, but only a few operate in areas where temperatures fall into the forties (°F) at night all the way through the summer. Whenever you breed for flavor, you are also automatically selecting for regional adaptation.

Professional breeders also usually breed for tomatoes with a lot of flesh and little juice. Such tomatoes ship much better than juicy tomatoes. They are also better for using as slices on a hamburger or in sandwiches; they don't make the bread soggy.

However, the acids and many other flavor components of tomatoes are mostly in the juice. I have yet to find any meaty tomato whose flavor I like as well as that of 'Early Girl'. And I don't care that a slice of an 'Early Girl' can completely liquefy a sandwich, because I don't eat many tomatoes in sandwiches. I eat them like the fruit they are, whole and out of hand. The juicy 'Early Girl' often drips or squirts juice on me, however, and it would be better if the tomato were smaller so I could get my mouth around it better.

What I want is a 'Baby Girl' tomato. It would have the juiciness, tenderness, and full sweet-tart flavor, as well as the earliness and cold resistance, of 'Early Girl'. But it would be smaller — perhaps even of cherry tomato size.

So far every cherry tomato I've tasted has been meaty (so they travel well) and either bland or sweet but lacking in tartness. Cherry tomatoes are convenient to eat, but they just don't taste as good as 'Early Girl'. My 'Baby Girl' would be commercially useless, of course, because it wouldn't travel well. But it wouldn't have to travel any farther than the distance from the vine to my mouth.

Like every gardener who has grown any specific crop for a while, I have my own individual priorities and preferences — my "wish list" of the characteristics of the perfect variety. Your ideas and your wish list are no doubt different from mine. Tim Peters, when he bred from 'Early Girl Hybrid', was selecting for larger, not smaller, size and for meaty tomatoes, not juicy ones.

There are a lot of professional tomato breeders in the world, but I can't count on any of them to want exactly what I do in the perfect tomato. In other words, if I want my 'Baby Girl', I'd better breed it myself.

I have saved the seed from 'Early Girl Hybrid', and each year I will plant a few seeds from that material in addition to the standard 'Early Girl'. I also have raided Tim Peters's trash can. I want an 'Early Girl' derivative that is almost the opposite of what he wanted. What did he do with all the material he eliminated during his breeding project? I asked him.

He had, as I suspected, saved it all, and he was quite happy to send me seed of lines from his inbreeding that he had discarded because the tomatoes were too small. Next spring I will grow some plants from that seed in place of my normal varieties and screen them for flavor. If one of them is suitable, it will be a number of generations closer to being a stable line than material straight out of 'Early Girl'. I think it's a good idea to save all the seed produced in a breeding project. Even if it doesn't meet your needs, it might, as in this case, give a head start to someone else who has different breeding goals.

I will probably also try some crosses of 'Early Girl' with varieties that produce smaller tomatoes. None of this will take any extra space, since I usually raise a number of varieties anyway and can eat all the tomatoes except the few involved in the crosses.

Flavor is dependent on many genes and how they interact with each other and with the environment. It is what we call a quantitative genetic trait — that is, a trait affected by many genes instead of just one or a few. Quantitative traits are not especially amenable to any of the new biochemical breeding approaches of genetic engineering. They are very amenable to traditional plant breeding methods — the methods that are fully accessible to the average gardener or farmer.

Other quantitative traits include size, yield, shapes and proportions of various parts of the plant, tenderness, earliness, lateness, cold hardiness, heat tolerance, regional adaptation, yield — in fact, most of the essential agriculturally important characteristics. Specific methods for breeding for quantitative as well as single-gene traits are covered in later chapters.

Selecting an established vegetable for other than the established use is one of the major ways new kinds of vegetables are developed. Professional breeders focus their attention on the most common uses of vegetables. When you do the breeding, you can focus on your uses.

Consider the lowly cabbage. It started as a plant that was grown for its edible loose leaves. Somewhere along the line, various people — amateurs all — decided they cared more about the central buds, the flower bud, the side buds, or the central stalk, and they bred for plants that excelled at producing those. So we have head cabbage, cauliflower, broccoli, Brussels sprouts, marrowstem kale, and kohlrabi, as well as collards and kale.

If you eat a bean as a snap bean, perhaps other breeders will satisfy your needs. Many professional breeders work on snap beans. But what if you prefer to use beans as shellies (shelled green seed) instead of eating them earlier as pods? I doubt that any professional breeder in the world is trying to develop better shelly beans.

And what if you prefer pole beans to bush types? American bean breeders have been concentrating on bush beans suitable for harvest by machine. For machine harvest, the pods should ripen simultaneously; should be held, if possible, above the leaves; and should be moderately tough — "firm" it's called. The homegrown bean can be as tender as possible. The commercial bean needs to have significant amounts of fiber in it so that it resists damage during machine harvest and shipping.

Flavor matters to the professionals, too, but it is usually not the top priority. If a bean bruises and then rots before it gets to the consumer, it doesn't matter how great it would have tasted had someone eaten it.

You might eat a vegetable raw, whereas others cook it, or vice versa. Cucumbers are a popular vegetable, and many American professionals work with them. But how many of them recognize the cuke as an excellent *cooked* vegetable and are breeding for better cooking cukes? You may eat a vegetable very young, whereas others eat it at a different stage. Or perhaps you like the stalk of the broccoli better than the top and would prefer more stalk and less top. You might like stuffed squash flowers more than squash and want a variety that produces a lot of huge flowers and doesn't waste its energy setting many fruit.

I grew scorzonera (*Scorzonera hispanica*) for the first time a few years ago. Scorzonera, or black salsify, is a root crop. It's a perennial that is grown as an annual. I had only a few plants at the end of the season — not enough to bother harvesting them and figuring out how to use them — so I just left them in the garden.

The leaves of the scorzonera plants were long and spearhead shaped. They lasted past the mild freezes and regrew quickly after the hard ones. That's when I noticed them and tasted them — and found that they have an excellent, mild flavor. They taste so much like lettuce that I think if you told someone they were eating lettuce, they would believe it readily.

When spring came, the scorzonera plants produced copious quantities of large, mild-flavored leaves long before the lettuce could even be planted. I stopped buying lettuce and filled my salads with "lettuce-salsify." Later in the summer the plants bolted and produced only small leaves, but by then lettuce was available.

Naturally, I am working further with this lovely perennial. I'll start by letting the original plants get bigger and seeing how they behave. The real characteristics and virtues of a perennial vegetable often don't appear until the plants are three years old or so. I'll also obtain and screen additional varieties for the size, quality, and cold hardiness of their leaves. It will probably be possible to lift the roots in fall and force them inside as well. And who knows? One of these days I might even get around to eating the roots.

Scorzonera is a minor crop. Professional plant breeders are doing little or no breeding work with it. I know of only one group working on it (in Europe), and they have been working with it as a root crop. In addition, part of what I would like is a plant that holds its leaves throughout the winter — a characteristic that is probably possible or advantageous only in mild-winter areas. If I want a better perennial lettuce-salsify, a salsify ideally adapted to my climate and for my purposes, I'd better get busy and develop it myself.

I may not need to breed a better lettuce-salsify, though. Perhaps all I need to do is explore and better understand the varieties that are already available. Developing new vegetable varieties includes not just plant breeding but also plant exploration, plant introduction, variety trials, kitchen trials, and experimental cooking. They all go hand in hand. They are all fun. The adventuresome gardener should be ready to do any or all of them — whatever a specific project requires.

Breeding for better regional adaptation of plants, including extending the boundaries of the areas in which specific vegetables can be grown, is of great interest to many amateurs and is very amenable to the breeding methods available to gardeners and farmers. Glenn Drowns wanted a watermelon that would grow in the mountains of Idaho, not a watermelon that would do better in the South, where most of the watermelon breeders are located.

Ken Allan, who lives in Ontario, Canada, is inordinately fond of sweet potatoes. The sweet potato is an important world vegetable, and many professionals work with it — but not in Canada. Yet sweet potatoes grow well for Ken. Some varieties have shorter seasons, but other, later types grow better when it's cold. So Ken is crossing short-season types with cold-growing types and plans to select for a short-season type that also grows well in the cold. In other words, he is breeding for a Canadian-adapted sweet potato. (For a description of his work, see The Vegetable Gardening Research Exchange, which is listed in Appendix D.)

Professional breeders for any specific vegetable tend to be concentrated in the areas where the vegetable is most important as a commercial crop — that is, where it already

does well. Professionals also are concentrated in temperate regions and work on vegetables that are important in those regions. Many vegetables that thrive in the tropics receive little attention. Yet vegetables are perhaps more important in tropical than in temperate zones. In many areas of the tropics, starchy vegetables, not grains, are the most important source of calories, the difference between starving and thriving for millions. There is a lot of room for amateurs to do important work with plants such as cassavas, sweet potatoes, and yams.

Other crops may be less important on a world basis but crucial in certain regions. Plantain (starchy) bananas are a major source of calories in some lands. Achira (*Canna edulis*), arracacha (*Arracacia xanthorrhiza*), oca (*Oxalis tuberosa*), ulluco (*Ullucus tuberosa*), and mashua (*Tropaeolum tuberosum*) are important root crops in certain areas of South America. (See the discussion of the book *Lost Crops of the Incas* later in this chapter.) Amateurs who live where there are special local crops are in an excellent position to make important contributions to their regional food base.

You don't have to live in the tropics to make an important contribution to your regional food base. Professional plant breeders are severely limited in the amount of time and effort they are able to devote to developing regionally adapted crops or cultivars. Large seed companies can usually afford to carry only varieties that sell in large amounts; this usually means cultivars that are widely adapted. Such cultivars often don't do quite as well anywhere as the varieties that are fine-tuned to specific areas.

Every region is unique and offers unique opportunities to the imaginative amateur.

Fava beans, for example, thrive in the maritime Northwest. Favas, also known as faba beans or broad beans, are *Vicia faba;* common green beans are *Phaseolus vulgaris*. Favas love mild, wet winters and cool summers. Only recently have people started to realize that favas could be a major crop here. The greens, the green pods, the green shelled beans, and the dried beans are all edible. The dried beans can be used in soup or turned into tofu or miso, just like soybeans, which don't grow well here. In fact, fava beans could become the soybean of the Northwest.

All the work on favas for the maritime Northwest has been done by amateur plant explorers and breeders and by people associated with the small, alternative seed companies. Alan Kapuler, of Peace Seeds, started obtaining and trialing favas about fifteen years ago. He currently offers about a dozen different varieties that do well in the Northwest.

Steve Solomon, founder of Territorial Seed Company, trialed large-seeded favas and introduced 'Aquadulce Claudia', a large-seeded, good-flavored bean that can overwinter here. The Territorial line is derived from two plants in a field of 'Aquadulce Claudia' that survived one winter when everything else died. More recently, Steve has been working with green-manure and small-seeded types. He has developed a good-flavored small type, he says, which will soon be introduced under the name 'Sweet Lorane'.

Dan Borman (Meristem Plants), who lives in maritime Washington, is especially interested in favas that will overwinter — that can be planted in fall and used for greens in the winter and spring as well as for green and dried seed later. He lives on Orcas Island, Washington, which is actually north of the Canadian border. He is screening varieties for overwintering ability and for flavorful greens, and he has worked out methods for making miso from favas as well. I've tried the miso, and it's excellent. Fava beans also can be roasted to make a rich, full-flavored, caffeine-free coffeelike beverage.

Farmers in the Skagit Valley of Washington and in southern Oregon have developed methods for large-scale planting and harvesting of favas, alone or in combination with grain crops. Seed size is so different that grain and fava bean seed can be harvested (combined) together and sorted mechanically.

My favorite fava is 'Aprovecho Select', which was bred by Ianto Evans, an architect by training. He grew up in Wales, where favas are a common vegetable. When Ianto arrived in Oregon, he looked around and asked, "Where are the fava beans?" The climate in western Oregon is quite similar to what he was used to. Fava beans ought to do well here, he thought.

He started collecting favas — from Britain and from wherever his various travels took him. He has collected many varieties from Latin America that were brought there by Europeans and grown in relative isolation. They may no longer exist in their original homelands.

Ianto bred 'Aprovecho Select' by selecting from a large-seeded European type he bought from a local feed store. He selected for large bean size, four or more seeds per pod, and good yield in the Northwest. He did not particularly select for flavor. The superb, full-bodied, rich flavor of the green shelled 'Aprovecho Select' is just one of those happy accidents.

Ianto put together an organization of people, the Fava Project, which publishes a newsletter on favas and fava research and runs an annual workshop for people interested in growing, researching, developing, processing, commercializing, or just eating favas. He also grows and sells about forty varieties. He and other members of the Fava Project continue to collect new germplasm, develop new varieties, and spread beans and information around to others. Those interested in the newsletter or workshop or in buying favas can contact Ianto at the address given in Appendix D.

Peace Seeds also carries an extensive line of favas, including 'Aprovecho Large', which Alan Kapuler developed from Ianto's variety by further selection.

You may grow a vegetable that is established in your area, but perhaps you grow it in a different way than it is normally grown. If so, you may want to select for a variety that does well under *your* growing methods. Or you may end up selecting such a variety accidentally.

I started breeding pole peas about four years ago. Pole peas are indeterminate varieties; bush peas are determinate, or, at least, more determinate. Pole varieties such as 'Sugar Snap' grow to about 6 feet high in most areas of the country. Bush types such as 'Oregon Sugar Pod II' grow to 2 to 3 feet.

Here in maritime Oregon, the peas all seem to grow about twice as tall as they do elsewhere. 'Sugar Snap' never quits at 6 feet; it reaches 10 feet readily if the trellis is tall enough. 'Oregon Sugar Pod II' is a handy 4 feet or so.

All I wanted, initially, were disease-resistant pole snap peas. 'Sugar Snap', the classic snap pea, is a wonderful cultivar. It's a tall, vigorous pole type with two pods per node, and the peas are large and of excellent flavor. But 'Sugar Snap' is susceptible to both pea enation mosaic virus and powdery mildew. Powdery mildew is common in most areas of the country. Pea enation mosaic virus is most common in western Oregon and Washington and is of little relevance elsewhere. Both diseases are so common here that one or the other always cuts production short in early summer. A variety resistant to both diseases might be able to produce all the way through our cool summers and, possibly, could be planted in summer for fall production. In other words, disease-resistant varieties might double or triple the length of our snap pea season.

One of the major accomplishments of modern professional pea breeders has been to incorporate genetic resistance to most of the major diseases into many of the most important commercial varieties. The commercial varieties, however, are bush varieties. I prefer pole types. They yield so much more per planted area. They produce bigger pods, and more of them, week after week after week — not a handful of dinky little pods all at once. In addition, the pods of pole types are clean and intact and can be used without washing. Bush types produce pods that, under my conditions, are muddy and slug-eaten and that often require considerable culling, washing, and trimming.

Professional breeders do little work with pole peas. Jim Baggett, at Oregon State University, is one of the foremost pea breeders in the world, and he has developed shelling and snow peas resistant to enation and powdery mildew. He is also working with snap peas. But he isn't doing anything with pole types. If I want to be able to grow the pea of my dreams, I'll have to develop it myself. And so I am.

I started by crossing 'Sugar Snap' with 'Oregon Sugar Pod II'. 'Oregon Sugar Pod II' is a dwarf snow pea bred by Jim Baggett that is resistant to verticillium wilt, enation, and powdery mildew. It bears two large pods per node, has excellent flavor, and can be planted so as to be harvested throughout the summer and into the fall.

Things went relatively smoothly. I dug back into the professional literature to find out how to cross peas. The project took only a few inches of row the first year and a foot or two of row the second year (to grow the F_1 plants).

The plants of the first generation after the cross (F_1) were all tall plants and had snow-type peas. This was what I had expected. The genes responsible for height and basic pod type are well known, so I was better able to make predictions for peas than when working with many other vegetables. (The genetics of pea height and pod type are covered in the section on peas in appendix A.) In the second generation after the cross (F_2), some plants were short but most were tall; some plants had snap peas but most had

snow peas. This also was as expected. Some plants had only one pod per node, however, instead of two. And some plants were earlier than others. All the peas, snow or snap, had excellent flavor.

Seven of the plants were the kind I was looking for — tall, vigorous pole-type plants with snap-type pods. I saved the seed from each of them so as to start seven lines. I am in the process of increasing the seed from each of the lines and evaluating and selecting for disease resistance.

The project as described so far went as expected. But an extra dimension developed because of the way I raise my peas.

The normal way to raise peas here is to plant them in early spring. For some years, though, I have been experimenting with fall planting, trialing various varieties and establishing planting dates and methods. I regularly plant 'Oregon Sugar Pod II' and 'Sugar Snap' in the fall. Both varieties overwinter well here.

To overwinter garden peas, I plant them in mid-November, the same time 'Austrian' field peas are planted here. It is important to plant them late enough so that by the time the peas emerge, the first early frosts have killed all the aphids. Aphids are sucking insects that don't seem to hurt the plants much themselves but that transmit enation and many other viral diseases. The plants freeze solid, with or without snow cover, off and on during the winter. Whenever it is above freezing, the plants grow. The over-wintered plants start flowering and yielding peas much earlier in the spring than spring-planted peas.

Naturally, I did my breeding work with

fall-planted, overwintered material. I had started the cross with parents that were especially winter hardy, and as I selected, the less winter-hardy peas were automatically eliminated. By the third year of the project, I already had material that overwintered better than either of the parents — peas that grow happily whenever it is above freezing, that can hold up 6 inches of snow for two weeks without the stems breaking, that can even withstand freezing rain, and that can be completely covered with ice without being damaged.

Freeze tolerance and cold tolerance are complex characteristics whose genetics are not understood. I did know, however, that varieties of vegetables that are especially cold tolerant are often especially heat tolerant as well, and vice versa. Often, apparently, it isn't cold tolerance from the plant's point of view; it's tolerance toward temperature extremes, perhaps even general tolerance toward environmental stress. So I hoped that my extra-cold-tolerant peas might be extra-heat-tolerant as well — which would make them all that much more able to last into summer and fall.

I made a midsummer planting of my pea lines in a field in Independence, Oregon, to see how they would do and to select specifically for summer performance. Some peas in my lines grew vigorously and yielded peas all summer. I ate a few of the snap peas, just to have done it, to have eaten Oregon-grown snap peas in the summer. I saved the seed from the best plants in late fall and then forgot them for a while. I expected them to get knocked out by the first freeze, which was imminent.

But they weren't. On November 9, after several freezes had occurred, I rediscovered the peas. They were still there, and many of the plants were still flowering and bearing peas. The peas were a little discolored, but they were still delicious.

These plants are apparently able to take freezing weather at all stages of the life cycle. That is something I did not, in my wildest dreams, ever imagine or expect. All I was trying to do was to breed a disease-resistant pole snap pea — a relatively straightforward genetic exercise. But I was selecting under my growing conditions, which are unique, and doing so produced a unique result.

The winter of 1990–91 was the worst winter for plants that we've had in the twelve years I've lived in Oregon. We had long periods of below-freezing weather, with temperatures frequently down into single digits (°F). Worse yet, there was at best only an inch or so of snow cover, and often none at all. Many plants that I never lost before were decimated. Most of my chickpeas, however, did fine — and they were growing in a wind-swept farm field, not in a pampered garden. They are, as far as I know, the first chickpeas to be overwintered in the Pacific Northwest.

Chickpea is the English word for the bean and plant of *Cicer arietinum*. Many Americans recognize them more readily by the Spanish name *garbanzo* or *garbanzo bean*. In India and Pakistan and throughout the Middle East, chickpeas are a major grain legume. They come in all sizes, shapes, and colors. In the United States, only the large cream-colored ones are used, and mostly only in salads. It's not obvious why.

Although the major uses are of the dry grain, the young green seeds and pods also can be used as a vegetable, as can the growing tips of the branches. Chickpeas also produce an acidic exudate that coats the plants and stems. It can be collected on cloths under the plants or washed off of them and used to make a drink similar to lemonade. The plants are a foot or two high, 1 to 3 feet across, and of various shapes depending on the variety. They have lacy foliage that is quite attractive.

Chickpeas fit especially well into sustainable, low-input agricultural systems. They produce a tasty, high-protein seed. The seed doesn't shatter; it stays in the pods until threshing. Threshing is easily achieved on a large scale with standard combines or on a small scale with the methods used for dry beans.

Chickpeas, being nitrogen-fixing legumes, don't require high-nitrogen fertilizers. In fact, they will not yield as well if the soil is too high in nitrogen. In addition, chickpeas are unusually well endowed with natural defenses against pests. The leaves and stems are fuzzy and covered with the sticky, acidic exudate. Insects and slugs seem to find the situation unappetizing. Friends tell me that deer and rabbits don't like the plants either.

I got interested in chickpeas by accident. It all started when I read *Lost Crops of the Incas: Little-Known Plants of the Andes with Promise for Worldwide Cultivation,* a book published in 1989 by the National Research Council of the National Academy of Sciences. One of the crops described was nuñas, which are a popping bean. They are the same species as ordinary beans (*Phaseolus vulgaris*),

but they are varieties that pop. Unfortunately, they are also short-day plants — that is, they don't flower until day length drops below a certain level. This means that in the United States, they flower, if at all, only in late fall. In addition, they also need cool, moderate weather to flower. The only place in the United States that one would expect to be able to grow them and get them to flower outdoors would be the maritime Northwest.

The nuñas would probably flower here, I thought, but not until about November. Sometimes we don't have freezing weather until December. The nuñas might make it in a mild year, at least well enough to raise a few plants and do some crosses. Perhaps I could transfer the genes for a temperate-adapted flowering pattern into the nuñas (using recurrent backcrossing, a method I'll discuss later).

So I called the Western Regional Plant Introduction Station at Pullman, Washington, which handles the U.S. Department of Agriculture (USDA) *Phaseolus vulgaris* germplasm collection, and found myself talking to Rich Hannan, the USDA horticulturist who is the curator for the beans. He also, as it turned out, handles the chickpeas.

Rich was willing to send me some nuñas, but he had had a lot of requests for them since that book came out; what *he* wanted to talk about was chickpeas. There are popping varieties of chickpeas, too, he said. A friend of his who had lived in Iran told him that in the Middle East, some types are popped in hot sand, just like nuñas. "Why is everybody interested in popping nuñas, which don't grow here," Rich asked, "and nobody is interested in popping chickpeas, which do?"

"Because I didn't know that," I responded cheerfully. "They weren't in the book. Which varieties pop?"

He didn't know. He had about four thousand accessions. An accession represents a collection event, an acquiring. If a plant explorer buys a pound of seed from a single basket in a market in Iran, for example, that seed and all its progeny are an accession. An accession may be the same as a variety, or it may not; it depends on whether the farmer had one variety in his field or a mix. (In the USDA-ARS National Plant Germplasm System, accessions have numbers that start with PI, for plant introduction. I discuss how to obtain germplasm from the USDA system later.)

Rich had a wealth of germplasm and information about where it was collected but no information at all on what it was good for. He had tried to get people interested in popping chickpeas, but to no avail. He talked about them with everyone who called him about the nuñas. He had himself tried to pop some chickpeas in hot sand, but he had chipped a tooth on them. Undaunted, he tried two more types and broke a second tooth. He decided he needed someone else to try the rest.

I didn't think I could handle four thousand accessions of chickpeas. We compromised. Rich sent me the nuñas I had asked for along with twenty accessions of chickpeas so I could test them and see if any of them popped.

Rich and I made some guesses about what kind of chickpeas might pop. That first batch were all #2s. (The smallest size is #1 — those that weigh less than 10 grams per 100

seeds. The #2s weigh 10 to 19 grams per 100 seeds. They're about the same size as popcorn. The largest size is #5 — greater than 40 grams per 100 seeds.) Most were black, but some were brown or cream.

I didn't happen to have any hot sand lying around. I also wasn't too enthusiastic about the idea of eating sandy chickpeas. So I did my popping the same way I pop popcorn — in a stainless steel skillet (which can be heated much hotter than cast iron without smoking) and a little safflower oil (which has a high smoking point). Peanut oil, sunflower oil, and canola oil also are fine.

All the chickpeas popped. I popped just five seeds of each variety. I heated a small amount of oil in the corner of the frying pan until it was just short of the smoking point. Then I added the chickpeas and stirred them with a fork until they all popped, usually within about five seconds. I put a little salt on them and ate them. They didn't expand very much, but they cooked completely and were tasty. They were delicately crunchy, like a potato chip, not crunchy and hard like corn nuts. There were obviously differences in flavor among the varieties, but I couldn't sort that out with such small samples.

I also tried to pop three larger commercial types of chickpeas I'd bought at my local food co-op. They popped, but they didn't cook. They were mostly hard and inedible and were a threat to my teeth and dental work. I did my evaluations with care.

I went back into my legume collection and started popping samples of everything. I found that just about any bean or pea will pop, but they don't cook, or don't cook completely. Adzuki beans are about the size of the #2 chickpeas, and they also pop, but they aren't made edible by popping. The small chickpeas really are different. As it turns out, there are large chickpeas that pop, too — chickpeas as big as nuñas, chickpeas big enough to make a fine candidate for a new, delicious, high-protein snack food. But I discovered them later.

James A. Duke's book *Handbook of Legumes of World Economic Importance* lists frying as one method of cooking chickpeas. I had initially assumed it was frying of cooked chickpeas, but now I guessed that it meant exactly what it said — stir-frying of little whole, dried chickpeas. A woman from India has since confirmed this for me. Dry, unsoaked small chickpeas are often stir-fried briefly to pop them, then cooked into various dishes. Of the small types I tested, some popped so quickly and uniformly that I suspect they have been selected specifically for stir-frying.

By this time, I had planted my twenty accessions of chickpeas. I knew they could be planted in early spring (February) here, even though we often get some of our hardest freezes in February. I figured that if chickpeas can be planted in February, it also might be possible to plant them in fall and overwinter them. If so, they would be of much more importance for the maritime Northwest.

Western Oregon gets all its water in the winter, so food plants that can be planted in the fall to grow during the winter are especially valuable. For a grain or seed crop, fall planting is especially valuable, because fall-planted plants can mature their seed in summer, which is the dry season. Spring-planted plants normally mature a crop in late summer at the earliest and are threatened by the fall rainy season, which starts as soon as August some years. Most fall-planted crops can mature without irrigation. If chickpeas will overwinter here, they might, like fava beans, be a major addition to the agricultural base of our area.

Chickpeas are traditionally planted in spring, though. Gardeners throughout the United States and much of Canada should be able to grow appropriate varieties.

I planted my first chickpeas in mid-November — just a short row (about eight seeds) of each of the twenty accessions. That was probably later than optimal. Sixteen accessions each had at least some plants that survived what was the worst winter in my experience. I think it should be possible to develop overwintering varieties quite easily by simple selection within available material.

Chickpeas were assumed to be unable to overwinter here. That they are drought resistant is well known, but they are also thought to be intolerant of wet conditions. If so, I think the intolerance must be associated with the mature stages. They don't seem to mind continual rain when young. Most of my accessions germinated well, even though they were planted as untreated seed in cold, wet mud, which actually froze solid at least once before they emerged. In areas with more severe winters, the frost-hardy types presumably could be planted as early in the spring as the soil could be worked.

After I found out that chickpeas would pop and that they could overwinter here, I became smitten with them. I began thinking about other aspects of chickpeas and how they might fit into the agriculture and diet of Oregon and America.

One way I find interesting projects is by bucking the trends. That is, I take note of what the professionals are doing and do the opposite. I work on pole peas when all the professional breeders work on bush types. If professionals are interested in bush-type squashes, I work with vine types. If everyone else is selecting under a spring planting regime, I select under fall planting. For thousands of years people have concentrated on developing annual vegetables. A mistake, I say. My major focus is on discovering, rediscovering, breeding, growing, and using perennial vegetables.

Most work with chickpeas in the United States has focused on large types, so I asked myself if I could think of any reason why little chickpeas might be more interesting than big ones. And, yes, I could. Little things often cook much faster than big things. Big chickpeas — the only kinds that are commercially available in the United States — must be soaked overnight, then cooked for two to three hours. That doesn't fit very well into the food preparation patterns of most Americans.

I knew that normal-size beans such as pintos also require soaking and cooking for hours. But little beans like adzukis can be cooked in about forty-five minutes without soaking. How small did chickpeas come? How fast would they cook using standard boiling? And what about a larger popping chickpea? I suspected there was one because Duke's book mentions that popping and eating out of hand is one of the ways chickpeas are used. The little chickpeas I had been working with popped, but they were too small to eat out of hand very easily.

I called Rich Hannan and reported on the stir-frying and overwintering, which was fun. He got rather excited about it. I asked for more chickpeas, and he promptly sent me about 120 more accessions — all the #1s and a good selection of every other size and shape, in various shades of cream, tan, brown, orange, black, green, and mottled and from India, Pakistan, Iran, Iraq, Israel, Turkey, Saudi Arabia, the former Soviet Union, Spain, Morocco, Ethiopia, and Mexico. I have been testing and growing them ever since.

All the smallest chickpeas I've tested — the #1s and #2s — pop readily and could probably be used for stir-fries. Many #3s or larger also pop, but not so quickly and reliably. I have found two large chickpeas, however, that pop very quickly and completely — within a second or two of hitting the hot oil. The first is a black and brown mottled #4. The other is an orangish brown pea type with a smooth skin. [They are, respectively, PI374085 and PI374090.]

Both popbeans have a wonderful flavor, and the flavors of the two are totally different from each other and from anything else, including boiled chickpeas. I mean really different — as different as asparagus and ice cream, for example. The black and brown mottled popbean has a wild flavor; wild rice or buckwheat groats come to mind, although the flavor is not like either of those. The orange popbean has a flavor like a roasted nut, though not any known one. Both are excellent.

I now have more than five hundred accessions of chickpeas, and I'm still evaluating them. I suspect there will be many additional popbeans among them.

My method of popping chickpeas, now that I have more practice, is to put a little oil (I'm using sunflower oil now) in a stainless steel frying pan and heat it to just below the smoking point. Then I dump in the beans, stir them with a spatula while they're cooking, and dump them out. They're actually just being stir-fried.

It's important to stir the popbeans during cooking and not to overcook them. Small, standard chickpeas may take several seconds (up to about five) to finish popping. Popping types, even large ones, pop and split open almost as soon as they hit the oil. If you cook them for more than two or three seconds, you'll probably burn them. Burned popped chickpeas taste awful. (Cream-colored types, for example, should still be cream after popping, not brown or black.)

Popping chickpeas expand just a little when they pop; they don't blow up like popcorn. (The same is true of nuñas.) The beans do split open, though, and they hop around a little in the pan while popping. Small ones occasionally leave and explore other parts of the room; big ones generally restrict themselves to jumping from one side of the pan to the other or hopping around in place.

I also did boiling trials. What I have found is that, so far, all the #1 and #2 chickpeas I have tried can be cooked without soaking. The #1s cook in about fifteen minutes without soaking. The #2s take about forty-five minutes. I think anything bigger than a #2 is best handled by soaking before cooking.

There are differences in both cooking and flavor within size classes, but defining those differences will require more work.

Meanwhile, I was also growing chickpeas. I spring-planted patches of about twenty more varieties in the field where my over-wintering chickpeas were growing. They germinated well, but then disaster struck. The recession hit the owner of the land my chickpeas were on, and he leased it to a neighbor. He assured me that the neighbor had no intention of using the field my chickpeas were in, but when I came back to harvest, there was nothing there but a field of corn. Forty varieties of chickpeas, the seed increase on all my own pea breeding work, and experimental plantings of thirty or so of my beloved perennial vegetables all had vanished. (I did have backup seed on everything. The germplasm, at least, was secure.)

I searched the field and found one small patch of nine chickpea plants in an area that the plow had missed. They were all nicely dried up and covered with pods, and they showed no signs of rot or damage, even though the patch had been irrigated with a sprinkler system. I grabbed the plants and took them home.

The variety that had been spared was a small cream-colored #1, PI359515. I threshed out the seed from each plant by putting the plant in a cloth bag and stomping on it. Since this type of seed was almost round, I separated it from the debris by dumping everything in a box and tilting it a few times in various directions so the seed rolled away from the junk.

The most productive plant had 834 seeds,

which weighed out to 128 grams (about 4½ ounces). The average for the nine plants was 55 grams (about 2 ounces). My plants were spaced somewhat erratically. They were originally planted about 4 inches apart in rows 2 feet apart. The rows can be closer; I used 2 feet for ease in evaluating the plants, not because that much space is necessary. Optimal spacing will vary with both conditions and variety. Big, bushy plants can grow to about 3 feet across, whereas other erect plants are only a foot or so across. Even the sprawling plants hold their pods well up off the ground.

My nine plants differed in size and shape, even though the seed all looked the same. The most productive plant was large and bushy. The second most productive plant was very erect and columnar. It might lend itself better to denser plantings and/or machine harvest. I saved the seed from each of those plants separately — that is, I started two lines in addition to the original unselected material based on single seed descent.

The seed I harvested was bigger than what I had planted. The accession was classified as a #1, and the seed in the package from the USDA weighed out as a #1. What I harvested was visibly bigger and weighed out as a #2. I am guessing that this has to do with the growing conditions.

For the first time, I had enough seed to boil up a real batch of chickpeas. I pooled the seed from the least productive plants and used part of that. The unsoaked seed took forty-five minutes to cook, typical for seed of that size. The flavor was just like that of large chickpeas but slightly more powerful — delicious. The chickpeas also popped readily and had a rich, nutty flavor.

I have been experimenting with the large, commercial cream-colored chickpeas, and I've found that because of their mild flavor, I can eat them as a staple, like rice or potatoes. They combine well with everything, even sweet things. I put them in soups, use them as the basis for curries and spaghetti sauce, combine them into rice dishes, grill cheddar cheese over them for a main course, add them to oatmeal in the morning, or mash them, drench them with maple syrup, and call them dessert.

As an experiment, I ate chickpeas every day (often more than once a day) for a month and found I didn't tire of them. But I definitely did tire of the presoaking and long cooking. I need the quick-chick types. I'm sure I'll make them a regular part of my diet once I have enough seed that I can eat it instead of saving it.

It's already obvious to me that different-colored chickpeas have different flavors. The cream-colored ones I have tried all have virtually the same flavor. I also have tried the three colored chickpeas (black, brown, and green) sold by Abundant Life. They taste very different from the cream ones — more like a wild grain and less like a nut. I boiled them after soaking, but I didn't especially like them just boiled. One friend tells me that he's grown the Abundant Life brown, and it makes especially good hummus. Another friend prefers the black chickpea 'Black Kabouli' (available from Peace Seeds and J. L. Hudson) to commercial cream ones. He says 'Black Kabouli' has a delicious "wild" flavor and that the creams are "tasteless." I tried 'Black Kabouli' just soaked and boiled, and I didn't especially like it. ('Black Kabouli' will pop, but not well enough or reliably enough to be considered a popbean.)

My Indian acquaintance told me that the green and brown chickpeas from Abundant Life look similar in size and color to the ones she is familiar with, but she isn't familiar with the black or cream types. The way they fix their brown or green chickpeas is to soak them, boil them, and then cook them into various spicy dishes. Maybe the problem is that I don't know those spicy dishes. I also don't know yet whether all the varieties of one color will taste the same or whether there will be differences.

Chickpeas can probably also be made into all the same kinds of fermented products for which soybeans are used. My local food co-op already carries a chickpea miso that I like far better than any soybean miso I have tried. Some varieties may be better than others for making these products. Chickpeas also can be used as livestock or poultry feed. Perhaps that's what some varieties were developed for. Chickpeas can be sprouted and used as a vegetable or in salads. Are there special varieties for sprouting? And what about the possibility of using the plants as a green manure crop? It's clear that we still have a lot to learn.

Often we manage to preserve or import the germplasm, the accessions or varieties of particular plants, but not the information about how to grow, harvest, prepare, or use them. Often we don't need to breed the plant so

much as to discover or rediscover its uses. In many cases we need to do both — in this case, for example, to evaluate accessions for cooking characteristics and uses, to identify accessions that grow well in specific areas of the country and under specific planting regimes and methods, and to select for high yield among the material that looks most promising.

One of the major schools of agriculture in the country is at Oregon State University, but no one there is working on overwintering peas, chickpeas, or favas. There are very few professional plant breeders in the world, and they can pursue only a minuscule number of all the projects that might be interesting. We can't expect them to do everything. If I want chickpeas for overwintering in the maritime Northwest, I'd better breed them myself. If you want chickpeas for your growing methods and area, you, too, should get involved.

When we look at the recent accomplishments of professional plant breeders, the tally is impressive. But it's nothing compared to what could be achieved if those few professionals were joined by an army of amateurs.

Rich Hannan, at the USDA Western Regional Plant Introduction Station, has done a seed increase on the rediscovered popbeans and on #1 chickpeas so that he can send small (thirty- to fifty-seed samples) to any researcher who wants them. A researcher, as far as Rich is concerned, is anybody who wants to do research with the seed and is willing to tell him what he or she finds out. This includes amateurs and backyard researchers. (See Appendix C for the address.) In addition, all seed companies, no matter how small, are welcome to request material.

While many small chickpeas will pop readily, the two accessions already mentioned, PI374085 and PI374090 remain the only large ones I've found that pop readily and become completely cooked just by popping. I think both have the potential for being developed into fine new snack foods.

The orange chickpea (PI374090) doesn't grow well here in western Oregon under organic farming conditions. I'm working further with the black-and-brown-mottled one (PI374085), developing lines for popping and other purposes. You don't have to wait until mine are ready, however. Both original accessions are available from the USDA.

My pea and chickpea projects are both interesting in part because I overwinter my plants — that is, I use other than the ordinary growing method. Any unusual growing method can produce special conditions for which specially bred plants may be most suitable. When you do the breeding, you select for plants that perform well using your growing methods.

My growing methods differ from those of most commercial growers in another way. I garden organically, and as a result, I probably now have the world's most slug-resistant pea lines.

There are a lot of slugs here. We call them garden slugs or gray slugs to distinguish them from the banana slugs found in the hills and coastal mountains nearby. Banana slugs grow to about 8 inches long. Ours, fortunately, usually don't exceed about 4 inches.

They're common, though, and they're active year-round, especially during the winter rainy season. And slugs are very fond of peas.

Many varieties of peas fail to overwinter here, not because they can't take the cold, but because slugs eat them. Dwarf types generally don't overwinter in my garden even in a mild year. They use up the energy from the seed in producing a bushy little plant whose apical tip is within reach of the slugs for a long time. Once the apical tip has been eaten, the seedling branches, producing more tips that are still within the slugs' reach and that, because they share the plant's energy, take even longer to grow out of the slugs' reach. The plant gets chewed down again and again, until its strength and ability to regrow are gone.

Peas from my lines shoot up first and concentrate on expanding their leaves and making a plant only after they already have considerable height. Slugs eat some of them anyway, but the peas are much more vulnerable when they're smaller. My peas are small for a very short time, so most of them survive.

Most commercial growers or professional breeders would use slug bait as a matter of course if they were doing pea breeding under such conditions. Therefore, they would not have selected for slug-resistant plants. Had I considered breeding for slug resistance, I might well have thought it impossible. But I didn't think about it. I didn't need to. Since I garden organically, I automatically selected for varieties that excel when grown organically.

*   *   *

More and more people are gardening organically. The desire to have wholesome, unadulterated food is a major motive for many of today's gardeners. In addition, many small farms, and, increasingly, even some large ones, are converting to more organic, more sustainable, or lower-input methods. Organic gardeners and farmers often have to grow commercial varieties, however. Such varieties often do poorly without pesticides, herbicides, and large doses of chemical fertilizers.

Every pattern of agriculture and every agricultural method is associated with certain crops and crop varieties that make it work and that work because of it. The varieties and the methods go together. To try to garden or farm organically with crops and varieties that were bred to excel under entirely different conditions is to assume a grossly unfair handicap. To use old-fashioned varieties that were developed originally under organic growing conditions is sometimes preferable, but older varieties often have been lost, lack resistance to modern diseases, or don't yield well enough to be competitive.

We need new varieties resistant to all the modern diseases but adapted to organic methods, new varieties to support a more diversified, more ecologically viable agriculture. We need entirely new crops, as well as new varieties of the old ones, to form the basis for a modern and competitive but fully sustainable agriculture. Those who envision and strive toward a different agriculture — a more diversified, more regionally based, or more sustainable agriculture — must be ready, willing, and able to do some of the breeding work.

* * *

You can breed plants because you have a vision of an agricultural future that you want to help bring about. Or you can breed plants because you just want to play around. Many gardeners will do some of both. I am strongly committed to increasing the biological diversity as well as the sustainability of our agriculture. I work with many new species, especially ones that lend themselves well to growing sustainably, and I encourage others to do likewise. But I also enjoy just playing around.

Can I get a leek to cross with a perennial bunching onion? That's what I wondered while looking at an especially large leek in the supermarket the other day. I bought the leek and stuck it into the ground in the middle of my bunching onion patch. Alliums are generally outbreeders — that is, the flowers are usually fertilized by pollen from other flowers. (Such information on various vegetables is presented in Appendix A and Table 1.) With just one leek surrounded completely by onions, perhaps some leek flowers would be fertilized by onion pollen and would give me a wide cross, a cross between two different species. And if they did, what in the world would the progeny be like?

If the cross is hard to do, if it requires special methods to overcome incompatibility barriers, I might not bother pursuing the project. But if the cross will go by itself — and, after all, many intergeneric crosses will — I'll have a lot of fun for just the amount of work it took to poke one leek into the ground.

I've already had a good bit of fun just thinking about it and, especially, just standing in the yard and looking at the leek. It's huge compared to the onions, its leaves are cut supermarket fashion, and much of the white stem is above the ground. It looks as if someone just stuck a supermarket leek into the ground in the middle of an onion patch. It's hard to walk by without noticing the leek, and the onion patch is adjacent to a much-used public sidewalk. The whole situation appeals to my sense of the absurd.

You can think as big or as small as you want. You can be serious or outrageous. You can merely try to transfer resistance to a specific disease into your favorite kind of pea. Or you can try to develop an entirely new species. You can even start from scratch — from an edible wild plant — and try to domesticate it. There are thought to be about twenty thousand edible wild plants. Only a couple of thousand perhaps have ever received any breeding attention at all from any human being.

And there is no law limiting you to the strictly practical. Ewald Eliason is dreaming of a potato that can be grown in the flower garden but will still give potatoes. Ken Allan wants, in addition to better Canadian sweet potatoes, a pole shelling pea with lots of peas in the pod, good flavor, and pretty purple flowers.

Amateurs often breed for unusual colors, shapes, or sizes — for novelty for its own sake, and for beauty. And why not? Having beautiful, unusual, or novel plants is part of the fun of gardening.

Glenn Drowns wants purple-podded snap peas, and so do I. I want purple-podded *pole* snap peas, of course, and they would have to be resistant to enation and powdery mildew. The purple pods would look beautiful on the vines. They would look beautiful sliced into salads. The project would take very little space and time. More important, I *like* snap peas. I like to grow the plants. I love to stand in the garden and look at them. And I love to eat snap peas, so I wouldn't mind eating the excess. In other words, I would enjoy the project. That is the best of all possible reasons for doing anything.

FOUR

Finding Germplasm

IN THIS CHAPTER and the next I discuss finding and evaluating germplasm and experimental material. Not every project starts with a search for and evaluation of germplasm. In some cases we have two tried-and-true varieties that are already our favorites and we simply want to combine the best characteristics of both; we start the project by crossing the two. In other cases a genetic accident of some sort turns up in one of our plantings and we decide to explore its possibilities. Or we may have a beloved heirloom variety that shows variability for a characteristic we care about, so we begin selecting for what we want. In such cases we already have the germplasm and are familiar with it. But many projects do start with a search for and evaluation of germplasm.

In some cases the search for and/or evaluation of germplasm *is* the project. For example, one or more varieties that would satisfy our requirements might already be available. We just don't happen to have them or know which ones they are. My initial work with chickpeas has been of this sort. Thousands of accessions are available; what is missing is information about what they are good for. It makes sense for me to start by finding out more about what there already is. Only then will it be clear what more is needed or where to start developing it.

However a project starts, it always requires evaluation during its course to choose what material to use. Evaluating germplasm, like maintaining germplasm and selection, is common to all plant breeding projects, as well as to all serious gardening.

42

Beginning gardeners are usually limited to the crops and varieties that are commonly available. As their interest grows, they begin trying different varieties and methods, which is what is known as doing trials. But they often don't know how to find the best material for their purposes; their trials are not designed, conducted, or interpreted properly; and the results are often incorrect or invalid. To a large extent, learning to find germplasm and evaluate it properly marks the difference between the beginning gardener and the sophisticated one.

Finding germplasm includes obtaining both the physical material and the information about it. Sometimes the germplasm is more readily available than the information about it. In this and the next chapter I cover everything from seed company catalogs and books to obtaining access to the collections of the USDA. I also consider the problems associated with the vague relationship between germplasm and variety names and its implications for the plant breeder.

My starting point for varieties and information about them is mail-order seed catalogs. I go to other, more specialized sources, too, but mail-order seed catalogs form my basic frame of reference. I have all the catalogs from my region and many from elsewhere. Some are free; some cost a few dollars.

A complete list of seed companies in the United States and Canada and their addresses can be found in Kent Whealy's *The Garden Seed Inventory* (see bibliography). For an excellent list of seed companies and other sources, both commercial and professional, throughout the world, see Stephen Facciola's *Cornucopia: A Source Book of Edible Plants* (see bibliography).

Companies that charge for their catalogs usually do so because the company is small or it handles largely rare or public-domain varieties. Often such companies have collections that include unique germplasm, and they produce an information-rich catalog that tells you more about the varieties and what they are good for than any other source. This information often is based on their own (unpaid) original plant exploration and variety trials. The catalogs that cost money are usually well worth it.

Some of the free catalogs also are excellent sources of information. I use the catalogs from Nichols Garden Nursery, Johnny's Selected Seeds, and Stokes Seeds (see Appendix D) to obtain information about disease resistance when I am designing breeding projects. When I lived in Minnesota, I learned how to garden from the Johnny's catalog. When I moved to the maritime Northwest, I learned how to garden from the Territorial Seed Company catalog.

I order regularly from about twenty companies. Most are in the Northwest or the Northeast, which shares some but not all climatic characteristics of the Northwest. You will develop your own favorites depending on your location and interests.

The second place I look for germplasm and information about it is the Seed Savers Exchange. Its annual *Winter Yearbook* lists thousands of varieties offered by members that are not available anywhere commercially. My favorite class is "Miscellaneous." That's where I find many rare perennial vegetables. The Seed Savers Exchange is open to anyone in the world. The Heritage Seed Program is an equivalent program in Canada.

I use the Seed Savers Exchange as a source of information as well as germplasm. Its book *The Garden Seed Inventory,* edited by Kent Whealy, lists all the commercial varieties available in the United States and Canada and includes a compilation of all the various catalog descriptions about them. I frequently find wonderful information there. In addition, since every variety listing includes all the seed companies that carry it, I also discover seed companies that are doing things I'm interested in. *The Garden Seed Inventory* is in its third edition; plans are to revise it about every two years.

The Seed Savers Exchange yearbooks list thousands of varieties that are not available commercially. There is some information with each listing — whatever the listing member volunteered. In some cases that's not much, but in others it's quite a lot. When something is listed by five or ten people, the total amount of information is often very useful, as are the different perspectives.

When I want additional information about a variety listed in the Seed Savers Exchange yearbook, I sometimes write or call the listing member and ask about it. I approached Dan Borman, who lives in Washington, over sea kale (*Crambe maritima*), for example. It turns out that he, too, is very interested in perennial vegetables. (He is, in fact, currently organizing a seed company, Meristem Plants, which will have a heavy emphasis on edible perennials.) Seed Savers Exchange members frequently list interests, projects, or questions in the yearbook to find others with specific information or similar interests.

The best single source of information about varieties (including rare species and varieties) is the book *Cornucopia: A Source Book of Edible Plants.* This amazing book, comprising 700 pages of fine print in a large format, lists edible plants and plant varieties and where to find them. It includes rare as well as common plants, cultivated and wild edibles, and is worldwide in scope. I have used it to find rare vegetables that have a single source in some other country, for example. I also have found information about how rare varieties are used. Yet I have only begun to explore the possibilities the book represents.

Cornucopia has made a major difference in my ability to find germplasm and information. I think all serious plant explorers should have a copy of it. For information about it, see the bibliography (under Facciola).

Organic Gardening and *National Gardening* magazines both have seed search or swap services that allow readers to exchange seed. Some seed companies also have seed search or trial programs. When they do, they describe them and issue invitations to participate in their catalogs.

When seed catalogs, books, and the Seed Savers Exchange don't provide me with the germplasm or information I want, I turn back to the mail-order seed companies, this time to their owners, founders, or managers. Most of them have much more than what they list in their catalogs. As Alan Kapuler (Peace Seeds) puts it, "I only list 900 varieties. I have thousands. If you don't find it in my catalog, ask. I probably have it."

Even when a seed company doesn't have what you want, it might be able to tell you where you can find it. The seed companies that specialize in preserving germplasm have extensive interactions with private collectors and knowledge about material that exists outside any official or commercial channels. Many such companies' catalogs actually include the offer to help you find something you haven't been able to get — in other words, they offer plant search services.

The right time to approach these seedspeople is after you are already thoroughly knowledgeable about the crop you're trying to grow and have tried all the standard varieties. You can go to them for varieties or information. ("Do you know of a muskmelon resistant to thus and so?" for example.) Realize, though, that many of these companies are largely one-person operations and that the person who knows the plants is usually very busy. In most cases you should write and be prepared to wait for an answer (until the heavy seed selling season is over or the harvest is in).

The basic operating rule is to give something back to those who go out of their way to help you. In some cases what is most valuable is information. In other cases a company may not list the species because it had only about three packages of seed. When a company gives you something it has very little of, it's nice if you can do a seed increase, even a small one, and give the company back more than you received. In some cases companies will bring that up as part of the deal.

Exercise some care in what you do with unique information or germplasm with which you've been entrusted. Don't undermine the commercial position of the company from which you got it. (If you are unsure about the extent to which you are free to pass the information or material along and to whom, ask.) These seedspeople are passionate about plants, and that passion usually matters much more than money. But there is always a bottom line; even dedicated, passionate, altruistic people have to survive.

The Agricultural Extension Service has a mission both to do research and to disseminate it to the general public — which includes you. The extension service often breeds and does trials of vegetables for your area. It frequently publishes the results in brochures written for the layperson that it will send on request. You often obtain the benefit of this work automatically because your regional seed companies use it in making their choices about listings. In my area, Nichols Garden Nursery and Territorial Seed Company make extensive use of Oregon State University and Agricultural Extension Service research and refer to it in their catalogs.

Various units of the Agricultural Extension Service also often provide a variety of special services. Mine, for example, has master gardening classes, a phone number you can call with questions, a plant disease clinic where you can take a sick vegetable and find out what is wrong with it, and people who can help identify wild plants, weeds, or garden pests. Some extension service people conceive of their service as intended primarily for commercial producers. Most, however, include gardeners in their concept.

University-based plant breeders and horticulturists, personnel at botanical gardens, and other professionals are important sources of information and experimental material for amateurs. You should realize, however, that their mission does not include disseminating material or information to the general public. They often will help, though, if you approach them properly with a reasonable request.

Once you are involved with a particular crop, you may want to approach the main breeders for that crop in the country or the world. Most of them are surprisingly open to requests from the knowledgeable amateur. Appendix A and Table 1 include references to the professional plant breeding literature. Use them to find the world's experts in any given specialization. As with seedspeople, the right time to approach the experts is after you have already grown all the commonly available varieties, are familiar with what is generally available, and have developed a clear idea of what you want to do.

Many university-based plant breeders make their experimental varieties available for prerelease trial to serious gardeners in the area. In Corvallis (Oregon State University), breeder Jim Baggett's new vegetables are announced in the local newspaper; anyone who wants to participate in gardener trials of them can drop by and pick up seed and report forms. At this stage the varieties have numbers, not names. They are named only

when the final decision to release them is made.

University researchers in your area also may have such a program. If not, don't let that stop you from participating in what your region's professional vegetable breeders are doing. Many researchers are glad to have serious gardeners and farmers try their material — if they are going to be informed of the results. Call the breeder who is doing work you're interested in and ask.

Regional seed companies usually credit regional vegetable breeders, so if you patronize your regional seed companies, you will automatically know the professionals who are doing work you're interested in. If you have catalogs from seed companies in many regions throughout the country, you will automatically know the names of all the professional breeders in the country who are doing work you're interested in.

In most of these relationships, feedback is the key. It is important to report back to the researcher on how the variety did. I suggest five basic rules for dealing with professionals or other amateurs.

1. Sometimes writing is preferable; sometimes writing doesn't get you anywhere, and calling is better. When you write, you may not include all the relevant information. When that happens, you usually don't get an answer. If you call, the expert can ask for the missing information. If you write, type the letter if at all possible unless it's very short. Always include a phone number if you have one. That gives the person from whom you are asking help the option of responding either in writing or by phone. Many people have strong preferences and will respond

graciously given the option they prefer, but not at all otherwise.

2. Do your homework. Don't take up a professional's time asking for varieties that are available commercially or for information that you can find in seed catalogs or books that are generally available. These professionals are not running seed companies or private beginning gardening lessons and won't appreciate your trying to use them that way. Exhaust the normal channels first. If the crop is a common one, get some experience growing the available varieties before you request rare material. Don't ask someone to go to special trouble sending you a rare broccoli, for example, before you've ever grown broccoli at all.

3. When you request help, start by explaining what you're trying to do and where you are with it. ("I've grown this and this and that, but they had thus and so problems. What I think I need is thus and so, or material I could use to develop it.") Plant scientists are nothing if not curious about plants. When you tell them about your project, they tend to get intellectually hooked and are much more interested in helping. Often they will have useful suggestions or criticisms or will be able to put you in contact with someone who is doing or has done what you're thinking about.

4. Keep requests modest initially. I limit my own requests for seed to a variety or two, for example. Even if I feel greedier than that and would like more, I exercise self-restraint. There will be other years to request and grow additional varieties.

5. Once the season is over, call or write your source and tell him or her your results.

After that the relationship is a two-way one. In subsequent years you can be more extravagant with requests for seed or information. You will have proven yourself to be someone who will give something back in return for what you get. You may find the professional calling you and offering material he or she would like to see trialed or asking for information. Most professionals have networks of people with whom they exchange material and information, and most of these networks include at least some serious amateurs.

There are many things professional plant breeders or other amateurs will help you with. But you should be aware that it isn't reasonable for you to ask a plant breeder (professional or amateur) for his or her mid-project working material. It's roughly equivalent to asking a writer for the first draft of her book for you to revise and publish under your own name; she would not be enthusiastic.

Breeders are much more likely to be willing to send you raw material or breeding material developed early in a project (such as an F_2) or finished varieties (even before formal release) than to send you material at later stages in the project. Once they have begun selecting and finalizing a variety, they usually want to perfect it and release it themselves. They don't want inferior or alternate versions of their work around. And they, like anyone else, want credit for what they do.

I often provide access to material that is at mid-stages in various breeding projects, but this is usually because I'm giving away or sharing the entire project, not just the material. I find it useful to start many more projects than I can complete, to start projects on a larger scale than I can handle, and to share or give most of them away as I find people who are interested. In this way I can help bring about far more work than I can do myself. I have some projects that are all mine, though. I don't make my mid-project working material available for those.

Three seed companies cater specifically to amateur plant breeders. All offer a wide range of breeding material as well as established varieties. By "breeding material" I mean material that has already been genetically manipulated so as to create a pool from which interesting new varieties can be selected.

In the past much of the potential of breeding material developed by either professionals or amateurs was lost. In a typical project a breeder might start by crossing the two most cold-hardy varieties of something and then raise the F_1 and F_2. Then he or she starts selecting in the F_2 for a super-cold-hardy type among a given choice of sizes, shapes, and flavors.

The result of this process is a single variety. That variety represents the breeder's single choice of plant type, flavor, and purposes. In addition, it has been automatically selected to perform well under the specific soil, climate, and growing conditions of that one individual breeder.

That same F_2 could be used by other people with other preferences to select other super-cold-hardy varieties adapted to *their* growing conditions and with *their* choices of colors, shapes, and flavors. That F_2 breeding material has the potential to be developed

into dozens of new and different varieties.

When companies sell breeding material as well as finished varieties, breeders who provide the material can watch their initial ideas and efforts being magnified by those of others. Those who receive the material will be able to save themselves a year or two or three on an interesting project. They'll begin with their own ideas and efforts and develop new varieties very rapidly.

Alan Kapuler, of Peace Seeds, offers a variety of breeding materials. His own breeding work has focused on dehybridizing hybrids. He has developed a number of public-domain open-pollinated lines from popular commercial hybrids. His interests include plant germplasm preservation, heirloom varieties, and vegetable varieties with especially high nutritional value. For information on access to his breeding material, see the listing for "Kapuler" in Appendix D.

Ecologist Dan Borman's company, Meristem Plants and Plant Research Supplies for Cool Climates, also offers breeding material. He is especially interested in perennial edibles (including vegetables). He carries a wide range of equipment for amateur plant breeders as well.

Tim Peters' company, Peters Seeds and Research, sells breeding material as well as finished varieties he has developed. He sells inbred lines and many varieties that are otherwise unavailable. He has an extensive list of breeding material of all kinds developed by himself and others. He has an unusual and extensive collection of breeding material for perennial grains.

Those who want to obtain or offer breeding material of fava beans should contact The Fava Project (see Appendix D). Those interested in perennial vegetables should consult the entry for "Perennial Vegetables" in Appendix D.

Finally, as of February 2000, I registered the domain name plantbreeding.net. I'm planning to use this website for one-time releases of my own finished varieties as well as breeding materials directly to farmers and gardeners. Through plantbreeding.net, I'll be able to sell breeder's-grade seed and foundation-grade seed (which is not ordinarily available commercially) as well as provide much more information about the material than is possible in a seed company catalog.

In addition, people with valuable collections of germplasm often approach me to find people who might be interested in it. I'll use plantbreeding.net to help people with plants to find those who want them. (See "plantbreeding.net" in Appendix D.)

Importing small amounts of seed from other countries — from foreign seed savers or from seed companies — is usually legal and relatively simple. (There are exceptions for certain crops.) Importing live plant material is a totally different matter. Because of the potential for importing new pests and diseases, live plant material needs to go through USDA inspection stations, and in some cases it must be quarantined. In addition, you usually need an import permit. You don't need any kind of professional status to obtain such a permit.

An excellent article on importing plants appears in the April 1992 issue of *Horticulture*. Refer to it for information about USDA rules, obtaining permits, preparing and shipping plants, and inspection and quarantine.

As a source of both germplasm and information, the USDA-ARS National Plant Germplasm System deserves special attention. (USDA-ARS stands for U.S. Department of Agriculture–Agricultural Research Service.) The role of the National Plant Germplasm System is to collect, maintain, and distribute germplasm and information about it to plant researchers. Some curators feel that backyard plant breeders and researchers are included in the mission; others don't. But even those who don't are still very likely to help you if your request is reasonable.

I know many amateurs who have obtained material from the working collections of the USDA. In fact, I don't know of any with a reasonable request who were turned away by a working collection unit. Do note, however, that one unit, the National Seed Storage Laboratory (NSSL), has as its mission long-term storage, not seed increase and distribution. Amateurs frequently approach the NSSL because of its name and then get discouraged when they are turned away. Usually, however, what they need isn't in the NSSL; it's in the working units.

There are many units of the National Plant Germplasm System in different areas of the country, and each is responsible for different crop species. In Table 1, I list the location of each vegetable in the system. Chickpeas, for example, are *Cicer arietinum*. (If you don't know the scientific name, you can find it by looking up the common name in the index.) In the listing for *Cicer arietinum* is the code "USA-W-6." This means that the species is handled by the W-6 branch of the National Plant Germplasm System. You should then look in Appendix C to find that W-6 is the Western Regional Plant Introduction Station in Pullman, Washington. The address, the name of the principal curator, and the phone number are given.

You can either write or call the unit. I usually call and ask the person who answers the phone who handles the crop I'm interested in. Then I ask to talk with that person. I don't ask for the director of the unit or the principal curator unless that person happens to be the one who takes care of increase and distribution for the crop I'm interested in. I always try to talk with the person who is most directly concerned with the specific plant, the one most likely to care about it.

The USDA system doesn't cover everything. I can find more different unusual perennial vegetables in the seed catalogs of Peace Seeds, J. L. Hudson, and Meristem Plants than in the USDA system. And the USDA collections usually concentrate on wild material and on land races instead of, for example, heirloom varieties. So you can often find things in the catalogs of alternative seed companies that specialize in heirlooms that you can't find in the USDA system.

In addition, the USDA folks don't necessarily send out everything they have. They usually do seed increases on material before they release it. If they haven't done a seed increase yet, they usually won't distribute the material.

If you need just an accession or two or five, you will almost certainly be able to get them if they are available, even if you don't know the accession numbers and need advice about what you need. On the popbeans, for example, I suggest you ask Rich Hannan what is available instead of just requesting the ones I've mentioned. I'll be discovering additional ones, I'm sure, and will be reporting back to Rich about it. That information then goes into the computer data base and becomes part of what you have access to.

If you want a lot of something — say, one hundred accessions of something instead of ten — I suggest that you start more modestly and establish a working relationship first. Directness works well. You can ask how much you should reasonably ask for, and you'll get a response that reflects where things are at that point.

If you get heavily involved in trialing a particular crop, you will probably want the USDA inventory for the species, if one is available. I have a copy of the *Cicer Inventory*, for example. It's an 81-page listing of the *Cicer* accessions and information about them.

Each order you receive from the USDA-ARS National Plant Germplasm System is accompanied by a packing sheet containing the lines, "This germplasm is being freely distributed. Continued reciprocal exchange will be appreciated." Each little seed pack is marked prominently with the words, "Report performance to the Coordinator of your

Region." You are sent the material free, even of postage, in exchange for a promise to give back that information.

In the 1990s the entire list of holdings of the USDA collections and plant introduction centers and all available information about the accessions was cataloged and put onto the Internet as GRIN (Germplasm Resources Information Network, see Appendix C.)

The search for and evaluation of germplasm is exacerbated by the fact that variety names don't necessarily correlate in a one-to-one fashion with germplasm sold under or associated with the name. It's common for different varieties to be sold under one name. There are dozens of different versions of 'Kentucky Wonder' pole beans, for example. Some have brown seeds; others have white seeds. Some are early; others are late. Some have strings; others do not.

In some cases, probably, mutations or crosses occurred and led to different varieties that were not given different names. In other cases different names might have been given that related to the original name — 'Early Kentucky Wonder', for example. But extra adjectives have a tendency to get lost, so pretty soon 'Early Kentucky Wonder' is being called 'Kentucky Wonder', with nothing to distinguish it from the original. Sometimes someone develops a new variety that resembles an established one, so he or she appropriates the established name as a way of describing or selling more of the new variety. Or the developer may call the new variety 'Improved Kentucky Wonder', but the "improved" gets dropped. 'Improved Kentucky Wonder' may be derived from 'Ken-

tucky Wonder', and it may not. Anytime I see "improved" as part of a variety name, I realize that this is a whole new variety that may be similar to the old one in the ways that matter to me or may not, that may be derived in part from the old one or may not.

Not only do many varieties get sold under the same name, but the same variety often gets sold under many names. 'Kentucky Wonder' pole beans, for example, according to the third edition of the Seed Savers Exchange *Garden Seed Inventory,* is also known as 'Old Homestead', 'Texas Pole', 'Egg Harbour', 'Kentucky Wonder Green Pod', 'Improved Kentucky Wonder', 'Kentucky Wonder 191', 'White Kentucky 191', 'Kentucky Wonder Brown Seeded', 'Old Homestead Brown Seeded', 'Kentucky Wonder Rust Resistant', 'Kentucky Wonder White Seeded', 'No. 191', and 'White Kentucky Wonder 191'. Some of these are clearly not synonymous with others, but if you order just plain old 'Kentucky Wonder' from somewhere, you could be getting any of them.

Anytime someone is maintaining a variety whose original name is unknown or lost, that person has to call it something, so alternate names proliferate. In addition, seed companies may give new names to old varieties so as to conceal their sources from other companies.

The situation is even worse when it comes to species vegetables. Many sources describe Good King Henry (*Chenopodium bonus-henricus*) as a delicious perennial substitute for spinach. My Good King Henry, though, is quite bitter. I've read a number of descriptions and accounts relating to Good King Henry, and no one mentions that it tastes horrible. Is all Good King Henry as unappetizing as my line? There aren't any variety names at all, so I can't tell whether the dozens of seed companies that sell Good King Henry are all selling the same material or not.

When a vegetable or herb is sold as a species plant, you have no idea where the original material was collected. Some lines might derive from wild material from southern Italy, and others might come from Siberia. One may be cold hardy enough to overwinter in your area, and the other may not be. Some lines may be primarily ornamental and of little use as a vegetable, whereas others might be, or at least might once have been, primarily an edible. There is absolutely no information in the name that lets you know. All you can do, if you care about it enough, is to round up as many samples as you can and try them out.

Seeds Blüm sells a sweet purple kale called 'Ragged Jack'. Peace Seeds, Nichols Garden Nursery, and others sell a kale called 'Red Russian'. Seedspeople consider the two the same thing. They are willing to buy 'Red Russian', for example, and sell it as 'Ragged Jack', or vice versa.

I know of at least three growers for 'Ragged Jack'/'Red Russian', and the way they grow the variety and maintain it is different. Each grower lives in a different climatic region, grows the kale differently, and is automatically selecting for a somewhat different line. The Peace Seeds 'Red Russian', for example, is planted in the fall and overwintered, so it is automatically being selected for overwintering ability under maritime conditions. It is undoubtedly different from lines

that are maintained by growers who plant in the spring.

Some retail seed companies buy the kale from two different growers, so even if you buy two packets of seed of the same name from the same company in one year, the seed in the packets may not represent exactly the same line. And anytime a retailer changes to a different grower for a variety, it probably also is automatically changing to an at least slightly different variety — although there will be nothing in the seed catalog to so indicate. Such changes can be good or bad, depending on your needs.

Whenever I receive a package of seed from anyone or anywhere, I label it with the source and the date, because it may not be the same as other packages from the same or other sources in the same or other years. When I do trials, I often obtain seed of (supposedly) the same variety from a number of companies and trial it as if it were all different. I have Good King Henry from half a dozen sources, for example, and plan on trialing them all in the hope of finding one that isn't bitter.

Any plant explorer or adventurer needs to be aware of the poor relationship between the names of varieties and the varieties. When something matters enough to you, be prepared to be stubborn. Don't get discouraged if one sample from one source doesn't give you what you want. Instead, look further.

In most cases you need to grow only a few plants of each sample from each source, so little gardening space or time is needed for comparisons of varieties from different sources. The major cost is the time involved in locating and requesting or ordering the seed.

Finally, just because someone else has tried something and says it isn't any good or that it doesn't work doesn't mean you shouldn't try it, too. In the first place, most people don't know how to trial things properly. But even if they have, they may have tried something that had the same name but was actually different material. In addition, your conditions are different, and you will probably use different methods and make different choices about some of the procedures. And you may be more stubborn or luckier, or you may have more of a knack for whatever is involved.

Evaluating Germplasm and Experimental Material; Variety Trials and Gardening Research

T O DO plant breeding, we need to be able to evaluate varieties, germplasm, and experimental material. Evaluation tells us what germplasm to work with and which material to go on with. In this chapter I speak in terms of evaluating varieties, but any lot of germplasm — such as material derived from our own crosses — is evaluated similarly.

There are two basic kinds of garden trials. Sometimes gardeners compare various varieties under constant conditions. Other times we're interested in how best to grow a specific variety or a crop in general, in which case we trial one variety under various conditions. Many projects combine both objectives, simultaneously trialing various varieties and various methods. Garden trials of vari-

eties or methods are central not just to plant breeding but to all good gardening.

Garden trials are scientific research. Good trials require good experimental design, execution, and analysis. But gardening trials also are just playing around, just trying things. Your trials are your curiosity incarnate. You conduct trials primarily because it's fun. In this chapter I explain how to get the most information from trials with the least space and labor — and without losing sight of the fun.

I love gardening trials. My own trials are almost never just trials, though. I eat them. Properly designed, a trial plot usually can produce nearly as much food as the same amount of space planted to a single variety. It yields information in addition to, not instead of, food.

A trial plot doesn't have to be devoted entirely to a single trial. Many different trials can be going on at once. Or a plot can be partly trials and partly for straight food production. A wide bed can have pole bean or pea trials planted down the middle and other crops planted along the edges, for example. I use this pattern all the time; it makes excellent use of a small amount of space and produces large amounts of both food and information.

Trial plots also can be ornamental. I sometimes interplant different-colored pole beans partly because I want to try them and partly to create beauty on a trellis alongside a much-traveled sidewalk. If I am trialing only green varieties, I plant an occasional 'Heavenly Blue' morning glory among them. The individual large blue flowers are truly glorious against the background of the green vines and beans.

Trials should be a routine part of gardening. Every planting can be a trial. All it takes is a few plants or a row of the trial variety among those of your standard planting. Two-variety comparisons where most of the planting is of your standard usually yield good information for very little work, and they are easy to design, interpret, and record.

You won't master doing great trials all at once, any more than you can learn to drive the first time you get in a car. It comes partly with experience. The more gardening research you do, the easier it gets.

There are many considerations when we design a trial, and they are all interrelated. The following section is an outline of points to consider when designing a garden trial.

The points are illustrated, explained, and tied together in the last part of the chapter with examples of some of my own recent trials.

Checklist for Gardening Trial Design

Recording

1. *Record things in some fashion that allows you to find all the information about one trial easily* either this year or in future years. I have a separate file for each kind of crop — mustards, for example. (It used to be a file folder; now it's a computer file. The principle is the same.) Anything I do with mustards is in the mustard file in chronological order. All the information for a trial is on a few pages, not scattered among records for everything I did that year. I can find anything I have ever done with mustards easily, even if I don't remember which year.

2. *Record only what's important, not everything.* If you try to record everything, your trials become too much work, and it's harder to find the information that matters.

3. *Dates often turn out to matter* for unanticipated reasons. Always record the date you plant. (If the new corn variety doesn't mature, it will be very significant that you planted much later than you could have.) Flowering and harvest dates also are often useful. Approximations are usually acceptable (for example, "mid-June," "first week in July," "for the last month or so," "April 10 or earlier").

4. *Record the sources of all your seed or material.* (As explained in Chapter 4, varieties by the same name from different sources may

be entirely different.) My records are full of abbreviations such as "AL89" or "J92" (Abundant Life 1989 and Johnny's 1992, respectively).

5. *Record information about the weather* in some general place where you can refer to it readily from year to year. It affects all your trials in any given year.

6. *Develop regular patterns that help minimize mistakes* (such as mixing up different varieties, skipping something, or planting two things in the same space).

I do everything left to right and front to back (as I face it). I arrange the seed on a tray left to right. I plant it left to right. I move each package of seed leftward on the tray as I put it back so there is a gap between it and the unplanted seed. I label my field marker (if there is one) and place it before I reach for the next variety. Even if I get interrupted, I always know where I am. (In complex trials, I label markers before I go out and order them left to right on the tray, too.)

I always place markers to the left of or directly in front of the varieties they go with (and record which). In my notebook I always list things with first to last representing left to right. In crosses I always list the female first.

What matters is that you have regular patterns ensuring that when your records say something is a given variety, it really is. Make sure that anyone who works for you or with you uses your conventions.

7. *Record what you plant. There are two kinds of records: those in the garden and those on paper or computer disk elsewhere. Markers in the garden are never enough by themselves.* Kids and birds pull them up, they fade even

though that kind of ink never did that before, or they vanish mysteriously, especially if you don't have records elsewhere. You always need something on paper or disk to help you figure out what is in the garden.

Markers in the garden may not be necessary, but they often add to the convenience and fun. I like to be able to tell what is happening in my trial as I watch the plants day by day.

8. If everything you are planting is distinguishable and you already know what each thing looks like, you may not need any records about what was planted where. If not, you will. *The more things you plant in one trial that you can't tell apart, the more complicated your experimental design and your labeling and records have to be.*

9. *Record anything that goes wrong as it happens.* Record anything that might limit the generality or validity of your results as it happens. ("Soil at left end has better tilth." "Ground froze right after planting." "Too much lime?") It's hard to remember later.

10. *Do elaborate prior planning if you want or need to; otherwise, don't.* I sometimes design a trial on paper and go out with a list of everything to be planted and the seed packets all arranged in the predetermined order. But sometimes I suddenly decide to start a trial right now, while the sun is shining and that cool breeze is blowing. So I look quickly through my seed, grab possible varieties, and toss them onto a tray with a piece of paper, a pen, some wooden markers, and a Magic Marker. Then I go out to the garden and sit on the ground in front of the space I'm going to plant and shuffle the seed packets on the tray as I think about them and what I'd like

to find out, eliminating some as I glance occasionally at the available space. Then I plant, labeling markers and placing them as I go. Finally, I record what I did and what's on the markers. Elaborate planning has its place, but so does spontaneity.

Designing and Conducting a Trial

11. *Figure out exactly what question or questions you are trying to answer.* What related questions don't you expect to be able to answer in this trial? Does your trial design address the question you're most interested in or only a related question? (See the discussion of corn trials later in this chapter.)

12. *What are the answers likely to be? Are you after qualitative answers or quantitative ones?* (That is, all-or-nothing kinds of answers or degrees of difference?) *How large a difference matters?* (You don't need many plants of each variety to get all-or-nothing answers. You need a lot more to see shades of difference.)

13. *Where should you plant?* Near the house or farther out where you tend not to notice things? Where irrigation is available or not? At home or in the space you're using at your friend's house? (How frequently will you need to tend the plot, harvest, or evaluate?)

What about diseases from former crops? What about volunteers? Is there seed already in the patch that you can confuse with one or more of the varieties you're planting? You may need to do the trial somewhere where you haven't grown that crop recently.

14. *How much space will you need for the trial, including space required to separate different varieties?* (You may be able to design the

trial so as to not need extra separating space.) *How long will you need the space for? When will the trial end?*

15. *How many varieties do you want to test? How many are you likely to be able to handle?* Weeding or watering may take more time than planting; harvest may take more time than weeding and watering; evaluation often takes more time than anything else.

Are you going to be able to evaluate the flavor of a hundred different tomatoes? You may want to know about a hundred, but planting a hundred may very well mean that you don't get around to evaluating any of them. If you do start more than you can finish, though, don't get discouraged and abandon everything. Instead decide what is most important and finish that part.

16. *The major difference between meaningful trials and useless ones is the inclusion of controls.* If you plant a new variety and it doesn't do well, that means nothing without controls. There might have been a freeze right after planting that you didn't notice. You might have used too much fertilizer (or walnut leaves in your compost) and stunted all plant growth. You might have so improved the soil that you now have too much nitrogen, causing excess vegetative growth but poor flowering and fruit yield or lower freeze tolerance.

You need something else besides the new variety in the same patch, something that you're already familiar with — a standard. *The simplest meaningful trial includes two varieties — the trial variety and the standard.*

17. *It's often useful to include more than one standard.* In elaborate trials, I usually include the popular standards as well as my own per-

sonal favorite(s) as controls. Doing so makes my results more useful to me and to others, and it makes it more obvious how they fit in with the work of others (who probably used at least one of the same controls).

18. *Whenever you grow out an F_1, an F_2, or other material derived from a cross, also grow out a little of each of the two original parents for comparison.* The material derived from crosses always includes plants with unexpected characteristics. You can best figure out what to do with the unexpected if you can infer the inheritance patterns involved. For that you need to be able to go back and compare with the parents.

19. *Whenever you evaluate the disease resistance of experimental material, try to include at least one variety known to be resistant to the disease and one variety known to be sensitive to it as controls.*

20. *When trialing different methods, use both positive and negative controls — controls that bracket the trial methods.* For example, if you are testing how much mulch to use, try to choose your least amount so that it will turn out to be too little and your largest amount so that it will turn out to be too much. If you are trying different planting densities, choose a closest spacing that you expect to stunt the plants and a widest one that is too wide (and will give no benefits over the second widest one).

21. *Consider plants of a different crop for some of the controls.* In trials of overwintering garden peas, I often include the 'Austrian' field pea. Garden peas aren't normally overwintered here, but that field pea is; it's planted in the fall and grown commercially.

If conditions are severe enough to kill 'Austrian', that is useful information. In the winter of 1990, my chickpeas overwintered in a field where 'Green Wave' mustard died. That mustard has never been winter-killed for me before; that adds perspective.

22. *How many replicas and/or how many plants do you need of each variety and control?* You don't have to have the same number for everything. The controls are the most important because without them, nothing else is meaningful. Where space is limited, it is often useful to plant more of the controls than of the trials.

More plants and repeats are not necessarily better. In some situations one plant and no repeats will do the job. In other cases a large number of plants or repeats may be needed. (See the discussion of my tomato trials later in this chapter.)

23. *How uniform is the environment?* Is one end of the trial patch shaded, or is the soil different in one end? If the growing area is nonuniform, you probably need one or more replicas of each variety and control planted in different areas of the plot so that you can avoid being misled by the environmental nonuniformities.

24. *Practice selective sloppiness.* Do things well enough to get the answers you need in whatever way minimizes the important costs to you (which is sometimes land, sometimes labor, and for me usually both). Don't do ten times as much as you need to do under the mistaken belief that the results will be ten times better. Don't do it ten times more carefully than you need to unless it doesn't cost any extra time. If you design the trial prop-

erly, ten times as much work will probably make the results only slightly better or not any better at all.

For example, don't waste huge amounts of time measuring everything so as to space it very accurately if small differences in performance aren't going to be meaningful. (They usually aren't in small trials.) Approximate spacing is good enough. My hand span is about 9 inches. Four inches is about half a hand. Your records should reflect how you measured. If you approximate, record the measurement as an approximation: "about 4 to 5 inches apart in the rows."

25. *Should you interplant, or should you plant the different varieties and controls in separate rows, beds, or patches?* Interplanting is better because it helps correct for the environmental erraticisms of small home plots — more shade at this end than that, slightly different soil, and so on. However, interplanting varieties that are not easily distinguishable requires more labeling and record keeping, so it can end up being a lot more work.

Interplanting is often the way to get good information with the least number of plants, and it may be essential where the conditions within the planting area aren't uniform. I often choose varieties for a specific trial that I can easily tell apart and interplant without elaborate labeling.

26. *Are you going to want to save seed from any of the material?* If so, you may need to arrange things with that in mind. For example, the special material might be at one end. Adjacent to it will be material that flowers at a very different time or that will be gone by when the special material flowers. Farthest

from it will be anything that would flower simultaneously and might contaminate it.

27. *What part of the experiment are you going to eat, and when? How can you tell it from parts you need not eat?* I've found that if I can't tell what to harvest and what to save *very easily* when I'm in the field, I usually don't harvest anything; the edibles get wasted. I go to some trouble to arrange my plot so that it is obvious what I can eat. (All the crosses are at one end, for example, or are elaborately marked not just with little white tags but with Christmas tinsel draped over them.)

28. *Does every seed or plant count, or can you afford to waste seed?*

29. *If every seed or plant counts, how can you minimize accidents* such as neighbors or visitors walking through the plot, kids or dogs running through, cats digging in it, deer or rabbits eating some or all of your plants, or birds pulling things up?

30. *Should you go to special trouble to start the seed or minimize damage from ordinary pests (such as insects and slugs)?* Or is ability to perform under ordinary conditions part of what you're testing?

31. *If every seed or plant counts, don't do anything new.* Of course, the more important the material, the more you'll be tempted to treat it specially. That's fine as long as the treatment is something you're already familiar with. Don't presoak seed for the first time when each seed matters, for example. There are almost always unexpected components of any new method. Learn on material that doesn't matter.

32. *If there is genetic heterogeneity in the*

planting material, which matters more, the fast-germinating seed or the slower one? The fast-growing plants or the slower ones? Often the fastest-sprouting seed and most vigorous plants are a result of an accidental cross. They are hybrids. If you are trying to preserve or "clean up" an open-pollinated variety, you may need to be careful to keep all the plants, not just the fastest, most vigorous ones. You may have to cull out all those fast ones to preserve the variety. If you instead cull out the slow plants (or don't plant the slowest-germinating pre-soaked seed), you may, by just that one decision in that one season, lose the entire variety. If you are trying to preserve germplasm, such as a land race of chickpeas, you want to preserve *all* the plants — those that do poorly under your conditions as well as those that do well.

33. *You don't have to find out everything in one year.* Often you can design a really good trial only after you know approximately what the answers are likely to be. *It's often most efficient to do a small, crude trial the first year, then do a much better one the following year(s).*

Whenever you are planning to evaluate many varieties and invest major effort, a small trial that is a subset of the major one can be used the first year to help shake down your experimental design and eliminate problems with both the principle and the practice.

34. *It's often better to do a number of small trials in different years than just one massive trial in one year.* Different years have different weather, and in different years you'll probably be using different plots of land and somewhat different methods. A variety may perform better than a standard one year but still be inferior to the standard, which performs reasonably well almost every year (which may be why it's the standard). Only if your work is spread over more than one year can you begin to see these kinds of differences.

In addition, the differences you see one year can be because of a flaw in your trial (that section got more fertilizer) or just because of chance. Many small trials in many years usually give you a more accurate view than one big trial done all at once.

35. *Can you use your normal plant spacing and layout for the crop, or will the trial require extra spacing or a different layout* so that you can tell one plant from another, get in and examine them, or make crosses?

36. *Blind trials are trials in which you arrange things so that you don't know which plants are which until after the trial is completed and analyzed.* They are most important when you're worried that your opinions or biases might influence the results or when what is being analyzed is very subjective. *Blind trials usually aren't necessary in gardening work* and are not always possible (as when varieties look different, for example). In addition, they spoil the fun of being able to watch and understand the trial as it is happening.

One case where I frequently use blind trials is for evaluation of flavor. I don't bother growing the plants blind, but arrange things so that I don't know which plant the material to be tasted came from until the tasting is done and the results recorded.

37. *What kind of labeling do you need?* Will

you need to be able to distinguish each individual plant or only groups of plants? How can you lay out the garden so that you can have a simple map that allows you to reconstruct what everything is after the markers mysteriously vanish?

38. *How can you best arrange the varieties within the trial plot so as to facilitate your being able to identify and evaluate the material?* If you are testing three pole beans that are to be grown on a single trellis, put the yellow one between the two greens. If the two greens have similar pods, you'll not be able to tell where one planting begins and the next ends. It takes about a running foot of empty trellis space to separate the two greens if you plant them side by side — a foot more space than is necessary.

39. *Where appropriate, use plants instead of empty space to separate varieties you're testing.* A few purple-podded pole beans can be used to separate a number of green varieties that you're testing. The space isn't wasted, since you eat the beans. I've separated green pole beans with 'Heavenly Blue' morning glory or climbing edible nasturtium plants. Fava beans can make good separators for early-planted material. A single fava bean can be planted between different pea varieties, for example.

40. *Don't cancel a trial prematurely.* When one treatment does better than another, don't rush in there and try to improve things for the other. The differences you see early in a trial may not hold up later. Further, seeing those differences was the point, not creating optimal conditions for everything. Finally, those differences, if real, are not the whole

story. (See the discussion of my potato trial later in this chapter.)

Evaluating a Trial

41. *In evaluating trials, take the environment into account.* The plant that is southernmost on a north-south row gets more sun than all the others; it may be biggest through no virtue of its own. The same is true if a plant is bigger or earlier than its neighbors but has more space because those next to it didn't come up. Don't plant in such a way that entire experimental groups shade others.

42. *The plants at the edges of patches are usually larger than those in the middle; you may need to take edge effects into account.* Suppose you have a number of 10-foot-square patches of something and are evaluating yield in preparation for planting whole fields. You should probably discard all the plants along the edges of your patches and harvest and evaluate only the plants in the middle. The yield of just the middle of the patch will give you the best approximation of what you could expect in a large field (where the contribution of plants along the edges is insignificant).

43. *Notice and record any evidence of genetic heterogeneity.* Do the plants that are supposed to be all the same variety look like it? Or do they appear to be a mixture of distinctly different types? It may be OK for the variety to have such heterogeneity — it may even be desirable — or it may not be.

44. *A general, subjective impression of performance is frequently adequate. General impressions can be misleading, though.* Impressions about yield can be especially mislead-

ing. For this reason, it's often necessary to take at least some measurements.

45. *Chance affects the results of a trial. If two varieties are identical, in any given trial, sometimes one will look better and sometimes the other.* Scientists use statistical methods to help evaluate the effects of chance. Most good gardeners develop an excellent feel for what kind of results are meaningful just as a function of experience.

46. *The fewer the plants in a trial, the larger the effects of chance.* A 10 to 20 percent difference in yield between two varieties based on, say, ten plants of each usually isn't reproducible. Even if the varieties are identical, you will often see differences that large just by chance. But a 10 to 20 percent difference between two varieties based on thousands of plants of each usually is reproducible and significant.

47. *If you want to be able to establish small differences, you need more plants than you need to establish larger ones.* To ascertain a 10 percent difference in yield between two varieties, you usually need hundreds or thousands of plants, not just a few. A 50 percent difference in yield between two varieties is usually obvious with just a few plants.

48. *When you see an apparent difference between two varieties and you are wondering whether it is meaningful, one trick is to look at the individual plants.* Is all the apparent difference associated with just one or two plants? If the difference would disappear if a single plant were not there, the difference probably isn't meaningful. Another trick is to divide your trial mentally down the middle (which turns it into two smaller trials) and ask whether both trials show the difference.

49. *Notice the range of both varieties.* Is nearly every plant of one variety better than nearly every plant of the other? If so, the difference between the varieties is probably meaningful, even if it is small. Are all the biggest plants of just one of the varieties? If so, that is probably meaningful, even if the averages for the two varieties are about the same.

50. *When I need quantitative data, I often measure and record certain plants instead of all of them.* In many cases I record just the best plant ("heights up to 7 inches"). *The performance of the best plant is especially important* because it represents what the variety is capable of under my circumstances. The worst plants aren't as meaningful because in many cases they were just unlucky (for example, that's where a cat dug).

51. *In small plantings the performance of the median plant is often a better indication of the true average than the average you would calculate by measuring or weighing all the plants.* The median plant is the one that is better than half of them and worse than half. To get the median, you have to choose and measure or weigh only one plant.

52. *When I don't need quantitative information, I usually don't collect it. I simply record my general impressions.* ("About the same as the standard." "A little less than the standard, but I'm not sure." "About ¾ as much as the standard, but still good enough to try again.") *When I do need quantitative data, I often record the best plant, the median plant, and the worst plant — that is, the median and the range.* I keep more elaborate records only if there is some reason for doing so.

53. *Gardening is both subjective and scien-*

tific. Acknowledge, appreciate, and value both the subjective and the scientific. There is no virtue in filling your notes with numbers and spending a lot of time getting them if the real result — that you do or do not want to try the variety again — is obvious and is all you really need. That conclusion may be subjective, but that is the result that matters.

The Bigger Picture

54. *When things go wrong, try to evaluate what is left without getting discouraged. If you've used good design, you almost always get at least some information* if you look at the situation and think about things before you plow up the plot in disgust. You usually get some of the information you want. You may end up getting some other information as well. Instead of finding out how five varieties performed under ordinary conditions, you may find out that when you go out of town and the friend who promised to water the plot doesn't, only a particular variety survives at all. That isn't what you wanted to know, but it isn't totally useless.

55. *The worst experiences seem to make the best stories.* You may not enjoy learning that you have done a trial to find out what happens when nothing gets watered, but your friends will undoubtedly love to hear about it. *Try to keep your sense of humor.* There will be other years.

56. *As you record and evaluate your trials, ask yourself whether the answers you got are to the questions you thought you were asking.* Sometimes they aren't, and it isn't necessarily obvious. Don't assume that they are without going back and reviewing. *Thinking about a trial afterward is often the most important part.*

57. *All experiments are flawed.* However you design and carry out your trials, there will be some flaws. Every trial also has limitations to how general and applicable it is. Some of these are inherent in the design, and others result from what happened along the way. Record any flaws or limitations that might have affected the results or that might have made the trial turn out differently than it might in some other year, on some other land, or under some other conditions.

What matters isn't to design and conduct an experiment with no flaws or limitations — which isn't possible — but to recognize and record what the flaws and limitations are. Examples of limitations I often list are that I planted later than I would optimally or normally, that part of the planting was more shaded, or that the summer was unusually cold or hot.

58. *Did anything unexpected turn up along the way? Did you find out something interesting that didn't happen to be what you asked? If so, record it.*

59. *Once you have your results, you need to decide what to do with them.* That isn't science; that's judgment. And it's subjective. *Your results don't dictate your judgment; they merely inform it.* They establish some of the facts that are relevant. Usually your judgment is to grow something again or not. Or to grow more of something or not. Or to grow it a different way. You put your trial conclusions together with everything else you know and care about, and then you choose.

The choices may not be permanent. You may think of ways to get around the problem your trial pointed to, for example. You only tried a specific case under specific conditions. You may decide that that didn't work but something similar might. So you do additional trials.

60. *In making your final judgments, consider the bigger picture.* If an open-pollinated variety that hardly anyone grows does almost (but not quite) as well as the fancy hybrid that everybody grows, maybe you should choose the open-pollinated variety. You can save its seed if you need or want to. In addition, when you buy the seed, you are probably supporting a small grower and a family seed company; when you buy the hybrid, you are probably supporting a multinational giant.

If a rare variety does as well or even almost as well as the common one, maybe you should grow it just because of its rareness. By growing it, you help to preserve it and the information about it. And by growing something different from your neighbors, you might not get their pest or disease when it sweeps over the whole area; your variety might be resistant. In addition, the pest or disease your variety might be most susceptible to will be less likely to sweep over your area, since your neighbors are growing something else. Whenever you grow something different from your neighbors, you are contributing to the agricultural biodiversity of your country, your region, and your neighborhood. This is part of the bigger picture.

Putting It All Together: Some Examples

Mustard Trials, 1991

I am inordinately fond of mustard greens. I plant in March for harvest a few weeks later, in July for fall and winter use, or in August for use in early spring. But so far I haven't found any way to have fresh mustard greens during the summer. To get summer mustard, I would need to plant in late spring; mustard planted then bolts without yielding much. One variety, 'Pak Choy', will not bolt when planted during that window. But 'Pak Choy' doesn't yield for me either when planted then. It's a mild type that is destroyed by insects and slugs that are especially efficient in late spring. I also don't much like the flavor.

Prior to 1991, various seed catalogs had been showing me pictures of beautiful new mustards, enticing me with suggestions about better or different gourmet flavors, and listing some as "for growing anytime." I wanted to see if "anytime" included my late spring; I wanted to try some of those new flavors. And I wanted a rough idea about vigor. How well would other varieties compare to 'Green Wave', my standard, which outgrows most weeds? If mild flavored, would they grow fast enough to keep ahead of the pests?

The space I had available for the trial was — well, none. Every square inch was already full of plants. I did have a compost heap that was more or less through cooking. It was located on a former hard-packed dirt driveway north of the garage and under a tree. I quickly shoveled one end of the pile into a flat section about 2 feet wide, 4 feet long, and 5 inches deep. That would be the mustard trial bed. It would support growth of the plants for a month or six weeks — long enough for me to taste everything and see what bolted and what didn't. It wouldn't support growth of full-size plants. But I wasn't expecting full-size plants; I was expecting premature bolting. (This is an example of how asking what answers you expect helps you design. In this case it suggested that I didn't need the land for very long — or even deep enough "land" for "real" plants.)

I would plant 'Green Wave' and 'Pak Choy' as my controls, along with some new types. I was expecting the 'Green Wave' to bolt quickly before producing much of anything, because that's what it always does when planted in late spring; that was the problem. The 'Pak Choy' would not yield because it would get eaten into the ground. Then there would be other mustards, new mustards. Since I was expecting all-or-nothing answers to one question (bolts or doesn't), nearly all-or-nothing answers to another question (gets decimated by pests or doesn't), only a rough idea about the third (vigor), and just a taste for a first impression of flavor, I wouldn't need very many plants per variety I tested.

The next problem in designing this trial was to note the extreme erraticism and nonuniformity of my growing conditions. The compost was pretty poor stuff. Some was loose; some was in not totally cooked clumps. Very large populations of sow bugs were in much of it. The entire patch was in the shade, and one end was almost in full shade — much more shaded than the other end. The whole bed was only 2 feet by 4 feet. What this meant to me was that I should interplant. I needed to have my controls and my experimental varieties spread out over the entire patch. Separate rows for each variety obviously wouldn't work. I would need

to plant at least two rows of every variety to correct for the variable conditions in the patch. Three or four would really be better. But there wasn't enough space for even three or four rows of each of the controls. Even with two rows of each variety, I would be limited to one or two new varieties.

If I mixed a number of varieties together with the controls and broadcast them throughout the patch, I would have all varieties and controls in all possible locations, and the nonuniformity of my patch wouldn't matter. If I interplanted in this way, I could test several varieties — if I could tell them all apart. In fact, since I didn't need many plants of each and I was expecting the trial to end with them still all quite small, if I interplanted, my patch would be a luxuriously adequate amount of land. I could easily try half a dozen varieties. The big problem would be telling them apart.

There were a lot of mustards I wanted to try, and I couldn't try them all at once, so when I went about ordering varieties, I deliberately picked ones that I thought I would be able to tell apart. (Color photos in the bigger catalogs help a lot.) My trial would include 'Pak Choi', 'Joi Choi Hybrid', 'Mei Qing Choi Hybrid', 'Tatsoi', 'Green Wave', and 'Red Giant'. The first four are *Brassica campestris* types. 'Joi Choi' is a hybrid pak (or bok) choy with green leaves and heavy white stems. 'Mei Qing Choi' is a hybrid Shanghai-type pak choy with light green, flattened stems. 'Tatsoi' is a pak choy with small, deeply intense green leaves and white stalks.

'Green Wave' is a *Brassica juncea* type. Its flavor raw is fiery, much too hot to eat in more than tiny amounts. I use small amounts raw, shredded fine in salads. But the real virtue of 'Green Wave' is as a cooking green. The chemical responsible for the hotness is destroyed instantly by cooking. What remains is a rich, uniquely flavorful green, and the broth it creates is also richly flavorful. 'Green Wave' is my favorite cooking green and soup green.

'Green Wave' also grows faster than anything else I grow. Slugs, insects, and other pests generally cannot eat it. When properly grown — in cool weather and in good soil so that it grows rapidly — the entire above-ground part of the plant is edible, including the succulent stalk and leaf stems. (With older or less optimally grown plants, the stems and stalks are stringy and bitter.) It also freezes well. I slice the whole plants into inch-long pieces and blanch them in unsalted water for a couple of minutes. Then I freeze the greens with the broth.

Somewhere along the line, we took a wrong turn with our breeding of cooking greens. Growing and thinking about 'Green Wave' made me realize this. We breed cooking greens that are the same as our raw-eating greens. A mistake, I say. Salad greens and cooking greens should be different plants.

Salad greens must be mild flavored when raw because we eat them raw. So do all the bugs and slugs. This makes them hard to grow. We end up fighting the pests or dumping pesticides on the plants so that there will be something left for us. But cooking greens don't have to be edible raw. We could raise them much more easily if they weren't. When planted at the right time of year, 'Green Wave' grows faster than the weeds.

I think 'Green Wave' represents a superior agroecological concept for a cooking green: *it can be eaten only by animals that know how to cook.* 'Green Wave' mustard fits superbly into low-input, sustainable agricultural patterns. I think we need more and better cooking greens than we have, and in my own breeding work 'Green Wave' is one of my models.

'Red Giant' also is a *Brassica juncea* type. The pictures in the seed catalogs made it look wonderful, but I had never grown it. I wondered whether it was just a different-colored version of 'Green Wave'. Would it have the same flavor or a different one? If different, was it a hot type or a mild type? Would it grow anywhere nearly as well as 'Green Wave'?

I mixed seeds of the different varieties together and broadcast it. I didn't bother labeling anything in the field. I did, however, record the date, varieties planted and sources, a few brief remarks about the compost, and a curse or two about the sow bugs — about three lines total.

The 'Green Wave', as usual, grew much faster than everything else. I thinned to cut back the 'Green Wave' and create a uniform stand of all the varieties throughout the patch. There weren't any weeds. I watered the bed every few days when I tended my ducks, since it was in the same area.

The 'Green Wave' bolted quickly before providing any significant amount of food, as I had expected. Nothing ate it. The 'Pak Choy' did not bolt but was eaten into the ground by pests before providing any significant amount of food, as I had expected. The 'Red Giant' was hot, but not nearly as hot as

'Green Wave'. And its flavor was entirely different. It was not simply a different-colored 'Green Wave'. It both bolted and was eaten into the ground by pests. Everything else also bolted and was eaten into the ground — except for one plant, which didn't look like any of the varieties I'd planted and which I'll return to later.

Only the 'Pak Choy' and the hybrid version thereof were bolt resistant under these conditions. That didn't do me any good because of their pest susceptibility. I didn't get to taste more than a tiny leaf or part of a leaf of each, but it was enough to tell me that as far as I was concerned, there was nothing new or special with respect to flavor that would make me want to grow the new varieties.

The 'Red Giant', though, was distinctly different in flavor. The plants were beautiful, too — intense purple-red tops of leaves and light green underneath. It was clearly not quite as vigorous as 'Green Wave', but it was pretty good.

None of my six varieties was successful at growing in the mustardless window. None of the B. campestris types had any remarkable new flavor. But I did find a new variety that I wanted to work with further — 'Red Giant'.

Of the flaws and limitations of the trial, the major one that might affect my conclusions was the abundance of sow bugs. Sow bugs aren't my normal pest, so the mustards that were decimated in this trial might behave differently in another trial with a different major pest. Shade was a concern initially. Had things failed to bolt, it might have been that they would have bolted with more

sun. That is, had the results been in the opposite direction, I would have listed shade as a limitation on the generality of the result. As it was, with everything bolting, shade seemed not to matter.

Also, my impressions about flavor have to do only with warm-grown material. It's quite possible that the flavors would be quite different under cool-grown conditions. I expect that they would be different once the plants were touched with frost.

Nothing suggested that the nonuniform, weird, ridiculous conditions of my trial bed were any problem. I had a good spread of plants of all kinds all over the bed. All the plants of the bolting varieties bolted, and none of the plants of the nonbolting ones bolted, wherever they were in the patch.

It didn't take me any time at all to figure out my trial design. I use this design all the time. I knew I would use this design when I chose what mustards to order from the seed companies. There are dozens I wanted to try, but I couldn't do them all, so I chose ones that I could tell apart readily. Once I've tried all those that work in this kind of trial, I'll have to use a different design to deal with those I can't tell apart.

Choosing four varieties that I could distinguish from 'Green Wave', 'Pak Choy', and each other took about an hour with the seed catalogs in the winter. Time required for building the bed and planting: about half an hour. Evaluation time: about half an hour.

The trial provided good basic information on flavor, bolting, and pest resistance, and rough information on vigor for four new mustards trialed against 'Green Wave' and

'Pak Choy' as controls. It required no real land and only two hours of work, one of which was lolling around in an easy chair indoors drooling over seed catalogs.

Interplanting is a powerful technique for getting the most information out of a small amount of land with erratic conditions. I commonly use a couple of other interplanting patterns. For example, if I want to trial a mustard that might look like 'Green Wave', I sow one end and the middle of a food production bed with 'Green Wave' and the other end and the middle with the test variety. That way, if I can tell the varieties apart, I have the advantages of interplanting in the midsection and can get a rigorous comparison. If I can't tell the two varieties apart, I find that out and at least get some information by comparing the plants at the two ends.

I would usually plant the bulk of the bed in 'Green Wave' and only one end with a little section of overlap in the new variety. That way I get almost as much food from the bed as I would without the trial, even if the new variety is worthless.

If I want to trial a number of varieties that I may not be able to tell apart but that I can tell from 'Green Wave', I plant the new varieties in separate rows and then broadcast 'Green Wave' over the entire patch. The varieties are separate from each other but interplanted with the standard. This design gives a good comparison of the growth and vigor of each variety as compared to 'Green Wave'. That is more important to me than how they do relative to each other.

Where different kinds of seed are the same size and shape, I usually mix them and broadcast all of them at once. If they differ substantially, I broadcast each separately. Afterward, I either rake or water in the seed. I usually try to sow carefully so that I don't have to thin too much. When I thin, I favor whichever plants I need to get an even stand with the appropriate proportions of all the varieties. When the varieties are in separate sections or rows, I use short little rake strokes within the section or row, not broad strokes that drag dirt and seeds between different sections.

I mentioned that there was one unusual plant in my mustard trial patch. I noticed the plant quite early because it was doing so well. It was growing as well as, or perhaps even a bit better than, the 'Green Wave'. After the 'Green Wave' began to bolt, the plant towered over everything else in the patch. It was not any more bug- or slug-eaten than the 'Green Wave', even though it appeared to be a mild-mustard type. And it wasn't bolting.

When I first noticed the plant, I gazed at it quite a while without touching it. Then I reached out a reverent hand to pluck a single leaf for tasting. *And it bit me!*

I withdrew my hand quickly. Then I got down on my hands and knees and *really* looked at the plant. It had bright green leaves and very light greenish stems. I reached out gingerly and touched a leaf. Hairs. It had hairs. Delicate little hairs all over both upper and lower surfaces. They weren't especially hurtful, if you were expecting them. Pests, though, apparently did not enjoy them.

I picked a leaf with its stalk, took it inside, sliced it, and dropped it in boiling water for a second or so. Then I withdrew the leaf and tasted it. All signs of the hairs vanished in even an instant of cooking, and the leaf had a mild flavor, which seemed to be something special. It seemed much more delicious than cooked greens of other mild mustards. But I might have been imagining the difference. I couldn't compare the plant's flavor with anything else, since nothing else produced enough edible greens to cook.

Aha, I thought happily, here, potentially, is another cooking green of the kind I argue is ideal for organic or sustainable, low-input gardening or farming — a cooking green that has strong natural defenses against any animal that doesn't know how to cook. And being both nonbolting and pest resistant, it could be planted in that mustardless window and give me mustard in midsummer.

I didn't know what the plant was or where it came from. It could have been a mutant in any one of the varieties I'd planted. It could have been a seed that resulted from an accidental cross in the field of the growers of any one of those varieties — a cross with a wild brassica of some kind or maybe even with a turnip. (Turnips are somewhat hairy, and the root on this thing looked somewhat like a turnip.) It also might have been a contaminant, something caught in the seed-processing equipment from whatever was processed before one of my varieties. But I don't know of any hairy 'Pak Choy'–type mustards. Since some of the varieties I planted were hybrids,

this plant also could have represented an accidental self-pollination of a parent variety, or it might have been from a seed from my own compost.

The unusual plant overwintered and grew to a huge size — 2 feet high, 4 feet across, and covered with flowers. I bud-pollinated some flowers in case it had an incompatibility system I needed to get around, but it wasn't necessary. The plant produced thousands of seeds, and the seed was nearly 100 percent viable. I have passed most of this seed on to Miriam Ebert, a friend and fellow gardener. Miriam and I will work together to select a new, insect-proof, hairy mustard variety. Should we develop something useful, we will offer third and further generations of breeding material to others, so that others can derive their own, locally adapted varieties.

Perhaps that original hairy plant will form the foundation for a new mustard that satisfies my needs during the hitherto mustardless summer. It may even be the foundation for a number of varieties derived by others to meet their needs and preferences. But whatever else, this plant has taught me something. It has focused my attention on hairiness.

I had not, until this plant suggested the way, recognized the full implications and possibilities of hairiness. Many wild mustards are hairy. It shouldn't be difficult to create hairy brassicas, having once decided that there is a reason to do so. The trick, of course, is the right sort of hairs. They have to disappear when cooked.

My trial wasn't aimed at finding out

whether hairiness could produce pest-proof but still edible brassicas, but that's part of what I found out. This is an example of what I mean by a trial producing answers to questions other than the ones you asked. The single hairy plant could turn out to be more important than all the rest of the trial, either by providing germplasm for future food or by providing food for future thought.

Corn Trials; Variable Spacing

I frequently try out new corn varieties. Sometimes I try sweet corns, sometimes field corns. Right now I am especially interested in old-fashioned roasting-ear dents.

When I trial corn varieties, I usually plant a 20-foot row of each. At one end of each row, I space the plants about 6 inches apart; that's close enough so that any variety will be stunted. Then I plant at wider and wider spacings until, at the other end of the row, the plants are 3 feet apart. That's far enough apart so that any variety will have more space than it needs. That way I see how each variety does when spaced optimally, and I find out how various varieties should be spaced under my conditions.

Let's suppose instead that I planted a small early variety and a large late one at a constant spacing of 8 inches apart in the row. The small variety would yield almost optimally; the large one would probably be so stunted that it wouldn't mature any ears at all. I wouldn't be asking how the varieties compared to each other. I would be asking how they compared when spaced 8 inches apart. The two are very different questions.

At a spacing of 2 feet, the large variety would probably yield more than the small. If I wanted to know potential yield per foot of row, I would need to evaluate at more than just one spacing.

With a 20-foot row with variable spacing, I can get a good idea of how each variety performs optimally, how each variety should be spaced so as to perform optimally, and how each variety performs with specific spacings. When I compare corn varieties grown in beds, I plant them 6 inches apart in all directions at one end, varying to 2 feet apart at the other end.

A practicality when you use variable spacing is that you don't want holes in the pattern. One way to generate a holeless pattern is to plant two or three seeds in a little cluster in every position where you are going to want a plant. Then thin to one plant.

When you are interested in how the varieties do, you need to ask yourself exactly what you mean by that. Is it total yield you care about? Yield in some part of the season? Yield of material only above a certain size or quality? Yield early in the season? And if it is yield you're after, is it yield per plant or yield per unit space?

In my trialing of different corn varieties with different spacings, I really care about yield per unit of garden space, not yield per plant. But it isn't total yield that matters to me; it is yield of ears of a certain size and quality. Usually there is some spacing that stunts the plants a bit and gives two or three smaller ears but gives the highest total yield per foot of row. And there is some other, greater spacing that gives less yield per foot but produces larger ears. I generally prefer larger ears. Little ones cost more labor in preparation per amount of food. I don't like shucking little ears and getting just a few bites of kernels. So the question I am most interested in isn't "Which spacing gives the highest yield?" but instead "Which spacing does the best job of giving me a reasonable number of *large* ears?"

To help yourself accurately formulate the question you really want to ask, it often helps to ask what you do *and do not* care about or expect to learn from the trial. In corn trials of the sort I have just described, I can get a good idea about earliness, size of plants, size and shape of ears and grain, and optimal spacing. I can get only a general idea about yield. And I can't get more than a rough idea about flavor. There are only a few plants at optimal spacing for each variety, and each variety has different characteristics that tell you when to pick the ear. I can't easily learn the tricks for telling when the ears of a variety are prime unless I have more plants of it.

If I am impressed by the performance of a variety in my initial trial, I plant an entire block of it, or two or three blocks, at the spacing suggested by the first trial. In the second year, I learn to harvest the variety in its prime and can compare it with other varieties harvested in their prime.

Tomato Trials, Spring 1992

Miriam Ebert and I are cooperating in some tomato trials. Altogether, we are trying about twenty cherry-type varieties and about twenty standard large ones. Miriam is a tomato fanatic. She normally plants twenty or more varieties and loves to try new ones. My major motive is related to my interest in creating 'Baby Girl' (see Chapter 3). 'Baby

Girl' would have the flavor of 'Early Girl Hybrid' and would be as early, but it would be open-pollinated and somewhat smaller. I may also want it to be a high-vitamin orange instead of a low-vitamin red (see Appendix B). I have a number of breeding projects aimed at creating 'Baby Girl'. But what about the flavor of the cherry and smaller tomatoes that already exist? Perhaps someone has already bred my 'Baby Girl' or has developed something that I can use as part of the germplasm for developing it. So I am interested in tasting as many cherry and smaller tomatoes as possible.

We are growing several plants of 'Early Girl', since that is the most important control. They are spread around the patch, interplanted with the other varieties. We're also growing a number of plants of certain other varieties we care a lot about. But we have only a plant or two of many varieties. There are reasons that even one plant of each would probably be all right.

One major reason has to do with the fact that tomatoes are such strong inbreeders. This means that it is likely that the seeds within each variety are close to genetically identical. Open-pollinated varieties of outbreeders typically show much more variability from plant to plant. (I explain why in Chapter 9.) With a strong inbreeder, an individual plant is likely to be typical; with an outbreeder, you usually need a number of plants to get an idea of what is typical.

A second reason that we can get away with just one or two plants for some varieties is that it wouldn't be a tragedy if we lost those plants. We would just try them again next year. It would be a tragedy if we lost the

'Early Girl', though; doing so would destroy the entire trial. So we need greater numbers of 'Early Girl' just to be on the safe side.

The major question in the trial is that of flavor. We'll be looking for plants that produce tomatoes that taste at least roughly as good as 'Early Girl'. We expect only an approximation. We will trial any variety that seems anywhere near as good as 'Early Girl' more thoroughly in future years. We also should be able to get a rough idea about earliness and tomato size, shape, and color.

We don't expect to be able to tell much about yields, though, since each plant will have different neighbors, which will cause variable degrees of crowding and shading. And each will be in a different section of the patch.

I often use an initial trial just to find out what a variety tastes like. The trial will be quick and dirty because I won't be doing it well enough (with enough different plants in uniform enough conditions) to evaluate yield. I don't care how much a variety produces of something that I don't want to eat. If I like the flavor, I'll trial the variety in additional relevant ways in future years.

Potato Trials; The Effects of Mulching

I was involved in a trial in which one bed of potatoes was mulched to see how that would affect performance compared with unmulched controls. Most of the planting was done normally. Just a single bed was mulched with an ample layer of straw. The mulched potatoes came up much more slowly than the others. That was no surprise, since mulched ground warms up later in spring than bare ground.

The other participants in the trial remarked on a number of occasions about how miserable the mulched potatoes looked. They asked me if we should take off the mulch. I said no without seriously considering it — or explaining. There was good reason for not even considering removing the mulch, but in retrospect, I should have explained.

The next time through the field, I wasn't there, and my coworkers unmulched the potatoes. It was already obvious that they weren't doing as well as the unmulched ones. They were only about half as high and were meager, spindly things compared to the bushes the controls had become. But removing the mulch at that point destroyed the trial. Here's why: The trial was of mulched versus unmulched. When the mulch was removed, it became a trial of temporarily mulched versus unmulched. That isn't a useful trial, since we would not normally do things that way. We destroyed a useful trial and substituted a useless one.

Second, the slower growth of the mulched potatoes wasn't the whole story. We weren't asking the question "Will mulching slow or not slow initial production of aboveground plant mass?" We don't eat potato plants. What matters with potatoes is the potatoes. Slower initial growth of the plant mass may not be disadvantageous. If the season is long enough, there may be just as high a yield of potatoes.

Later maturation may avoid important pests or diseases. Later maturation might be better for storage. Mulching may create better moisture conditions that more than make up for cooler soil and slower initial growth.

Or it might make the patch more drought-proof. The mulched potatoes may yield more even if they start slower.

But even if mulched potatoes yield less, mulching might be preferable to us. The potatoes may be bigger, cleaner, easier to harvest, or more disease free. All these factors need to be considered.

The initial question was "How will it work if they're mulched?" That is a subjective question, but that's OK. It was the right question, the relevant question, the question we really cared about. But we couldn't get the answer to that question without going all the way and producing the potatoes.

In general, you shouldn't cancel a trial just because an experimental treatment shows a difference from the controls in a given direction at a given time. Nor should you try to rearrange the experimental design in mid-course. There may be exceptions, though. If most of the labor is still to come, for example, you may prefer canceling a trial that has already developed serious problems. It may be more efficient use of your time to restart a new and improved version and invest the evaluation time in it instead of in a badly flawed initial attempt.

The Pea Patch, Spring 1992

My southernmost raised bed lies between the sidewalk and a busy street. It is about 4 feet wide and 18 feet long. All my pea breeding work and much else has been done in it. (I did plant peas on a farm in Independence, Oregon, one year, but that material was tilled under and lost.) The peas occupy a wide row down the center; the sides of the patch are full of other plants.

The patch is irregular in shape. One side is straight; one has a decided dip. One end is square; one end isn't. The supports for the trellis are at irregular intervals. They are not especially straight. Two-by-two-inch boards run across the bed every few feet, and they are at irregular intervals, too. Some go directly across the bed, and others go more diagonally. I like asymmetry.

At the moment the bed with my pea breeding also contains trials of ten different daylily varieties for flavor of the flowers; several 'Ragged Jack'/'Red Russian' kale plants for my salads; a few brassicas (a 'Green Wave' mustard and a 'Red Giant' mustard, a couple of 'Marrowstem' kales); some new brassica, whose identity I don't know, that came from a seed saver in Denmark but that is flavorful and very vigorous (perhaps I should start thinking about finding out what it is); three different wild mustards I might want to make crosses with; two plants of a mutant onion that showed up in my onion patch last year; one chickpea; one fava bean; and a garlic chive and two huge French sorrel plants at the end.

The fava bean is one representative of a planting I have at a friend's house. The chickpea is the most cold-hardy orange pop-bean from last year's overwintering trials. When I have plantings elsewhere, I often transplant one representative to my home garden, where I can better watch it and notice things. But that may be an excuse. One chickpea growing in my home garden provides me with more pleasure than a thousand growing elsewhere. I'll be moving the garlic chive and sorrel plants this year so I can expand the pea planting area.

They're holdovers from a larger perennial planting.

At the north end of the patch is a pile of leaves I dumped there last winter. Underneath the leaves, undoubtedly, are a lot of large night crawlers (8 inches or more long) busy softening the soil. When I expand the patch later this year, they will have done most of the work.

I planted the daylilies last spring and have been redigging and planting around them since. This year they should all flower, and I'll taste them all. Then I'll lift the winners, move them to another place, and discard the rest. This bed is not a good one for perennials because it is between two trees whose roots move up into the patch if it is not completely redug every year or two. The redigging usually doesn't happen with the whole bed at once, though.

Last fall I removed the pea vines and planted a combination of brassicas over most of the bed (around the daylilies). The bulk of it was of 'Green Wave' mustard and 'Ragged Jack'/'Red Russian' kale, because the primary purpose was food production and those are my favorites. The 'Red Russian' I use mostly in salads. It's a kale of extraordinary sweetness and flavor — unlike any other; it has no cabbagey or bitter flavor. Its blue-green leaves and purple stems also are beautiful. The rest of the brassica planting included trials of various unfamiliar brassicas, along with small plantings of a few other familiars. I ate most of the brassicas in the winter and early spring, leaving just a few red kales for salads and a few other individual plants I might want for various crosses.

The daylilies I selected are almost all intensely yellow or intensely red, because these are the colors I most want to add to my salads. Three of the varieties flowered last year. Of them, one small red was much more beautiful to me than my standard tawny but had pretty much the same flavor. A large yellow was more beautiful than the tawny but had an inferior flavor. A huge, beautiful, red tetraploid had a fine, lovely, sweet flavor far better than that of the tawny, and its petals are thick, with a crisp texture. It is something special.

I don't expect to be able to evaluate the productivity, or even the earliness, of the daylilies under my conditions. Some are under the tree, and others aren't. They are all being invaded by tree roots to various extents, and I disturb their roots to various extents as I dig around them to plant other things. My trials are directed only at two questions: flavor of petals and beauty. Once I have a number of winners in flavor and beauty, I'll evaluate productivity and other characteristics under other circumstances. I want to know about size, flavor, and production of tubers, too, for example. I also want to know which varieties can naturalize here — that is, which grow and produce vigorously in an unwatered, untended bed.

As I mentioned earlier, the peas are planted in a wide row down the center of the bed. I redug that area and trimmed back the daylily roots before planting. In most sections the rows are five peas across, spaced about 1½ inches apart in all directions. In some sections the row contracts to fewer peas because of the daylilies (more lovely asymmetry). Down the center of the bed near the ground run five wires, which evolved from one wire.

I used to use one wire down the middle as the bottom support for the trellis. But our neighborhood has a lot of cats, which relish digging in fresh gardens. For planting unique material, excluding cats is essential. I began adding extra wires to exclude cats. Five wires running the entire length of the pea row at a 1½-inch spacing is effective in discouraging them. (The wires are about an inch above the surface of the ground.) Initially I was string-ing the additional cat-suppressing wires after planting, but I soon discovered that it made more sense to string them before planting and use them as guides. And I no longer use any of them as support for the trellis. They are now strictly cat suppressors and planting guides.

With the five wires down the center of the patch and wooden stakes at intervals along it, I have many short rows laid out on a grid. This grid allows me to identify every single pea seed by location. That is very helpful in a situation where every plant is likely to be different and where I will be evaluating and making records about each. In this case it requires little extra work, since something has to be done about cats anyway. The five wires work better than covering the whole area with a strip of chicken wire because they are easier to string and you can just leave them there as the peas come up. In addition, they don't interfere as much with weeding.

In my neighborhood, it's useful to have a trellis down the center of beds like mine even if you don't have anything that needs one. Without it, grown-ups walk and kids and dogs run through. Some obvious barrier is necessary. For this reason I put the trellis up as soon as possible after planting, to pro-tect the bed from gratuitous footprints and rearranging.

There is a modification to the bed to take account of my need to get at the peas for evaluation and breeding — the 2-by-2-inch boards running across the patch. They rest on the edge supports for the bed, not on the ground; they run under and are not attached to the trellis. They're new for me this year. They are a footrest, a place where I can put one foot up and into the patch so as to reach things better.

I presoak my pea breeding material indoors, so that it emerges very quickly and is less vulnerable to slugs. In addition, I can avoid taking up planting space with dead seeds.

The bulk of this year's (1992) pea planting is aimed at evaluating the disease resistance of the material segregating out of different crosses. In addition to the segregating mate-rial, the planting includes both parent vari-eties as controls. One parent, 'Sugar Snap', is sensitive to both enation and powdery mil-dew; the other, 'Oregon Sugar Pod II', is resistant to both. I have some other known sensitive and resistant varieties in the patch as well (a few seeds here and there). I don't expect to be able to evaluate the earliness or yield of my peas from this planting, given the variable amounts of shade and competi-tion with tree and daylily roots.

I have other pea projects in the patch.

There is an F_1 of a cross between a purple shelling pea and an edible-podded variety. That's the beginning of an attempt at getting purple snow and snap peas. Then there is a yellow-podded snow pea, 'Golden Sweet'. It has good-quality pods that are on the small side, and it is not disease resistant. The plants are yellowish green, the pods yellow, the flowers two-toned maroon. The entire plant is beautiful, especially when distributed among standard types with green plants and pods and white flowers. As far as I know, 'Golden Sweet' is available commercially only from Peace Seeds. Alan Kapuler says it is one of the varieties that Mendel used.

I love the beauty of 'Golden Sweet', but I would prefer better disease resistance and bigger pods, so I'm planning some crosses with it. I also used 'Golden Sweet' as a separator between some of the other plantings. It makes an especially good separator because it is easy to distinguish the plants as well as the flowers and pods from all others. I added a 'Golden Sweet' seed here and there throughout the rest of my planting, too, wherever I felt the extra touch of color would be appropriate.

Then there is a seed increase and evaluation on another purple-podded variety. And there are some plants that represent the start of some crosses to obtain tall forms of certain shelling pea varieties.

The plantings of the controls is material I get to eat and need to be able to distinguish especially easily, so they are set off clearly. In some cases tall varieties alternate with short ones. In other cases snow peas alternate with snap peas. In yet other cases a row of 'Golden Sweet' serves as a divider. The purple variety for seed increase is at one extreme end of the patch.

I have about eight pea breeding projects going on in this one patch, a little bit of pea food production, the daylily trials, various plants for other projects, and a good bit of red kale, mustard, and sorrel for my salads. And then there are the aesthetic values. The bed runs along a busy sidewalk. Small trials I've done with yellow and purple peas have aroused intense interest from passersby. And I've responded to that. This pea patch, I am determined, will be both my best and my most beautiful.

To develop disease-resistant snap peas, I have done crosses followed by inbreeding and selection. To combine yellow or purple pod color with snow and snap peas with large pods and disease-resistant plants, I'm doing an initial cross followed by a very powerful technique called *recurrent backcrossing*. The wild mustards will be used in *wide crosses* with domesticates to try to develop whole new crop species. The mutant onions are an example of noticing and using *happy accidents*. These projects and much more are all going on in one tiny 4-by-18-foot bed between a sidewalk and a busy street.

SIX

Genetics and Plant Parenthood

WHY SHOULD we bother learning any genetics when the amateurs of the past accomplished so much without it? Early amateur plant breeders had no idea of the genetic basis for what they were doing. But there were thousands of them operating over hundreds or thousands of years. Generations of those amateurs developed new crop varieties. Most individual breeders probably achieved nothing. In addition, when significant advances did occur, they were undoubtedly more a result of luck than of skill. Nevertheless, crops were developed and improved, if only because there were so many breeders and so much time.

This book is not about how thousands of amateur plant breeders can develop new varieties given hundreds or thousands of years. It is about how individual amateur plant breeders can develop new varieties in one year or decade. To be able to do that, we need to know more about what we're doing than our ancestors did. We need to know something about the genetic basis of plant breeding.

What we need, however, differs from what is usually presented in beginning biology or genetics courses in high school or college. Those courses present the facts and concepts that are most important to our overall understanding of biology, not those that are most relevant to the practical business of breeding plants. In addition, they present nothing in enough depth so that we can actually go out and breed plants much more efficiently.

This book does the opposite. It extracts out of modern genetics only those concepts that are the basis for plant breeding. It covers those few essentials well enough so that we can go right out into the garden and use them. With this background the individual modern amateur should be able to accomplish more in a few years than thousands of amateurs of yesterday put together could achieve in their entire lifetimes.

Only ten methods have accounted for more than 90 percent of the accomplishments of plant breeding, both ancient and modern. They are finding and evaluating germplasm, maintaining germplasm (seed saving and clonal propagation), selection, making F_1 hybrids, using F_2s, inbreeding, outcrossing, backcrossing and recurrent backcrossing, using new mutations, and doing wide crosses (to generate entirely new species). You can do them all. This book covers them all. Of these methods, maintaining germplasm and selection are the foundation for everything else. They, in turn, are based on understanding the essentials of plant reproduction. This chapter introduces genetics, plant reproduction, and selection.

The Basic Observation

The reason that those thousands of amateurs of yesterday managed to accomplish anything at all is because they were doing something right. They had noticed something important, even though they didn't understand it. I call it the *basic observation*.

This basic observation is that offspring tend to resemble their parents. For thousands and probably tens of thousands of years, people both noticed and applied it.

They kept puppies from their favorite dogs, saved seed from their best plants, and were suspicious if children were not at least vaguely similar to their parents. Long before people knew anything about the physical basis for inheritance, they noticed that "like breeds like," and they used that observation to improve their domesticated plants and animals.

When we save seed from the best plants so as to maintain or improve an old variety or to develop a new one, we call it *selection*. We are selecting the best possible parents for the next generation of seed. We do it because we believe in the basic observation; we believe that offspring from the earliest parents are more likely to be early, those from the highest-yielding parents are more likely to be high yielding, and so on.

Until recently, crude selection — selection based on nothing other than the basic observation — was the only deliberate method of plant breeding. Combined with accidental crosses, occasional random new mutations, and a certain amount of luck, however, it created our domesticated food crops, nearly all of which have existed in essential form for hundreds or thousands of years. Most vegetable crops have been improved significantly in the past two centuries, but nearly all were originally developed hundreds or thousands of years ago by amateur plant breeders who relied exclusively on "like breeds like" as their only genetic fact and framework.

The basic observation is only statistically true, however. Offspring do tend to resemble their parents, but many times they do not. Like does breed like — sometimes. Selection, alas, works only in certain situations; it is often useless. In addition, other plant breeding methods are superior to selection in certain circumstances, or they must be used along with it to achieve maximum effectiveness.

The basic rule of selection — to keep seed from plants that have the most desirable parents — is still at the core of plant breeding. But saving seed from the best parents is sometimes more complicated than it seems, because with plants it isn't necessarily obvious who the parents are.

Plant Parenthood

A profound advance in plant breeding came about with the realization that most plants, like most people, have fathers. With plants, however, as with people, it may not be entirely obvious who the father is. Flowers are the reproductive organs of the plant. The plant that produces a seed is clearly the mother of the seed. You can watch the seed-containing fruit develop from the flower on its mother. That the seed has a father isn't necessarily obvious.

Human understanding of plant reproduction was undoubtedly complicated by the fact that plants are much more sexually diverse and versatile than people. Humans come in two and only two morphologically distinct sexual types — male and female — and they have only one pattern of reproduction. A male and a female can mate to produce offspring. Neither can reproduce by himself or herself.

Spinach plants and humans have a lot in common. Like people, spinach plants come in two sexes that are morphologically distinct and have different reproductive roles. Human females produce eggs, which remain inside their bodies and are fertilized by sperm. Human males produce sperm, which they deposit inside the reproductive tract of the female. The sperm swims its way to the egg and unites with it, thus combining the genetic contribution of the mother with that of the father to produce a fertilized egg (the zygote). The zygote develops into a baby human inside the mother's body.

Spinach females produce only female flowers — that is, egg-bearing flowers. They don't have male flowers — that is, pollen-producing flowers. The eggs of a spinach female remain in the ovary of the flower and are fertilized by pollen from spinach males. Fertilized spinach eggs develop into spinach embryos in the female plant.

Spinach eggs are specialized for staying where they are and producing an embryo after fertilization. Male spinach plants produce flowers that look very different from those on spinach females. The male flowers don't produce eggs; they produce pollen, which, in spinach plants, is dustlike grains specialized for being blown around by the wind. Would-be spinach fathers produce a lot of pollen, because only chance determines whether any individual pollen grain will land on your car or on the ground as opposed to on a stigma of a female spinach flower. When a spinach pollen grain lands on a receptive stigma, it grows into and through it, down the style and into the ovary, where it fertilizes the egg.

Spinach eggs and human eggs are essentially equivalent. Both are very large cells that contain the genetic material of the female along with all the machinery and raw materials necessary to translate genetic material into an embryo. Sperm and pollen also are similar. Both are very small cells that contain the genetic contribution of the male. They contain some of the information for making an embryo but none of the machinery and raw materials required. Both specialize in mobility. The sperm has a tail and mechanisms for swimming. Pollen has specializations that permit it to be transported by wind, insects, or other means.

Plant seeds, however, are not really equivalent to human babies. A seed is more than just a plant embryo, a baby plant. It is an embryo carefully packaged with a sophisticated food supply adequate to allow development up until the embryo is able to survive on its own. It often contains mechanisms that hold the embryo in a state of suspended animation until circumstances are right for its development.

Plants such as spinach that have morphologically distinct sexes are known as *dioecious* plants. Some other domesticated dioecious plants are asparagus, yam, date palm, papaya, pili nut, carob, persimmon, fig, hop, mulberry, pistachio, jojoba, hemp/marijuana, and muscadine grape. The dioecious pattern of reproduction is not very common among crop plants, however familiar it seems to us. Most plants are *hermaphrodites.* They have both female and male sexual parts, produce both eggs and pollen, and can function as both mothers and fathers.

Some hermaphroditic plants, such as bean, pea, tomato, lettuce, and cabbage, have hermaphroditic (perfect) flowers — flowers that have both male and female parts. Other plants, such as squash, muskmelon, watermelon, and corn (maize), are hermaphroditic as plants but have separate, unisexual male and female flowers. We refer to such plants and such a flowering pattern as *monoecious.* Glenn Drowns's watermelons, for example, had separate male and female flowers, but each plant had both. Corn has tassels (the pollen-producing male flower) at the top of the plant and female, silk-producing flowers that become corn ears lower down.

Dioecious plants are the extreme case of outbreeding plants. Like humans, the only way they can produce offspring is by a mating of two different individuals, a male and a female. Pea plants, with their perfect flowers, are a good example of the opposite breeding pattern. A pea flower has both male and female parts and produces both pollen and eggs. In peas, however, the stigma of the pistil usually becomes receptive and pollen is usually released and lands on it before the flower opens. So pea flowers nearly always pollinate themselves. A pea still has both a father and a mother, but usually the same plant is both. Pea is thus known as an *inbreeding* plant. Pea, bean, tomato, and lettuce are among the vegetables that are primarily inbreeders — that is, primarily self-pollinating.

Some hermaphroditic plants are primarily *outbreeders,* even though they can function both as male and female. In cabbage, kale, and many other brassicas, for example, an incompatibility system usually prevents self-pollination. Any individual flower functions both as a male and a female and produces both eggs and pollen. But any given seed usually has two parents, not one. Onion also is a hermaphrodite, but it is *protandrous* — that is, the male parts of the flower develop first and release pollen long before the female parts finish developing and become receptive. Onion is primarily an outbreeder, but it is self-compatible. Even though a flower is normally pollinated by some other flower, it can be another flower on the same plant.

Squash, cucumber, and melons, with their separate male and female flowers, are primarily outbreeders. But as with onions, nothing prevents a female squash flower from being pollinated by a male flower on the same plant.

A final major mode of reproduction is important to many plants and alien to humans. Many plants can clone themselves. They can reproduce a new plant vegetatively by sending up a shoot from a root fragment or cutting, or by sending out runners (like strawberries do). Such clones are genetically identical to the plant from which they were derived. In a sense, they have neither mothers nor fathers, since they do not arise by sexual reproduction. Any plant normally grown by planting tubers (or tuber pieces) or cuttings or started from runners is being reproduced vegetatively instead of sexually — that is, it is being cloned. Examples include potato, sweet potato, yam, strawberry, and most fruit trees.

In summary, many plants can clone themselves as well as reproduce sexually, most plants can function as either mothers or fathers, and the seeds of many plants may have either one or two parents.

Basic Selection

Only for the past two centuries or so have people understood what pollen is and what it does. Before then, there was no such thing as deliberately crossing two different plants to combine their desirable characteristics. Any crossing that happened was a function of accident. There was, however, isolation of different varieties to keep them pure. The necessity for it must have been established empirically. The American Indians, for example, grew different varieties of corn in separate fields. They had come to recognize the need for isolation in preserving different corn varieties, even though they did not understand what the corn pollen blowing all over the place had to do with it.

Selection practiced without any understanding of plant reproduction often isn't very powerful, because if you don't understand the role of pollen, when you try to save seed from the best parents, you are actually only choosing good mothers. Each seed has a father, too. In the species that are largely inbreeding, when you select good mothers, you are automatically selecting good fathers. But for species that are largely outbreeding, when you select good mothers, all you have is good mothers. You don't know who the fathers are; they might be undesirable.

Suppose, for example, that you have a

patch of melons of a variety whose seed you are saving. The variety is one of your favorites, but you would prefer a somewhat larger fruit size. So you try to select for large fruit by saving seed from the largest melon in the patch. When you grow the seed out the next year, the melons aren't particularly big. In fact, they don't even look like the same variety. Something has gone wrong.

Actually, several things have gone wrong. You were trying to use the basic rule of selection by saving seed from the best, but you made four important but common mistakes. First, you assumed that the melon is the offspring. It isn't. It doesn't result from fertilization with the pollen; it is just part of the mother. Second, you assumed that the seed would carry genes that would generate melons like the one they started out in. But they don't necessarily, and they didn't. Third, you didn't choose the best father; you didn't know who the father of your seed was. Finally, you didn't necessarily even choose the best possible mother plant.

The flower of a plant is part of the mother plant. It develops into a fruit that is exclusively maternal tissue except for the seed inside it. The fruit — the melon, tomato, or whatever — is part of the mother plant, just like the leaves and roots are. The appearance of the fruit is determined only by the genes and environment of the mother. It is the same, whatever the characteristics of the plant whose pollen fertilized the flower from which it developed. Pollination affects only the seed inside the fruit, not the fruit.

Actually, parts of the seed also are maternal tissue. It is the embryo itself (within the seed) that is the new individual. So if you pollinate a melon flower with pollen from a male of a different variety, the melon will look the same as it would have if the flower had been self-pollinated. In most cases the seed also looks the same. You can't tell anything by looking at the melon or the seed. When you grow the seed up, the plants that result are derived genetically from both the mother and the father. If there has been a cross, the plants will probably look different from the mother and produce melons that are different from the ones they came from.

In some cases in some crop species, the seed may look different if there has been a cross to a genetically different parent. In peas and beans, for example, crosses sometimes show up as differences in the first generation of seed. This is because the huge cotyledon leaves of legumes are the main food storage organ of the seed, and the cotyledons are part of the embryo. Not all crosses show up in the seed even in legumes, however.

In the example I gave earlier, you saved seed from your largest melon, but the next generation produced fruit that looked totally different from the mother. The flower that gave rise to that fruit was probably pollinated by a bee that had just visited a flower from a completely different variety — one growing in your neighbor's garden, for example.

Whenever we try to save seed or do selection, we need to choose both the female and the male parents. This means that we need to control pollination. In strongly inbreeding crops such as pea, we don't actually have to do anything, because left to her own devices, the mother plant also contributes the pollen. So if we choose a good mother plant, she is also the father plant. Selection is very easy with inbreeding plants; we just save seed from the best plants.

With outbreeding plants, however, whenever we maintain a variety, practice selection, or do any kind of plant breeding, we have to deliver the desirable pollen to the plant and exclude unwanted pollen. This means that before we can maintain any crop variety or develop any new ones, we need to know whether the plant we're working with is an inbreeder or an outbreeder. That information is included in Appendix A and Table 1 for all the common vegetables. In chapter 9 I explain how to figure it out for yourself in cases where no information is available.

The final mistake you made in trying to select for large melons by saving seed from the largest melon was that you didn't necessarily even choose the best female parent. The size of a melon is affected very much by whether it is the first melon on the plant or a later one, for example, and by how many melons there are all together. Suppose your plant had difficulty setting fruit under your climatic conditions and only managed to set one fruit. That fruit would be much larger than fruits from sister plants that set several. But would you want a variety that produced so few fruits, whose ability to set fruit at all was so marginal?

When we practice selection, we need to consider the entire plant, not just the fruit on the plant. We would choose a plant whose overall performance came closest to what we wanted and then either self-pollinate it or cross it to another, superior individual of the variety. (Which we do depends on considerations I'll cover later.)

When we act positively and save seed from only certain plants and eliminate the rest, we call it selection. But sometimes we do the opposite. We save all the seed except that from the worst plants. This is selection also, but it is usually called *culling* or *roguing* and refers to eliminating the rogues, or bad plants. Normally there are inferior plants in every generation. When we maintain a variety, we try to eliminate seed from the inferior plants. In an outbreeding crop, we do the roguing before the plants flower, if at all possible, so that the inferior plants don't contribute to the next generation as either maternal or paternal parents.

In a home garden, we usually don't discard the rogues; we usually eat them. In a corn patch, for example, we don't have to cull out whole plants and lose their production. Instead, we can remove the tassels from the inferior plants so that they cannot produce pollen; then we eat their ears instead of saving them. In such cases, we always need to consider contamination from other corn patches. Sweet corn pollen, for example, travels long distances. For this reason most home gardeners who save their own corn seed must hand-pollinate it.

There are other aspects of selection. Selection doesn't always work. It depends on the existence of genetic variability for the characteristic in which we are interested. If there is no genetic variability for fruit size in our melon variety, we can select year after year and achieve absolutely nothing. We would need to use other methods.

Selection is both the simplest and the most sophisticated of genetic methods. In this chapter I've outlined the essentials. I'll take up selection again later. At that point you will learn when to use selection and how to use it most effectively. You also will be able to do power selection instead of just basic selection. In addition, selection is just one of the ten major plant breeding methods. Most of the other methods, however, depend on having a better understanding of genes and how they behave — which is the subject of the next chapter.

Sex and the Single Gene; Mendel's Genes

PEOPLE HAD the idea that there was such a thing as heredity long before they knew anything about the physical basis for it. They knew that offspring tended to resemble their parents, and they supposed that there must be some hereditary material that offspring got from their parents that caused the resemblance. (The terms *heredity* and *genetics* are equivalent.) Early ideas about the nature of the hereditary material seem to be that it was some kind of fluid.

It's easy to see how the fluid concept of heredity arose and why it still seems so intuitively reasonable to the nonspecialist. People probably noticed that when a person with dark skin married one with light skin, their children's skin color was generally somewhere in between that of the two parents —

a result that seems analogous to what happens if you mix two different shades of inky water, for example. Likewise, the offspring of a tall person and a short person are most often somewhere in between the two in height. And when two dogs from breeds with different conformations are mated, the pups usually have a conformation that is quite obviously an average of that of the two parents. Early theories about the mechanism of inheritance were based on a model that involved the blending of the genetic fluids from the two parents. Words such as *half-breed, half-blood,* and *pure-blood* are based on that ancient assumption that the hereditary material was some kind of liquid or blood, perhaps even associated with the blood.

Once scientists understood the role of pollen in plant reproduction, they began to try to decipher the mechanisms of inheritance by doing deliberate crosses of different varieties of plants. Sometimes the results seemed to fit in with the concept of blending inheritance. The offspring sometimes did seem as if they were some kind of blend of the two parents. When a tall variety was crossed with a short variety, for example, the hybrid often was in between in height. When an early variety was crossed with a late one, the hybrid often was intermediate in timing. When white-flowered varieties were crossed with red-flowered varieties of the same species, in some cases the offspring were pink. These particular cases fit in with the concept that the genetic material was a fluid of some kind and that a hybrid between two pure breeding varieties would blend the genetic fluids and therefore be intermediate between the parent varieties.

But hybrids weren't always intermediate between the two parents. For every case where a cross between two varieties gave an intermediate hybrid, there were other cases where the hybrid resembled just one of the parents. And where there were a number of different characteristics involved in a cross, the hybrid might be like one parent for some characteristics, like the other for others, and intermediate for still others. In some cases the hybrid was even more extreme than or entirely different from either parent. Sometimes a cross of two varieties with white flowers would result in offspring that had purple flowers, for example.

In addition, in later generations after a cross, characteristics of the parents that had not appeared in the hybrid could suddenly reappear in some of the progeny in the later generations. These progeny were called *throwbacks*. The word is meaningless now, but it reflects the fact that, according to a blending theory of inheritance, there was no way to explain the sudden appearance of progeny with characteristics that were not present in the immediate parents.

The scientists of the day simply could not explain or understand what was going on when they crossed different plant varieties given the conceptual framework they had for the nature of the hereditary material and the mechanisms of inheritance. This did not mean that they were eager to hear alternative explanations from amateur-scientist monks, however.

Gregor Mendel was an Austrian monk. He is considered to be the founder and father of genetics, although, as far as I can tell, he had exactly zero impact on the field. He was just too far ahead of his time. The theories he proposed were correct, but there was no basis for understanding why they were correct. The framework of facts and concepts that would have made his assumptions acceptable and his theories logical didn't exist yet. Not enough was understood about mitosis and meiosis (asexual and sexual cell division).

Forty years later, after people had better microscopes, could see chromosomes, and had come to understand cell division, there was a physical mechanism to explain and justify Mendel's assumptions. At that point, in the early 1900s, three other scientists did work that was essentially equivalent to Mendel's. None knew anything about Mendel and learned about his work only during the literature searches they did as part of publishing their own work. So Mendel, the father of genetics, had no influence on the course of any science, including the genetics he is considered to be the father of. The three independent discoverers of Mendel's laws had no trouble getting their work accepted, since the framework for understanding it existed by the time they came along.

Mendel did a lot of breeding work with peas. It was a fortunate choice because peas are almost totally self-pollinating. He had a number of different pure breeding varieties — tall ones, short ones, ones with yellow pods, ones with green pods, and so on. He made a number of crosses between varieties that differed by just one characteristic. In each case he collected and planted the F_1 seed and examined and recorded the appearance of the F_1 plants. F_1 stands for "first filial" generation. We use the term F_1 or F_1 hybrid to refer to the first-generation hybrid made by crossing two different pure breeding varieties.

Mendel took things one step further than many others. He also considered the F_2 generation. That is, he planted seed from the F_1 hybrid plants, the F_2 seed. And he examined and recorded the appearance of the F_2 plants. The F_2 is the "second filial" generation, or the second generation after a cross between two pure breeding varieties. It is obtained by crossing two F_1 plants. In the case of a self-pollinator such as peas, you get an F_2 generation just by letting the F_1 plants produce seed.

Mendel had two pure breeding varieties that differed in height, for example. The tall variety always produced tall plants, and so did their offspring. The short variety always produced short plants. Mendel did crosses between plants of the two varieties and saved and planted the seed (the F_1 seed). He recorded the appearance of the F_1 plants and allowed them to set seed (the F_2 seed). Then he planted the F_2 seed and recorded the appearance of the F_2 plants.

Mendel found that the F_1 hybrids between the tall and short variety were all identical to each other. Furthermore, they were all tall — every bit as tall as plants from the tall variety. It was as if the short parents made no hereditary contribution to the F_1 hybrid at all. (According to a blending concept of heredity, the F_1 plants should have been intermediate in height.)

When Mendel planted the F_2 seed and examined the plants, he found further unexpected results. The F_2 plants were of two distinct types. Some were tall, and some were short. There was nothing in between. There were 787 tall plants and 277 short ones. That is, about three-quarters of the F_2 plants were tall; about one-quarter were short. The tall plants were just as tall as plants from the pure tall variety, and the short plants were just as short as those from the pure short variety.

Sometimes Mendel did the original cross using a tall female and a short male. Other times he did the reciprocal cross (short female × tall male). The results were exactly the same for all crosses, in whichever direction they were done.

Mendel proposed that inheritance was, basically, an either/or sort of thing. Tallness and shortness stayed separate, segregated, from each other. They didn't seem to blend or influence each other at all. (This phenomenon is often referred to as the Law of Segregation, or Mendel's First Law.) The hereditary material can't be fluid at all, Mendel said. It must come in units, or factors — lumps, so to speak — and these factors could exist in two different forms (*alleles*) that were inherited as alternatives to each other. Today the factors are usually called *genes*. The word *factor* is still used in certain contexts, though.

Mendel proposed that his two original varieties differed by one gene that influenced height. That is, the two varieties had different alleles for a gene involved in determining height.

The F_1 hybrid plants looked exactly like the plants of the tall variety, but it was clear to Mendel that they were not genetically identical. They couldn't be, since what they could pass on to their offspring was different. The tall variety was pure breeding for tallness. All the offspring were tall, and all the offspring of those offspring were tall. But the tall F_1 wasn't pure breeding for tall. Some of the offspring of tall F_1 plants were short. So the tall F_1 must be carrying genetic material that could determine shortness, even though it wasn't expressed in the F_1.

The numbers of the two different classes of the F_2 also were very interesting. About three-quarters of the F_2 plants were tall, and about one-quarter were short.

Mendel did six other crosses that produced similar results. When he crossed a green-podded variety with a yellow-podded variety, for example, the F_1 all had green pods. Then the F_2 fell into two distinct classes that resembled the two parent varieties. There were 428 green-podded plants and 152 yellow-podded ones. Again, the F_1 resembled just one of the parent varieties, but there were two discreet classes in the F_2 that resembled the two parents. And the numbers approximated three-quarters of the total for one class and one-quarter for the other. Furthermore, whatever the F_1 looked like was the majority class in the F_2. The same kind of results, with the same disappearance of one of the parent classes in the F_1 and same reappearance of it as one-quarter of the F_2, occurred in all Mendel's other crosses as well.

Mendel apparently knew enough math to realize that there is something special about the number $1/4$; it's what you get when you multiply two $1/2$'s. And $1/2$ is really special. It's the probability associated with a random choice between two alternatives that are equally probable. (*Random*, in this context, means "by chance.")

Example: If you flip a balanced coin, it is just as likely to land heads as tails. That is, there are two possible results, and the choice between them is random, so the probability of getting a head is 1 in 2, or $1/2$.

Suppose you flip two coins. What's the probability of both being heads? Well, the first one will be heads half of the time, and of that half of the time, the second coin also will be heads half of the time. So the probability of getting two heads at once is ($1/2$ × $1/2$) or $1/4$. The generalization is called the Product Law of Probability, and it comes up over and over again in genetics. It states that if you know the probability of two events occurring separately, and they are independent events, then the probability of their occurring simultaneously is the product of the probabilities of their occurring separately. Example: If you toss a coin and one die, what's the probability that the coin will be heads and the die will be a three? Answer: The probability of a heads is ½. The die is equally likely to land on any of six sides, so the probability of getting a three is ⅙. The probability of getting a heads and a three simultaneously is one-half times one-sixth, or (½)(⅙), that is, ¹⁄₁₂.

We also use the Product Law of Probability to describe all possibilities and their probabilities. When you toss one coin one time, all the outcomes and their probabilities can be represented as (½H + ½T), where H stands for heads and T stands for tails. The expression (½H + ½T) shows that there are only two possible results from one toss of one coin and they are equally probable.

If we toss a second coin, it, too, can be expected to land as (½H + ½T). If we want to know all the possible outcomes of the toss of two coins, we can simply multiply the two expressions. That is, all the possible outcomes of the toss of two coins and their probabilities will result when we multiply (½H + ½T) by (½H + ½T). That multiplies out to (¼HH + ¼HT + ¼TH + ¼TT). In most cases the two middle terms look identical, because we don't care which coin landed heads or tails, only that we got one of each. So the expression becomes (¼HH + ½HT + ¼TT). That is, one-quarter of the time we get two heads, one-quarter of the time we get two tails, and one-half of the time we get one of each.

In case your math is rusty, expressions such as (¼H + ¼T) or (a + b) that have more than one term are called polynomial expressions. You multiply two polynomials just by multiplying each term in one by each term in the other and adding all the resulting products. So (a + b)(c + d) works out to (ac + ad + bc + bd). Some other examples: The product of (a) and (b + c) would be (ab + ac). The product of (d + e) and (f + g) would be (df + dg + ef + eg). The product of (h + i) and (j + k + l) would be (hj + hk + hl + ij + ik + il). And (m + n + o)(p + q + r) would multiply out to (mp + mq + mr + np + nq + nr + op + oq + or).

I don't use much math in this book, but you often will want to manipulate polynomials to figure out the probability of getting progeny of some specific desired type from a given cross or series of crosses. This tells you how many offspring (seeds) you need to plant to get what you want. In other cases you can use such estimates to evaluate whether a specific plant breeding approach is likely to work at all. The two are often related. If you calculate that you need to grow ten thousand of something to get one of what you want and you only have room for twenty, you know that you will need to use some other approach.

To go back to coin tossing, if we toss two coins, what's the probability of getting at least one heads? We can see from the (¼HH + ½HT + ¼TT) distribution that it is ¾. That is, the distribution can be collapsed to (¾H_ + ¼TT): three-quarters of one or more heads plus one-quarter of no heads, or three-quarters of one outcome and one-quarter of the other. *These are exactly the same numbers that were showing up in Mendel's pea crosses.*

Mendel supposed that there was some unit of heredity, some factor, some gene, associated with the difference in height between the two varieties, and it had two alternate forms that were different in the two varieties. He further supposed that every pea plant had two alleles for the height gene in question, and they could be the same or different. He yet further supposed that during formation of an egg, a plant passed on just one of its two alleles, and it was chance as to which one it passed on. Likewise, during formation of pollen, the plant would pass on only one of the two alleles it had, and it would be random as to which.

When an egg was fertilized, the resulting zygote would get one allele from the egg and one from the pollen, and would have two alleles. Fertilization was equivalent to the simultaneous occurrence of two random and independent events. Which of two different alleles went into an egg would be like whether one coin landed heads or tails. Which of two different alleles went into the pollen grain would be like whether another coin landed heads or tails. And which combination of egg and pollen came together in a fertilization would be like considering two coins tossed simultaneously.

Mendel used uppercase and lowercase letters to represent different alleles of genes involved in his crosses. Using this convention, we can name the gene involved in his tall × short crosses the *T factor* and the two different alternative alleles involved in his cross *T* and *t*. The tall variety would have the genotype *TT* and the short variety genotype *tt*. The original varieties were both pure breeding. Another way of envisioning this is that they could pass on only one kind of gene associated with height. We refer to the condition of the original varieties as *homozygous* — that is, each had two identical alleles. The F_1 hybrid would get a *T* from the tall parent and a *t* from the short parent, so it would have the genotype *Tt*. It would be *heterozygous* — that is, it would have two different alleles.

The F_1 was tall, however. It looked the same as the original tall variety, but it was genetically different. (It had to be genetically different from the parent variety because it could pass on shortness to some offspring, but the tall variety couldn't.) Mendel used the word *genotype* to describe the actual alleles present and the word *phenotype* to describe what the plants looked like. Apparently, said Mendel, the genotype *Tt* looks just like the genotype *TT*. And he coined some terms. The *T* allele was *dominant* to the *t* allele; the *t* allele was *recessive* to the *T* allele. The F_1 was phenotypically identical to the tall variety, but it was genotypically different. (The terms don't explain the whys of it; they just describe the situation.)

Given the rules Mendel set up, what kinds of genotypes would we expect in the F_2, and with what frequencies? Well, the F_2 seed is the progeny of a self-pollination of the F_1 pea plant, which is of genotype *Tt*. The F_1 plant has two different alleles, so it can pass on either *T* or *t* to its eggs. Mendel assumed that chance determined which allele the egg got, so half the eggs would get *T* and the other half *t*. Likewise, the F_1 pea plant could pass on either allele to its pollen. Then either kind of egg could combine with either kind of pollen. So one could get four different kinds of fertilization, and they would be equally probable.

To calculate all possible genotypes that can result from the cross and their probabilities, we multiply the egg distribution by the pollen distribution. All possible genotypes of eggs and their frequencies are represented by the expression $(\frac{1}{2} T + \frac{1}{2} t)$. All possible genotypes of pollen grains and their frequencies are represented by that same expression. So the expression for all possible genotypes of zygotes and their frequencies is $(\frac{1}{2}T + \frac{1}{2}t)(\frac{1}{2}T + \frac{1}{2}t)$, which gives $(\frac{1}{4}TT + \frac{1}{4}Tt + \frac{1}{4}tT + \frac{1}{4}tt)$, which reduces to $(\frac{1}{4}TT + \frac{1}{2}Tt + \frac{1}{4}tt)$. In terms of phenotypes, this is $\frac{3}{4}$ tall $+ \frac{1}{4}$ short.

According to Mendel's theory, the short plants should all be pure breeding. He confirmed this by self-pollinating a number of them to obtain an F_3. His theory also specified that the tall offspring weren't all identical. Some should be homozygous and pure breeding. Others should be heterozygous and should be able to pass on either tallness or shortness. By inbreeding a number of tall plants, Mendel showed this, too, was true. In this fashion, he showed that the $(\frac{3}{4}$ tall $+ \frac{1}{4}$ short) he obtained in the F_2 was actually $(\frac{1}{4}$ pure breeding tall $+ \frac{1}{2}$ talls that carried short $+ \frac{1}{4}$ pure breeding short). He did the same work with his crosses involving other characters and obtained the same results.

Mendel also did another kind of cross, which we now call a *backcross*. He crossed tall F_1 plants back to the short parent. This cross can be represented as ($Tt \times tt$). The genotypes Mendel's theory predicts and their frequencies are given by the expression $(\frac{1}{2}$ $T + \frac{1}{2} t)(1\ t)$. That would be $(\frac{1}{2} Tt + \frac{1}{2}$ $tt)$, that is, one-half tall and one-half short. This is what Mendel obtained. Furthermore, Mendel's theory implied that all the tall offspring of the backcross should be heterozygous, and he showed that this, too, was true.

Mendel did another experiment. He took the tall offspring from a backcross and backcrossed again. This cross is again ($Tt \times tt$). It again gave one-half tall and one-half short. According to a blending concept of heredity, these talls were only quarter bloods, so to speak, with respect to the tall variety. But they were every bit as tall as the original tall variety. And when backcrossed yet again, the offspring were still one half tall and one half short. These talls were still as tall as the original variety, even though they were now eighth bloods; and the shorts were still as short as the short variety. There is no way to explain the results of Mendel's serial backcrosses with a blending concept of inheritance.

Mendel also did various crosses involving two different traits at once, and he found simple, predictable results in these cases also. When he crossed a tall variety with green pods to a short variety with yellow pods, for example, the F_1 plants were all tall and green podded. The F_2 plants were ($\frac{9}{16}$ tall green $+ \frac{3}{16}$ short green $+ \frac{3}{16}$ tall yellow $+ \frac{1}{16}$ short yellow).

Mendel realized that his results were what you would expect if the inheritance of the two traits were independent. In other words, a two-factor cross was just two one-factor crosses occurring simultaneously. If so, to obtain the classes and frequencies of both traits simultaneously, you just multiply what you would expect of each alone. The F_2 with respect to height is ($\frac{3}{4}$ tall + $\frac{1}{4}$ short). The F_2 with respect to pod color is ($\frac{3}{4}$ green + $\frac{1}{4}$ yellow). So the F_2 considering both simultaneously is ($\frac{3}{4}$ tall + $\frac{1}{4}$ short)($\frac{3}{4}$ green + $\frac{1}{4}$ yellow). That multiplies out to ($\frac{9}{16}$ tall green + $\frac{3}{16}$ tall yellow + $\frac{3}{16}$ short green + $\frac{1}{16}$ short yellow).

To portray these results in terms of genotypes, let's assign the letter G to represent the allele associated with green pods. So the green-podded variety is $GG,$ and the yellow-podded one is $gg.$ The cross of the tall green variety and the short yellow one can then be represented ($TTGG \times ttgg$). The results expressed in terms of genotypes are ($\frac{3}{4}$ $T\_$ + $\frac{1}{4}tt$) ($\frac{3}{4}G\_$ + $\frac{1}{4}gg$), or ($\frac{9}{16}$ $T\_G\_$ + $\frac{3}{16}T\_gg$ + $\frac{3}{16}ttG\_$ + $\frac{1}{16}ttgg$).

Mendel performed all possible two-factor crosses with his seven traits, and in all cases he found the same pattern. In the F_2, $\frac{9}{16}$ of the plants displayed the two dominant traits and resembled the F_1; $\frac{3}{16}$ had one dominant trait and one recessive; $\frac{3}{16}$ had the other combination of dominant and recessive; and $\frac{1}{16}$ had the two recessives.

In each of the two-factor crosses, four classes of F_2 progeny were obtained. Two classes resembled the two original parent varieties. But there were two new classes, two classes in which the traits of the parent occurred in different combinations. The cross (tall green × short yellow) produced an F_2 that included both the parental classes, tall green and short yellow, and also the recombinant classes, tall yellow and short green.

Mendel's explanation of how two different traits are inherited in two-factor crosses is referred to as independent assortment or recombination. (The Law of Independent Assortment is Mendel's Second Law.)

The phenomenon of independent assortment or recombination of phenotypic traits in the F_2 forms the basis for much of plant breeding. Independent assortment allows the breeder to cross two varieties that differ in a number of ways and to derive new varieties with different combinations of the characteristics of the parents, or even completely new phenotypes.

EIGHT

Modern Genes

T HE GENETIC material of plants and other higher organisms is made out of DNA, a long, chainlike, linear molecule composed of four different kinds of subunits. DNA has two profound functions that are essential to its role as the hereditary material. First, it self-replicates. Given the machinery in the cell, a DNA molecule can synthesize a duplicate of itself, using itself as the template. No other biological molecule can do so. Second, DNA encodes information. The four kinds of subunits can occur in all possible orders along the DNA chain; the order of the subunits encodes the information required to make proteins.

Proteins are one of the most important classes of substances in living things. Structural proteins form much of the basic struc-

ture of cells. Other proteins, called enzymes, are responsible for catalyzing nearly all the chemical reactions that take place in cells — all the reactions that allow cells to grow, divide, develop, and do all the things that cells do. In other words, DNA, by encoding the information to make proteins, specifies everything living things need to be and do to be living things.

Today quite a lot is known about the structure and function of specific genes. But the genes that have been most studied and that are best understood are mostly those in bacteria, viruses, yeast, and other laboratory organisms such as mice, as well as those of medical importance in humans. The structure and function of most genes in most plants is unknown. Peas, tomatoes, and corn are the only vegetables that have been studied much genetically. Most other vegetables, even common ones such as squash, are nearly complete genetic unknowns. Even with peas, tomatoes, and corn, most of the agronomically important genes are still unidentified.

In most practical breeding of vegetables, we manipulate and recombine genes without knowing specifically what they do. We usually don't even know what genes we're working with. It doesn't matter, though. The manipulation and rearrangement of genes into new varieties using classic plant breeding methods does not in any way depend on knowing the genetic or biochemical identity of the genes involved. All it depends on is an understanding of the general principles.

The other major profound characteristic of DNA, besides its ability to encode information, is its ability to replicate. If DNA is to be the genetic material, it must be able to replicate so that each time a cell divides and becomes two cells, both can have a copy of all the genetic information they need.

In this book I don't go into the biochemistry of how genes code for information or how they replicate. More about that can be found in any beginning biology or genetics textbook. One aspect of replication does matter to plant breeders, however — the fact that it isn't perfect.

Mutation

DNA replication is usually very exact. If it were totally exact, we would not need this book. There would be no plant breeding. Plant breeding involves selecting for genetic differences, and genetic differences are created by new mutations, or changes — mistakes in the replication of DNA. We also wouldn't be here to read the book. Evolution requires selection based on the existence of genetic differences. Fortunately, DNA replication is very accurate, but not too accurate.

Occasionally when a gene replicates, a mistake is made. The frequency of spontaneous mistakes — that is, spontaneous mutations — varies with the gene and the kind of mutation. Mutations of specific types in specific genes commonly occur between 1 per 10,000 replications and 1 per 100,000. Mutations are rare when we think in terms of the replication of specific genes. But they are common when we consider how many genes a complex organism such as a plant or animal has and how many different ways they can mutate. Humans, for example, are thought to have 50,000 to 100,000 genes. Most humans probably contain at least one or more brand-new mutations — altered genes that neither of their parents carried. And so do most plants.

Once a mutation occurs, the mutant gene self-replicates just as the original gene did, so the change is passed along to offspring as an altered gene. In many cases new mutations are lethal. They cause dramatic enough alterations of function so that the organism cannot survive. In other cases mutations are viable, but the enzyme involved is altered. There may be much less of it, or it may be much less active than the original form of the enzyme. In many cases mutations cause an alteration in one or more characteristics of an enzyme for which they code. Two enzymes that differ from each other by a single amino acid might have somewhat different temperature optima, for example. That could translate into a mutant plant that thrives best at a temperature higher or lower than the original strain.

Mutation is a random process. If we want peas that grow better in the heat and so grow an entire field of peas in the heat, we are no more likely to obtain a new mutation that improves heat tolerance than with a whole field of peas grown under moderate conditions. In the hot field, however, we would be better able to identify any mutants with improved heat tolerance, because they would look different (better than) the heat-sensitive plants. Under moderate conditions the heat-resistant plants would not necessarily be distinct. Growing the plants in a hot field doesn't cause mutations to heat resistance, but it does allow us to recognize any that might already be present in the seed we planted. The occurrence of the new mutation that confers greater heat tolerance is a chemical accident to the DNA and is in no way caused by or related to whether the pea plants are growing under hot conditions.

Scientists using simple organisms such as bacteria and viruses have shown that there is no such thing as directed or purposeful mutation. If I have a tall pea variety and want a bush form, I might chop off a pea plant so that it bushes out and then collect the seed. But my efforts to obtain a bush variety in that way would fail. When I chop off the top of the pea plant, I don't make any change in the DNA contained in all the cells of the plant. I have turned the individual plant into a bush, but that bushiness isn't inheritable. The seed of the bush would still be genetically pure breeding for tall, even though the parent was phenotypically bushy.

If I want taller pea plants, there is no way I can treat or grow the plants to create a mutation that will make the plants taller. That specific mutation isn't likely to occur unless I have hundreds of thousands or even millions of plants. But it is often the case that the mutation we need occurred at some time in the past and exists within some variety of our crop somewhere. We learn how to find such genes, get them out of the variety where they occur, and get them into new varieties of our own.

Genes and Alleles

We don't know the exact identity of the gene that was associated with the tall versus short growth pattern in Mendel's peas. But this pattern in crop plants is, in a number of cases, associated with differences in the synthesis of the plant hormone gibberellin. The "tall" version of the gene is usually the form that is found in the wild. The "short" version, in many cases, has a less active form of one of the enzymes involved in synthesis of the hormone, so the plants are shorter. We refer to two genes as alleles of each other when they are inherited as alternatives to each other. In molecular terms, alleles are different forms of the same gene. There can be more than two alleles of a gene in a population of organisms. But any given individual has only two alleles at the most.

Shorter plants usually cannot compete with the taller forms in the wild; a short mutant in a patch of tall plants would be shaded out. That problem isn't relevant when a human plants a patch or field with nothing but short plants. And short plants may be earlier than tall ones, or less subject to lodging (falling over) in rain or wind. They also may have a much higher proportion of grain compared to the rest of the plant. So shorter plants can be advantageous as cultivated crops. Specific mutations or alleles are not good or bad in and of themselves, but only within a certain context. An allele that promotes better growth in hot weather may promote inferior growth in cold weather, for example.

Many systems are involved in determining the size and shape of a plant. Probably dozens of different genes in peas are capable of mutating so as to create a tall versus short difference, and hundreds of genes probably influence plant height to some extent. When we say that tall versus short was a one-factor difference in Mendel's crosses, that doesn't mean that only one gene is involved in determining height in peas. All it means is that the two varieties that Mendel crossed differed by only one gene involving height. The other dozens or even hundreds of genes involved in determining height must have been the same in the two varieties he crossed.

Genes, Chromosomes, and Cell Division

Most plants and animals are diploid organisms. *Diploid* means that they have two copies of every gene in each cell of their bodies. Genes in higher organisms such as plants and animals are arranged in structures called *chromosomes*. Each chromosome contains the DNA for many genes, in most cases hundreds or thousands of genes. Each cell has two chromosomes of each kind. Gametes (eggs and pollen) are an exception; they have only one chromosome of each kind.

The pea genome, for example, is organized into seven chromosome pairs. (*Genome* refers to all the DNA there is in an organism.) Each cell in a pea plant, except for the gametes, has two chromosomes of each kind for a total of fourteen. Each pea gamete has only one chromosome of each kind for a total of seven. The letter n is used to indicate the number of different chromosomes a species or variety has. The expression "$2n =$ whatever" is used to indicate the total number of chromosomes, or the diploid chromosome number. Peas are $2n = 14$; tomatoes are $2n = 24$; lettuce is $2n = 18$.

The two chromosomes of a pair are said to be *homologous* and are referred to as *homologs*. What makes two chromosomes homologous is that they have the same basic genes in the same order. Some of the genes are exactly identical to the corresponding genes on the homologous chromosome. Other gene pairs aren't exactly the same; they are slightly different — that is, they are different alleles of the given gene.

There doesn't seem to be much relationship, at least on a gross level, between which chromosome a gene is on and its function. Three major genes influencing height frequently show up in crosses of garden peas, for example. One is on chromosome 2, one is on chromosome 3, and one is on chromosome 4. (The numbers are just arbitrary names for the specific chromosomes.) Likewise, genes involved in flower color and pod color are scattered among the seven chromosomes. There are two major genes involved in the difference between the pod characteristics of shelling peas and those of edible-podded snow-type peas. One is on chromosome 4 and the other on chromosome 6.

The word *chromosome* refers to the fact that the structures absorb certain stains that made them easily seen once people began to use microscopes. The behavior of chromosomes during cell division is dramatic and corresponds exactly to what one would expect for structures that make up the genetic material. By the time Mendel's work was rediscovered, people knew chromosomes behaved in a way that was exactly analogous to the patterns Mendel outlined. The behavior of chromosomes in cell division explained Mendel's laws completely.

Two kinds of cell division occur in diploid organisms — mitosis and meiosis. *Mitosis,* also called *somatic* cell division, is the kind that occurs as the body of a plant or animal grows. Somatic cells are ordinary cells, as opposed to gametes, which are the sex cells. All vegetative growth — growth of shoots, leaves, stems, roots, and tubers — is somatic growth and uses mitosis as the mechanism of cell division. The other kind of cell division, *meiosis,* is restricted to sexual reproduction. In fact, it is restricted to just certain cells within the reproductive organs — those that divide to give rise to the gametes.

The purpose of mitosis is to produce two identical cells where before there was just one. A mitotically dividing cell gives rise to two daughters that are identical to it and to

each other. Each daughter cell has the standard quota of two of each kind of chromosome (and, therefore, two of each kind of gene), just like the cell that divided to give rise to it.

The purpose of meiosis is to eliminate one chromosome out of every pair so that the resulting gametes have just one chromosome of each kind (and, therefore, one gene of each kind). This is done by a special chromosome-pairing process that is unique to meiotic division. During this process, each chromosome physically pairs with its homologous chromosome. Then one chromosome of each pair migrates to one pole of the cell, the other chromosome migrates to the opposite pole, and a membrane forms down the middle. The result is two daughter cells each having one chromosome of each pair. Meiosis is sometimes referred to as *reduction division* because it reduces the total amount of genetic material.

Any beginning biology textbook presents sequential pictures of the behavior of chromosomes during mitosis and meiosis — usually in such detail that the important differences get obscured. For our purposes we don't need the mechanical details. All that matters is the overall result.

Somatic cells are diploid; gametes are haploid. When two somatic cells unite as a result of fertilization, the resulting zygote is diploid again. Since somatic cell division generates genetically identical cells, when we take cuttings from a plant or grow a plant from a tuber, the new plants are genetically identical to the mother plant and to each other. They are clones.

Gametes, however, are not identical to the

mother cell from which they arose. They have only one gene of each pair instead of two. In addition, in most cases gametes that arise from a single mother cell aren't identical to each other either. To understand why, let's go back to Mendel's laws. When we follow what happens when we do a cross, it's easy to see exactly why all Mendel's laws are the way they are and what is actually happening.

Mendel crossed, for example, a tall pea variety with a short one. We represented the tall variety as TT and the short variety as tt. We represented the F_1 as Tt, and we said that the genotypes and ratios expected in the F_2 correspond to ($\frac{1}{4}TT + \frac{1}{2}Tt + \frac{1}{4}tt$).

Let's suppose for purposes of illustration that the T gene is on chromosome 2. The tall parent has two chromosome 2s, and each has a copy of the T gene. As the tall parent grows, mitosis occurs in the somatic cells, giving rise to more cells with genotype TT. When the tall parent produces eggs and pollen, they will have just one copy of chromosome 2, and their genotype will be T. Since both chromosome 2s had identical alleles for the T gene, all the gametes are identical with respect to the T gene.

The short parent, likewise, has two copies of chromosome 2, each carrying t, which is allelic to (a slightly different form of) the T gene. The short parent's gametes also will have one chromosome 2, but their genotype will be t.

When we pollinate a plant from the tall variety with pollen from the short variety, we end up with a zygote that has two chromosome 2s, one carrying T and the other carrying t. In other words, the F_1 is genotype Tt.

Now we grow the F_1 plants. As they grow from small plants into big plants, the kind of cell division involved is mitosis, or somatic cell division. Through all the hundreds of cycles of somatic cell division, the genotype Tt is replicated exactly. All the cells in the F_1 pea plants are identical to one another and have genotype Tt. But now the pea plant starts making flowers and forming eggs and pollen — conducting meiosis. In meiosis, the two chromosome 2s pair, and only one or the other winds up in each gamete. Half the cells get the chromosome 2 with the T and the other half get the chromosome 2 with the t. In other words, it is complete chance as to which of the chromosome 2s any specific gamete inherits. The result is that each gamete gets either T or t, and either is equally likely. Half the gametes carry T and half carry t, and none carry both or none.

To get the F_2, the F_1 plant's eggs are fertilized with its own pollen. Half its eggs carry T. If all the pollen is equally viable and functional, then an egg carrying T will be fertilized by pollen carrying T half the time and pollen carrying t half the time. That is, all the zygotes and their probabilities are ($\frac{1}{2}T + \frac{1}{2}t$)($\frac{1}{2}T + \frac{1}{2}t$), which produces the familiar ($\frac{1}{4}TT + \frac{1}{2}Tt + \frac{1}{4}tt$) zygotes. It is the pairing of homologous chromosomes in meiosis and the inheritance of only one chromosome of each pair that is the physical basis for Mendel's Law of Segregation.

Mendel's Law of Independent Assortment also is easy to understand. Let's consider Mendel's cross of a tall, green-podded variety with a short, yellow-podded variety. As mentioned, three genes that commonly appear in pea crosses can be responsible for tallness, and we don't know which Mendel was working with. The three are on chromosomes 2, 3, and 4, respectively. For purposes of illustration, let's suppose that Mendel's tall gene was the one on chromosome 2. Mendel's gene associated with yellow pod color was on chromosome 5.

We can represent the cross of the tall green variety and the short yellow variety as $TTGG \times ttgg$. The F_1 is tall and green and is represented as $TtGg$. Now let's consider what happens during meiosis of the F_1. As we've seen, the two chromosome 2s pair; half the gametes inherit the one carrying the T, and the other half inherit the one carrying the t. But the two chromosome 5s also pair, and half the gametes inherit the one with the G and the other half the one with the g. The chromosome 2s pair separately and independently from the chromosome 5s, so the inheritance of T and G are independent. Of the gametes that have T, half have G and half have g. Of the gametes that have t, half have G and half have g. Mendel's Law of Independent Assortment is based on the fact that genes that are on separate chromosomes physically assort independently.

But what if two different genes are on the same chromosome?

Linkage

Suppose, for example, that genes T and G are both on chromosome 2. Our cross of the tall green variety with the short yellow would still be represented as $TTGG \times ttgg$, and the F_1 would still be represented as $TtGg$. But now, when we consider what happens during meiosis in the F_1, only one chromosome pair, pair 2, is involved. One of the 2s carries both T and G; the other carries both t and g. Any given gamete gets only one chromosome 2 or the other, so half the gametes would be TG and half tg. The F_2 would be $(\frac{1}{2}TG + \frac{1}{2}tg)(\frac{1}{2}TG + \frac{1}{2}tg)$, which works out to $(\frac{3}{4}T\_G\_ + \frac{1}{4}ttgg)$. That's very different from the $(\frac{9}{16}T\_G\_ + \frac{3}{16}T\_gg + \frac{3}{16}ttG\_ + \frac{1}{16}ttgg)$ we get when G and T are on different chromosomes. We recover only the parental classes (those that resemble the original parents); there are no recombinant classes. There is no independent assortment of T and G, for the simple reason that they aren't independent if they are physically on the same chromosome.

Linkage affects our ability to do breeding work. Let's suppose we cross a tall green with a short yellow, with the objective of getting a tall yellow, for example. If tall and yellow are determined by unlinked genes, we can recover tall yellows very easily; three-sixteenths of the F_2 plants would be tall yellows. But if T and G are tightly linked together on the same chromosome, we won't get the desired class at all. With only seven chromosomes, and with crosses that involve dozens of genes that matter agronomically, linkage of some of the genes we care about is common.

Fortunately, the effects of linkage on the recovery of recombinant classes is usually not as extreme as I have portrayed, because a phenomenon called *crossing over* occurs. When homologous chromosomes pair during meiosis, breakages and reunions can occur, so genes that started out linked can become separated or recombine even though they are on the same chromosome. How likely this is depends on how far apart the two genes are on the chromosome.

If T and G are very close together, a crossover between them will be rare. With no crossovers between the genes, the gametes will all be parental types — that is, TG or tg. We will get $(\frac{3}{4}T\_G\_ + \frac{1}{4}ttgg)$ and no T_gg (tall yellow). If T and G are far enough apart to allow some crossing over, we will get some tall yellows, though not as many as we would expect from independent assortment.

If T and G are located far apart on a chromosome, crossovers between them are likely to occur. They will stay together whenever no crossovers or an even number of crossovers occur; they will recombine whenever an odd number of crossovers occur. Half the time, the number of crossovers will be even and half the time odd. So the gametes expected will be $(\frac{1}{4}TG + \frac{1}{4}tg + \frac{1}{4}Tg + \frac{1}{4}tG)$ — that is, exactly the same as we get when the two genes are unlinked.

Thus genes on different chromosomes assort independently because they are actually physically independent. In addition, most pairs of linked genes act genetically as if they were independent even though physically they are not. Only closely linked pairs of genes show their nonindependence by giving us smaller frequencies of the recombinant classes than we would expect.

When we do crosses to develop new varieties, linkage can either aid or complicate our efforts. Mendel worked with only seven genes, and none of them displayed linkage. It is now thought that three of his genes were on separate chromosomes, another two were far apart on another chromosome, and the other two also were far apart on yet another chromosome. In addition, we believe that the other two chromosomes in peas don't have any of Mendel's genes.

One other basic genetic phenomenon is very common, but Mendel failed to run into it with his small sample of genes. This phenomenon is codominance.

Codominance

Mendel worked with only seven different pairs of alleles, and each displayed complete dominance. That is, in a cross of a tall and a short variety, the heterozygotes (heterozygous individuals) resembled one of the parents. The cross was $TT \times Tt$. The F_1, which is Tt, was just as tall as the tall parent, even though it had only one copy of the T gene instead of two. We use the term *dominance* to describe such situations. Not all allele pairs display simple dominance, however. In some cases the appearance of the heterozygote is in between that of the two parents. The classic example involves a cross between two pure breeding varieties of four-o'clocks, one with red flowers and one with white flowers. When the red and white varieties were crossed, the flowers of the F_1 plants were pink. The F_2 was $\frac{1}{4}$ red-flowered + $\frac{1}{2}$ pink-flowered + $\frac{1}{4}$ white-flowered.

In this example, we can portray the red variety as RR, the white variety as rr, and the F_1 between them as Rr. It is clear from the appearance of the F_1 that Rr is pink. So it is no surprise that the F_2 plants with genotype Rr also are pink. Where alleles display codominance, the F_1 is distinguishably different from both parents, and there are three phenotypic classes in the F_2, with the largest class having the same phenotype as the F_1.

In most randomly chosen crosses, most genes — at least 75 percent, I would say — display simple dominance instead of codominance. But codominance is common enough that you'll be likely to run into it fairly often.

Purple color in plant leaves and stems frequently shows codominance, with the hybrid being purple, but not as purple as the parent. A smooth as opposed to a deeply indented leaf margin also is frequently codominant. Hybrids usually show an intermediate amount of indentation. Many gene combinations associated with flower color show codominance.

One important aspect of codominance is that the phenotypes that show up as a result of codominance are heterozygous and are inherently not pure breeding. With the four-o'clocks, for example, you could not obtain a pure breeding pink strain from the cross of the red and white. You could inbreed pinks from this cross for the rest of your life, but you would still never get a pure breeding pink. Understanding about the existence of codominance allows you to recognize the cases where that work would be futile. If you wanted a pure breeding pink, you would use some other cross to get it.

If we do a cross involving two different genes, and one displays codominance, this affects the number and kinds of classes we get in the F_2. To use an abstract example, let the cross be $AABB \times aabb$. If both genes display dominance, the F_2, as we have seen, is $(\frac{3}{4}A\_ + \frac{1}{4}aa)(\frac{3}{4}B\_ + \frac{1}{4}bb)$, which works out to $\frac{9}{16}A\_B\_ + \frac{3}{16}A\_bb + \frac{3}{16}aaB\_ + \frac{1}{16}aabb$. But if the A locus dis-

plays dominance and the B locus codominance, then the phenotypes of the F_2 would be calculated from the expression $(\frac{3}{4}A\_ + \frac{1}{4}aa)(\frac{1}{4}BB + \frac{1}{2}Bb + \frac{1}{4}bb)$, which works out to $\frac{3}{16}A\_BB + \frac{3}{8}A\_Bb + \frac{3}{16}A\_bb + \frac{1}{16}A\_BB + \frac{1}{8}aaBb + \frac{1}{16}aabb$.

Exceptions and Apparent Exceptions to Mendel's Laws

There are exceptions to Mendel's laws — some only apparent, some genuine. Exceptions are common enough that you will run into them fairly often in practical breeding work, even if you operate on a small scale.

Lethals, Detrimentals, and Steriles

Some genes are recessive lethals. In other words, the plants that are homozygous for them don't survive, but the plants that are heterozygous for them do. For instance, the plant with genotype Aa is perfectly OK, but the plant with genotype aa doesn't survive. The seed may abort or fail to germinate, or the seed germinates but the seedling dies when it runs out of the stored food in the seed. A *detrimental* is a gene or genotype that is sometimes lethal or causes a loss in fitness or vigor.

Where the effect of a genotype is to cause the individuals that carry it to vanish, you don't see the expected Mendelian ratios. In a cross of $Aa \times Aa$, for example, we expect $(\frac{1}{4}AA + \frac{1}{2}Aa + \frac{1}{4}aa)$. But if aa is lethal, we get only two classes. The modified ratio is $(\frac{1}{3}AA + \frac{2}{3}Aa)$. (There are twice as many Aa's as AA's.)

Alternatively, the genotype aa may be severely detrimental or lethal only in some circumstances, so the *aa* class appears, but with a lower frequency than expected. In addition, if you are following an ordinary gene that is closely linked to a lethal or detrimental gene, this can skew the ratios of the ordinary gene.

Some genes are lethal in the gamete stage. Pollen lethals are especially common. In this case the lethal allele displays normal segregation in formation of eggs but isn't transmitted via pollen. In other cases the gene may be associated with poor gamete performance. For example, a gamete carrying allele *s* may grow down the style of the female flower more slowly than a gamete carrying *S* and fail to be recovered with the expected frequencies in certain crosses.

In addition, some genotypes may be viable but be associated with the sterility of the plant as a male or a female or both. Some genes are viable and have a dominant phenotypic appearance as heterozygotes but are recessive lethals as well. For example, *YY* plants may have normal green foliage, *Yy* plants yellow-green foliage, and *yy* plants no chlorophyll at all. Thus, the *yy* plants die as small, yellow seedlings.

Genes also can be lethal or detrimental only within specific contexts. The genotype *aa* may fail to survive in cold weather, for example, but have no visible effect in hot weather. Or it might even outperform *A_* in hot weather; in that case it would be advantageous in hot weather but lethal in cold weather.

Lethals, detrimentals, and steriles of various kinds are especially likely to show up when you save seed from a plant that is nor-mally propagated vegetatively (where such genes are not selected against in each generation), They also are likely to appear when you forcibly inbreed a plant that is normally strongly outbreeding — by, for example, doing special tricks to bypass its incompatibility mechanisms. Simple, commonplace acts such as saving seed from a potato or bud-pollinating a broccoli can produce progeny that include a number of weird types.

Aberrant Segregation

The crosses we have been discussing are crosses involving relatively close relatives. In these crosses both parent varieties contribute chromosomes with the same arrangement and order of genes on the chromosomes. If we cross two quite distant relatives, their genomes may have different chromosomal arrangements. They have evolved away from each other a bit. The block of genes that is at the left end of chromosome 2 in one parent, for example, may be on the right end of chromosome 5 in the other, and vice versa.

In such cases we can still cross the two varieties and obtain the F_1 without difficulty. Each gamete contributes one and only one copy of every gene in the genome, so the F_1 plants have two copies of each gene, as they should, and they grow and divide normally. Chromosome pairing doesn't occur during mitotic division, so the fact that the chromosomes aren't homologous doesn't matter.

However, problems occur during meiosis in the flowers of the F_1 plants. Homologous chromosomes can't pair. For chromosomes 2 and 5, there are no proper homologous chromosomes. Each is homologous partly with chromosome 2 and partly with chromosome 5, and the appropriate parts attempt to pair with the parts of the chromosome with which they are homologous. When cell division occurs, many chromosomes get torn apart, tangled around each other, or left out altogether. Many gametes are inviable. In addition, many that do function contain major deletions or excessive numbers of genes or sections of chromosomes that give rise to aborted or inviable seed or abnormal progeny. In short, the meiotic process that is responsible for causing each gamete to have one and only one copy of each and every gene — and which produces Mendel's laws — depends on chromosomal homology. Varieties that are distant enough to have evolved away from each other somewhat in genome arrangement don't have that homology.

Crosses between two varieties that are only distant relatives are some of the most interesting crosses to do. But often we just don't see "normal" Mendelian segregation in such crosses. We may fail to recover all the expected classes or to recover them with expected frequencies. The crosses are what we geneticists call "messy."

The practical approach is to go ahead and do whatever crosses we want and look for the class or classes we are interested in. If we can find the appropriate individuals at all, that is good enough to continue. At later stages of the project, chromosomal homology recurs as various nonhomologies are eliminated through pollen and plant lethality. So in later generations of the project, we begin

to see normal Mendelian segregation for characteristics that were messy initially.

Incomplete Penetrance and Variable Expressivity

So far in this discussion, I have been presuming that the relationship between a genotype and a phenotype is consistent. This is not always true. Many genes are variable in their expression. They may be affected in important ways by even minor differences in the environment, or there may be an underlying stochastic (chance-associated) mechanism involved in the degree to which they express or whether they express at all.

The individuals of a pure breeding variety that carries a gene for resistance to a disease often show considerable variability in their resistance, for example. We may be able (using methods described later) to establish that the variety is, indeed, pure breeding for the genes involved in disease resistance, but some plants are very resistant and others less resistant. And some plants of the variety may even be fully as sensitive as plants from sensitive varieties.

We use the term *variable expressivity* for the concept that the degree to which the character is expressed can vary even when the genotype is constant and the environment is as constant as we can make it. When the expression of a gene is so variable that sometimes the gene seems to have no effect on phenotype at all, we say the gene is not fully *penetrant*.

If a gene doesn't always express itself, this affects the ratios we obtain in crosses. If expected classes are ($\frac{3}{4}A\_ + \frac{1}{4}aa$) but *aa* can be distinguished from $A\_$ only part of the time, then the ratios we observe are skewed more toward $A\_$ and less toward aa than we would expect.

Genes with variable expressivity are undoubtedly much more common than we tend to think, because geneticists almost never choose them to work with unless they are deliberately studying variable expressivity (and hardly anybody does). Genes with large and consistent effects on phenotypes are the ones that are easiest to work with. However, genes with small and/or variable effects will be present by the dozens in many of the varieties we work with and the crosses we do.

Variable expression of genes matters a lot in plant breeding. In the first place, we may need to select for a gene that has variable expression. Many genes of great agricultural importance are variable in expression or not fully penetrant. A gene with variable expression that is associated with disease resistance, for example, may be adequate to allow you to grow the crop; it may not matter that some plants succumb to the disease as long as most do not. The gene may be messy to work with but very practical to have and use anyway.

In the second place, if we have a variety that is pure breeding at a given locus but the gene is variable in expression, the variety will display variability that is not genetic. When we look at it, our first inclination will be to think that the variety is not pure breeding. But it may be. If we try to select for the phenotype associated with the expressed gene, we will be wasting our time; we can select successfully only if genetic differences are present. We could spend a lot of time trying to improve the variety via selection and make absolutely no progress. So it's useful to be able to distinguish when phenotypic variability is caused by genetic variability and when it isn't. I'll explain how to do that when I come back to selection.

Maternal Inheritance

The pericarp, the outermost layer of the seed, is maternal, not embryonic tissue. So its phenotype reflects the genotype of the mother, not of the embryo the seed contains. Some other genes whose products operate at very early stages may also show maternal inheritance — in other words, the phenotype for the grown plant is established by the genotype of the mother. Maternal inheritance of this sort is Mendelian, but the expected ratios appear one generation late.

The Cytoplasmic in Inheritance

More than 99 percent of the genes in plants are arranged on chromosomes located in the nuclei of cells, and they are inherited according to Mendel's laws. However, some small fraction of genes are located on organelles in the cytoplasm, and they aren't inherited in this way. Even though there aren't very many *extrachromosomal* or *cytoplasmic* genes, they have a disproportionate importance in plant breeding.

The inheritance of genes in cytoplasmic organelles is completely non-Mendelian. It is almost always exclusively maternal. The egg is a big cell with a lot of cytoplasm. The pollen is a very small cell with little cytoplasm, which is not normally transferred to the egg during fertilization. Nuclear DNA comes equally from both parents, but cytoplasm comes only from the maternal parent.

Once in a while you can do a cross where one of the characteristics you are interested in is associated with a gene that is cytoplasmic instead of nuclear. Suppose you do a cross between a variety that has a mottled coloration pattern on the leaves and one that does not. If you cross a mottled female to a normal male, some or all of the F_1 plants are likely to be mottled. But if you cross a mottled male with a normal female, none of the F_1 plants will be mottled, nor will the offspring in any future generations.

The cytoplasm is relevant to us in additional ways. The cytoplasm of a variety and its nuclear genome have evolved to be compatible. When distant relatives are crossed, sometimes the cytoplasm and the nuclear genome of the F_1 or of some of the F_2 progeny aren't compatible. In extreme cases lethality results. In less extreme cases the plant may be sterile or pollen-sterile. *Male-sterile cytoplasms* have, in fact, been useful in the production of hybrids. Such cytoplasms aren't absolutely male-sterile; they are sterile only in the presence of one or more nuclear genes.

Anytime we are crossing two fairly distant relatives, we need to realize that the reciprocal crosses might be genuinely different. Both will give rise to the same nuclear genotype, but they will set it down in the middle of different cytoplasms. This could have a dramatic effect on the phenotype of the F_1 and subsequent generations.

Many modern crops trace their nuclear genomes to only a dozen or so sources. But it is common for vast fractions of a modern crop to have only one cytoplasmic genome. The Southern corn blight in the United States in the early 1970s was caused by the blight sensitivity of a particular cytoplasm — a male-sterile cytoplasm that was in virtually the entire commercial crop, whatever the nuclear genome involved.

Whenever we do a cross using a particular variety as the maternal parent, we are preserving its cytoplasm and discarding that of the male. I think we should pay more attention to preserving and increasing the diversity of the cytoplasms in our food crops. All we have to do to preserve a particular cytoplasm is to make sure that we use it as the female parent when we do crosses.

Gene Interaction

Often more than one gene is involved in determining a trait. Consider an abstract example in which A and B are genes that code for different enzymes involved in the biosynthesis of the plant pigment anthocyanin. Let's also assume that A is dominant to a and B is dominant to b, and that the A locus and the B locus are not linked. Let's also imagine that the a allele makes a completely inactive form of its enzyme, and so does the b allele.

Now let's consider four different pure breeding varieties representing all possible genotypes with respect to the two loci. Variety 1 is AABB and has purple flowers. Variety 2 is AAbb and has white flowers. Variety 3 is aaBB and also has white flowers. And variety 4 is aabb and also has white flowers. Variety 2 would have white flowers because it has no good enzyme for one step in the synthesis of the purple pigment. Variety 3 has good enzyme for that step, but has no good enzyme for some other step that is also necessary to synthesize pigment. And variety 4 lacks any good enzyme for either of two of the essential steps in pigment biosynthesis.

If we cross variety 1 (purple) with variety 2 (white), the cross is: $AABB \times AAbb$, and the F_1 would be $AABb$ (purple). And the F_2 would be $\frac{3}{4}AAB\_ + \frac{1}{4}AAbb$; that is, $\frac{3}{4}$ purple $+ \frac{1}{4}$ white. This is just a one-factor cross. That is, there is only one factor — the B gene — *that differs between the two strains.* Information about the other gene, the A locus, is only included for later reference.

If we cross variety 1 with variety 3 we obtain similar results. The cross would be $AABB$ (purple) $\times aaBB$ (white) and the F_1 would be $AaBB$ (purple). The F_2 would be $\frac{3}{4}$ $A\_BB + \frac{1}{4}aaBB$, that is, $\frac{3}{4}$ purple $+ \frac{1}{4}$ white respectively. This is also a one-factor cross, but the factor involved is different from the one involved in the cross of variety 1 and variety 2.

Now let's consider a cross of variety 2 with variety 3. The cross is $AAbb \times aaBB$, *and is a cross of two white-flowered plants to each other.* The F_1 would be $AaBb$, *and it would be purple.* Each locus displays dominance, that is, one "good" copy of each gene makes enough enzyme to perform its step. The F_1 plants have one good copy of each gene, and thus make enough good enzyme for both steps in biosynthesis of pigment.

These sorts of situations — where a breeder crosses varieties of identical color or phenotypes similar in some other regard and obtains an F_1 of a different color or phenotype — are quite common. In some cases they have caused great anguish because the new phenotype was undesirable or inedible compared to the parents. In other cases the new color or phenotype was useful itself.

But let's carry the cross one step further. The F_2 genotypes would be $\frac{9}{16}A\_B\_ + \frac{3}{16}$ $aaB\_ + \frac{3}{16}A\_bb + \frac{1}{16}aabb$. And all the classes except the first one would be white. So the phenotypic classes would be $\frac{9}{16}$ purple $+ \frac{7}{16}$ white.

The final possible cross, that of the first variety to the fourth, would be $AABB$ (purple) $\times aabb$ (white) and would give a purple F_1. But the F_2 would be exactly the same as in the cross of varieties 2 and 3 and would reduce to phenotypic classes of $\frac{9}{16}$ purple $+ \frac{7}{16}$ white.

In a cross of a pure breeding soybean with black pods to one with tan pods, the F_1 was black. The F_2 was $\frac{12}{16}$ black $+ \frac{3}{16}$ brown $+ \frac{1}{16}$ tan. Where did the brown come from? Well, we generate two new genotypes in the F_2 of a two-factor cross, so one or two new phenotypes shouldn't be surprising. In this case, the cross could be symbolized as $EEFF$ (black) $\times eeff$ (tan), and the F_1 would be $EeFf$, which was also black. The F_2 genotypes would be $\frac{9}{16}E\_F\_ + \frac{3}{16}E\_ff +$ $\frac{3}{16}eeF\_ + \frac{1}{16}eeff$. The E_ff and the $eeF\_$ are new genotypes. One of them must look black and get lumped with the $E\_F\_$ group. The other must look brown.

Two different genes that affect the same trait can interact in all kinds of ways. The basic ratios in the F_2 of 9:3:3:1 can become 15:1, 9:7, 9:6:1, or 9:4:3 depending on what phenotypes are generated by the various genotypes. New phenotypes can appear in the F_1 or F_2 depending on the specific cross and the interactions between the genes. And that is all with just two genes, both of which show dominance.

In a cross with just one factor and codominance, one new phenotype appears in the F_1 and F_2. In a two-factor cross where both loci show codominance, each factor generates three genotypes, and the F_2 has nine different genotypes. That means up to seven new phenotypes could appear.

There are two major genes, p and v, involved in the difference between snow (edible-podded) peas and shelling peas. Both are responsible for reducing the amount of fiber in the wall of the pod; they are apparently involved in different steps or aspects of production of the fiber. Both genes are simple recessives. And the two loci are on different chromosomes; p is on chromosome 6, and v is on chromosome 4.

Shelling peas are generally of genotype $PPVV$. Most modern varieties of snow peas are genotype $ppvv$, and the pods are edible even when they are quite large. Some older snow-pea varieties are $ppVV$ or $PPvv$, and are not as fiber-free as varieties that are $ppvv$, or are fiber-free only when the pods are small.

If you crossed a semi-snow of genotype $ppVV$ to another semi-snow from a different variety that happened to be $PPvv$, the F_1 would have tough, shelling-type, inedible pods. The genotypes in the F_2 would be $\frac{9}{16}P\_V\_ + \frac{3}{16}P\_vv + \frac{3}{16}ppV\_ + \frac{1}{16}$ $ppvv$. The phenotypic classes we would

obtain would be $\frac{9}{16}$ shelling $+ \frac{6}{16}$ semi-snow $+ \frac{1}{16}$ snow. (We would not be able to distinguish the two different classes of semi-snows, but we could distinguish them from the true snows and the shelling types.)

So you could cross two semi-snow type peas and, in some cases, recover a much better snow pea than either parent as $\frac{1}{16}$ of the F_2. This is probably how our modern $ppvv$ varieties were developed.

Note, however, that not all crosses of two semi-snows would have the potential for generating a better snow. Suppose we had only semi-snow type snow peas, for example, and you wanted a better snow pea — one that stayed tender even when it was a bigger pod. You wouldn't know anything about the p or v genes, and you wouldn't know that some semi-snow varieties are $PPvv$ and others are $ppVV$. But as a general principle, you would cross two varieties that were among the best available with respect to the characteristic that you cared about — pod type, in this case.

If you happened to cross two varieties that were both $ppVV$ or two varieties that were both $PPvv$, then the F_1s would still be semi-snows, as would all of the F_2s. In other words, you would have just one phenotype in the F_2, and it would be the same as that of the parental varieties and the F_1. There would be no variability for pod type. Even if you grew only twenty or so F_2 plants, which might not be enough to reliably recover a member of a class expected only $\frac{1}{16}$ of the time, you would still expect to see at least two classes. So if you didn't, you probably wouldn't bother growing more seed from that cross. Instead you would try different crosses.

If, on the other hand, you crossed two semi-snow varieties and obtained an F_1 that was a shelling type, you would know right away that you had generated genetic variability for pod type with that cross. If you then grew up only twenty F_2 plants and found two classes, shelling and semi-snow, but failed to find any snow types, you would probably try more seed from the same cross before giving up on it and going to a different cross. If you plant only twenty seeds, much of the time you will miss a class you only expect $\frac{1}{16}$ of the time.

The concept that different varieties may have different genes contributing to similar phenotypes is a major part of practical plant breeding. Let's suppose we want a variety that is more cold hardy than anything available. We would make one or more crosses between the most cold-hardy varieties we could find. We would be hoping that two varieties would be cold hardy for different reasons and that from them we could produce a new variety that was more cold hardy than either parent.

It's also possible to have more than two genes involved in the inheritance of a single trait in a cross. With flower color, for example, the cross between two varieties may involve three, four, five, or even more factors. By the time we have four factors involved, even if all loci show simple dominance, we will have up to sixteen phenotypic classes in the F_2. The cross may produce all kinds of colors we never imagined, with some genes affecting the kinds of pigments produced, others affecting the amounts, and still others affecting the distribution of the pigments in the parts of the flower.

Once a number of factors that affect a trait are involved in a cross, people sometimes use the terms *major genes* and *modifiers*. Two major genes may account for most of the genetic variation in the F_2, for example, so there will be the expected discreet classes in the F_2. But there may be additional variation within each class caused by the segregation of other genes that, in the context of the particular cross, have smaller effects on the phenotype. If the cross involves color, for example, there may be four basic colors in the F_2, but various individuals may have different shades of them. We refer to the genes that segregate to generate the main classes as the major genes and those that are involved in creating the shades as minor genes, modifier genes, or modifiers. The terms are merely subjective, though, as one of those minor genes could be a major one in another cross.

Quantitative Inheritance

In many crosses many different genes affect the characteristics we care about. General plant vigor, for example, is often influenced by many genes, each having effects that are too small to sort out individually. Sometimes there are signs of dominance at some loci; often there is codominance. Where many genes contribute to a characteristic, we don't see discreet classes when we do crosses. If we cross a high-yielding variety with a low-yielding one, for example, it is likely that the F_2 will include plants that cover the entire range from low to high, with no discreet classes obvious. The effects of the individual genes appear to be *quantitative* instead of discreet. Each gene or allele can be thought of as adding an increment to the phenotype.

Size, growth rate, vigor, and yield show quantitative inheritance in most crosses. They also can be affected by single genes that have large enough effects to show discreet classes. When I make crosses involving tall and short varieties of peas, for example, I obtain discreet classes of tall and short. But within each class there is considerable variability for height. In other words, in most of these crosses one major gene affects height, but a lot of other genes have less dramatic effects.

Codominance is common with quantitatively inherited characters. A common model is that several genes are involved whose effects are too small to tally individually, and there is often codominance. If we cross *AAbbCCdd* and *aaBBccDD,* for example, where the large-letter alleles all contribute to higher yield and are codominant, then the F_1, *AaBbCcDd,* will have about the same yield as the parent varieties. Each parent has four of the large-letter alleles, but among the F_2 plants will be some with genotypes having more of the large letters than either parent. These will average much higher yields than either parent. There also will be individuals that have fewer large-letter alleles than the parent varieties and have lower yields.

To consider another example, imagine a low-yielding variety of genotype *aabbccdd* and a high-yielding one with genotype *AABBCCDD.* If we cross the two, the F_1 will be intermediate between the two parents, and the F_2 will include individuals of every

yield from as low as the low parent to as high as the high parent. Because the effects of each allele are small and the environment also influences each plant, there will be no discreet classes. *It will seem exactly as if inheritance were based on the mixing of some kind of fluid.*

In fact, we now know that even in these cases, inheritance is caused by discreet genes. But when there are many genes and their effects are small, we cannot decipher what is going on without investigating it specifically, without understanding Mendelian genes based on how they behave in simpler crosses, and without sophisticated statistical methods.

Whenever we cross two varieties that are distant relatives, many genes differ, so many characteristics show quantitative inheritance. These are the kinds of crosses that most pre-Mendel scientists were trying to learn about heredity. It wasn't possible to understand such crosses, however, until Mendelian patterns were understood and there were better statistical methods.

Predictions and Actuality

It's useful to be able to figure out what classes and frequencies we expect when we do crosses, because this helps us decide what to do. That is, the predictions help us evaluate whether a project is possible with a specific approach given the time and space we have. If we calculate that the class we want is expected to occur with a frequency of $1/1{,}024$ in the F_2, we know we will probably not get any at all if we raise only twenty F_2 plants. If the class we want has a frequency of $1/16$, we have a good chance of getting it with just twenty plants. But how good?

How many plants would we need to be reasonably sure of getting at least one of the type we want? If we expect a coin to land heads half of the time and we toss it twice several times, we won't get exactly one heads every time we do it; chance matters. Likewise, if we expect a 9:3:3:1 ratio and we raise exactly sixteen plants, we won't ordinarily get exactly 9, 3, 3, and 1 plants of the respective classes. Sometimes, for example, we will get none of the class with the expected frequency of $\frac{1}{16}$, and sometimes we will get two.

What we need to know is how many of something we need to grow to be reasonably sure of obtaining at least one of the class we want. This is something that we can calculate statistically. The table in Appendix H represents the results of these calculations for the probabilities that come up most frequently in genetics. To use the table, look up the probability of the class you want; the table tells you how many plants you need to be either 95 percent or 99 percent sure of obtaining at least one of the desired class.

If the desired class is expected $\frac{1}{4}$ of the time, we can look up "$\frac{1}{4}$" in the table and see that we need eleven plants to be 95 percent sure of obtaining at least one of the desired class. If we want to be 99 percent sure, we will need sixteen plants.

If the class we need has a probability of $\frac{1}{16}$, we should grow forty-six plants if we want to be 95 percent sure of obtaining at least one of the class. This means that if we raise forty-six plants every time we need to obtain at least one of a class with a probability of $\frac{1}{16}$, 95 percent of the time we will succeed, and 5 percent of the time we will not. If we want to be 99 percent sure, we will have to raise seventy-one plants.

But what if we don't have enough space, time, or seed for forty-six plants, let alone seventy-one? We may then choose a different project, a different approach to this project, or a different kind of cross (backcrossing, for example). Or we may just go ahead with the project and the cross, raise however many plants we can, and figure on obtaining what we want in more than one step. These options are aspects of breeding project design, which I discuss in chapter 10 in the context of specific examples.

NINE

The Genetic Basis of Seed Saving

SUPPOSE YOUR grandmother had a special bean variety. You remember it and would love to grow it, but as far as you know, the seed was lost. Then you find a single seed of the variety caught in a crack in a chest of drawers that belonged to your grandmother. You plant it, and it germinates. Can you restart and maintain your grandmother's variety from just the single seed? The answer is probably yes.

Now let's suppose that you go back and examine all the cracks and crevices in grandmother's chest of drawers, and you also find a broccoli seed. And you remember that Grandmother had her own broccoli variety, too. Can you restart and maintain her broccoli from just a single seed? The answer is probably no.

Beans are an inbreeding plant. Chickpeas,

peas, tomatoes, and lettuce are other common inbreeders. Broccoli is naturally outbreeding. Corn, spinach, most of the brassicas, and most of the cucurbits are among the common outbreeders. Varieties of inbreeders and varieties of outbreeders are essentially different — so much so that it is misleading that the word *variety* is used for both.

The genetic structure of a population of plants of an inbreeding variety is usually very different from that of a population of an outbreeding variety. In addition, many outbreeders display inbreeding depression. Some also have incompatibility systems. These factors affect every aspect of how inbreeding and outbreeding varieties are both created and maintained.

To understand how inbreeders and outbreeders are different, and to make the characteristics of each work for us, we have to understand a bit more about what inbreeding actually does.

Inbreeding and the Genetic Nature of Inbreeding Crop Varieties

Imagine that we make a cross of two different varieties of an inbreeder that generates heterozygosity at the A locus, so that the F_1 is Aa. Then let's allow the plant to inbreed one generation to obtain the F_2 generation. We know the results: ($\frac{1}{4}AA$ + $\frac{1}{2}Aa$ + $\frac{1}{4}aa$). What's interesting to us at this point is that one generation of inbreeding eliminates half the heterozygosity. The F_1 population was 100 percent heterozygous at the A locus. One generation of inbreeding gives us an F_2 population in which half are heterozygous (Aa) and half are homozygous (either AA or aa).

Now let's allow all the plants of the F_2 population to inbreed another generation. And let's assume that each produces equal amounts of seed (or that we save equal amounts from each deliberately). What will the F_3 population be like?

Well, one-quarter of the F_2 population is AA, and the one-quarter of the F_3 population these plants give rise to will all be AA. Similarly, one-quarter of the F_2 population is aa, and the one-quarter of the F_3 population these plants give rise to will be aa. The half of the F_2 population that is Aa will give rise to ($\frac{1}{4}AA$ + $\frac{1}{2}Aa$ + $\frac{1}{4}aa$), and since these plants contribute half of the F_3 population, their contribution to F_3 is ($\frac{1}{2}$)($\frac{1}{4}AA$ +

$\frac{1}{2}Aa$ + $\frac{1}{4}aa$), or ($\frac{1}{8}AA$ + $\frac{1}{4}Aa$ + $\frac{1}{4}aa$). When we add their contribution to that of the other genotypes, we get an F_3 population that is ($\frac{3}{8}AA$ + $\frac{1}{4}Aa$ + $\frac{3}{8}aa$).

The percentage of the population that is heterozygous went from 100 percent to 50 percent when we inbred one generation. Then it went from 50 percent to 25 percent when we inbred another generation. If we inbreed further, the percentage of the population that is heterozygous at the A locus will continue to drop by 50 percent with each round of inbreeding.

The reduction of heterozygosity with inbreeding happens at all loci. This means that we lose, on average, half of the total heterozygosity in the population with each generation of inbreeding. This has profound implications for the genetic nature of inbreeding crop varieties.

In an inbreeding crop variety, whenever an event creates genetic heterogeneity — an occasional cross or a new mutation, for example — sequential rounds of inbreeding promptly eliminate the heterogeneity and sort it out into homozygous genotypes. For this reason, when there is variability in an inbreeding crop variety, it is because it is a mixture of slightly different lines, each of largely homozygous individuals. In other words, if a population of inbreeders contains variability at the A locus, it is usually because it contains individuals with genotype AA and individuals with genotype aa. Little of the variability is in the form of individuals with genotype Aa.

In other words, individuals of inbreeding varieties tend to be highly homozygous.

There are three important practical implications of this genetic structure of inbreeding populations. First, since any given seed or plant is generally very homozygous, offspring generally resemble their parents very closely. They are, in fact, very close to genetically identical to their parents unless some mutation or accidental cross has occurred quite recently in their pedigree. The common expression "as like as two peas in a pod" is based on an important component of the nature of reproduction in peas and other inbreeding plants.

The second important practical aspect is that when we choose a superior pea plant, it is usually the case that the plant is already homozygous for the genes that make it superior. If I find a chickpea that yields twice as well as the others of the variety, *there is a good chance that I can derive a new variety in one step just by saving its seed separately from the rest.* If the plant is genuinely superior to the rest genetically, it is likely that all its offspring will be also. (The exception is if a new mutation or accidental cross occurred recently in the superior plant's pedigree.)

When I pick one superior plant out of a land race of chickpeas and save its seed separately, however, I discard most of the variability that was present in the land race. That could include natural resistance to diseases that might appear in future years. But it also includes, say, variability for yield, which I'd prefer not to have. I would like uniformly high yield.

Variability is not necessarily either good or bad in itself. Variability for yield tends to make the average yield low, so the crop isn't worth growing at all. But variability for plant color, for example, won't matter at all if the plants all yield well. Variability for plant shape often is acceptable for hand-harvested chickpeas, but erect plants are better for machine harvest. Simultaneous maturation is essential for machine harvest of garden vegetables. For the home gardener who eats the vegetable fresh, variable maturation over a period of time often is preferable.

The third practical implication of the genetic nature of inbreeders is how the preservation of variability relates to the number of plants from which we save seed. With chickpeas, farmers grow a lot of plants and save seed from them all, since the dry seed is the crop. Saving seed from many plants maximizes the probability that all the various related sublines will be preserved. With garden plants such as peas, however, you have to sacrifice the food production of some plants in order to save seed. In addition, you don't grow as many plants as a farmer with a field of chickpeas.

The farmer who saves seed from thousands of chickpea plants is probably saving seed from every genotype and subline that existed in the variety the generation before. But if I save peas from ten plants in each generation, I can save only a few of the sublines that might have been present originally. This often is adequate, though, since the sublines I save are likely to be those most suitable for my conditions. In addition, I am not usually the only gardener with a variety. Different seed-saving gardeners are usually preserving slightly different lines. Even if each person is maintaining only a single subline, a good amount of total heterogeneity is preserved.

If I start a new chickpea variety by saving seed from one superior plant, my new variety has almost no genetic variability, so little will be gained by saving seed from many plants in each generation. If, however, I want to preserve the original land race, it's better to save seed from many plants.

You can do a pretty good job of maintaining most of the genetic variability in a land race or other heterogeneous material by saving seed from a hundred plants each generation. You can do an excellent job with seed from a thousand plants per generation. But you can maintain much of the genetic variability in heterogeneous material with just twenty plants per generation. In the latter case, you tend to preserve the more common sublines and lose the rarer ones.

You can do a good job of maintaining a relatively uniform inbreeding crop variety (a subline) by saving seed from five or so plants in each generation. You can do an excellent job and give yourself plenty of margin for error with ten or twenty plants.

The only reason you need to save seed from more than one plant to preserve your subline of an inbreeder is that new mutations happen all the time, and so do accidental crosses. If you save seed from just one plant and it happens to be carrying a deleterious mutation or be the product of a contaminating cross, you will lose the entire variety. In other words, you can maintain a subline of an inbreeding crop variety by saving seed from just one good plant per generation. You save seed from somewhat more than that as a margin for error.

Remember that when you choose one, five, twenty, or however many good plants, you need to look at the entire plant, not just the fruit. You would not save seed from the plant with the largest pea pod, for example, if that was the only pod on the plant.

Note that if you obtain seed from someone who generated it from a single seed, most of the variability is gone. Even if he or she sends you a hundred seeds, that doesn't mean you can maintain much variability by saving seed from a hundred plants in each generation — not when you started with just one subline. The one-seed generation represents a genetic bottleneck that eliminates virtually all the variability.

Grandmother's bean variety might have been maintained by saving seed from just a few plants each generation. If so, it probably had very little variability. By finding and planting that single viable seed, you will, indeed, be regenerating Grandmother's variety.

If Grandmother saved seed from dozens of plants in each generation and created no recent genetic bottlenecks in her maintenance of the variety, her variety might have contained considerable genetic variability. The single seed you found would represent just one typical subline; the others would be lost. You could regenerate one subline of her variety, but not all of them. You could then preserve the basic line, but not the broader line with all its variability.

If you save seed from a couple of dozen or more plants each year, mutations will accumulate, and more variability will be created. You'll have your grandmother's basic line, but with a new repertoire of variability. This is quite possibly what Grandmother did in the first place. Heirloom garden beans are often passed on by the ones and twos and half dozens, not by the pound. Varieties are often passed through bottlenecks, as someone finds or is given just a single precious seed. The basic types of inbreeding crop plants can usually be reestablished from just a single seed.

Heirlooms

Various people have various definitions of an heirloom. Loosely speaking, an heirloom garden cultivar or variety is just like an heirloom anything else. It's something that has been passed down from generation to generation, preserved, and cherished. And like other heirlooms, heirloom garden varieties often have a flavor of romance.

Some people consider a cultivar an heirloom only if it was, at least at some point, maintained and distributed informally by gardeners or farmers. Others don't care how the variety was maintained but feel that there should be a cutoff point with respect to age — say, at least forty years old. Whatever else, heirlooms are always open-pollinated — that is, nonhybrid — varieties. Many people, myself included, use the word fairly loosely to mean any open-pollinated variety that has been around for a while.

The word *heirloom* also suggests excellence and value. Heirloom garden varieties have survived for decades in the absence of patents or anything else that makes distributing them especially profitable. They have done so very often because they are genuinely excellent. They are the varieties our ancestors' friends or relatives were most likely to give them personally, because they were the best these people had. They are, of all the varieties that have been developed and grown, the ones that gardeners refused to do without. Some heirlooms have broad regional adaptation. Others are fine-tuned to a specific region or climate and don't do well elsewhere.

Many heirloom varieties — beans, for example — contain a lot of genetic variability, but many others do not. It depends on how the seed was saved. With garden varieties of inbreeding crops, you can maintain vigorous varieties by saving seed from just a few plants each year, as long as they are good plants that are typical of the variety. When a variety is so maintained, it usually has little genetic variability. But this usually doesn't matter. If you need more variability, you can get it by obtaining seed from a number of other people who maintain it and then pooling the seed. Usually a single, excellent subline of an heirloom variety of an inbreeding crop is all you need for your home garden.

Outbreeding and the Genetic Nature of Outbreeding Crop Varieties

Not all heirlooms and other open-pollinated varieties are naturally inbreeding. Many are outbreeding. To understand the genetic nature of outbreeding crop varieties, we need to understand what happens when plants of an outbreeding variety sit in a field open-pollinating (that is, crossing however they want or however the bees fly).

Let's imagine that we make a cross between two varieties of an outbreeding crop and start with an F_1 population that is 100 percent heterozygous at the A locus, or Aa. Let's imagine an extreme outbreeder, one with a tight incompatibility system that prevents self-pollination. If we plant the F_1 plants in a patch and leave them alone, none of them will self-pollinate, but they will cross with each other. That is, they will do the cross $Aa \times Aa$ to get an F_2 generation. That generation will be ($\frac{1}{4}AA + \frac{1}{2}Aa + \frac{1}{4}aa$), just as was the case for the inbreeding population we considered at the beginning of this chapter.

What about the F_3 generation? That gets a bit complicated. With inbreeders only three kinds of crosses (self-pollinations for each of the three genotypes) can occur in the F_2 population. With outbreeders there are three genotypes, but there also are many more than three possible kinds of crosses. This is because each genotype can cross with plants having any of the other genotypes. So nine kinds of crosses are possible.

In the outbreeding F_2 population, the $Aa \times Aa$ cross causes a loss of heterozygosity, but the $AA \times aa$ cross creates heterozygosity. For every $AA \times aa$ cross, the heterozygosity of the parents is 0 percent, but that of their offspring is 100 percent. In an outbreeding population, the frequencies of the various genotypes shift with additional generations until the amount of heterozygosity lost by the $Aa \times Aa$ crosses is exactly balanced by the heterozygosity that is gener-

ated by the $AA \times aa$ crosses. An equilibrium then exists, and the same frequencies of genotypes occur in future generations. *In an outbreeding crop variety allowed to pollinate spontaneously, heterozygosity is not lost — it is maintained.*

In a population of outbreeder that contains genetic variability, the variability is usually both within and between individuals. One important practical implication is that the typical individual of an outbreeding crop plant contains a lot of heterozygosity. This is why we say "as like as two peas in a pod" but don't say "as like as two broccoli seeds in a pod." The seeds from an individual broccoli plant of a variety with standard levels of uniformity are often fairly different from each other and from the mother. This means that if I find a superior individual in a field of broccoli, it's fairly likely that it is not homozygous for whatever genes make it superior. Thus only some of its offspring will likely inherit its superior characteristics. I can still use it to begin deriving a new variety, but doing so will probably take longer and be more complex than with an inbreeding plant.

Most plants are not totally inbreeders or totally outbreeders; they are somewhere in between. The genetic structure of their populations also falls somewhere in between, depending on just how much inbreeding and outbreeding they do under the given circumstances.

Many outbreeding or partially outbreeding plants are characterized by two other phenomena that complicate our efforts to create or maintain them: inbreeding depression and self-incompatibility.

Inbreeding Depression

Many varieties of outbreeders tend to deteriorate in size, vigor, and yield if inbred too much. Deterioration as a function of inbreeding is called *inbreeding depression*. The converse is *hybrid vigor* — the tendency of many hybrids to be larger, more vigorous, and better yielding than either parent.

Apparently, genetic heterogeneity makes an important contribution to vigor, yield, and many other traits in these plants. The classic example is corn. If you cross two open-pollinated varieties of corn, the F_1 hybrid is usually dramatically more vigorous and high yielding than either parent, and the effects can be great. Certain combinations of parents will produce F_1 hybrids that yield more than twice as much as either parent. This hybrid vigor is the basis for the modern hybrid seed industry.

The other side of the coin is inbreeding depression. If you try to develop a new variety of corn by saving seed from the best plant in each generation, it won't work. Even if you are selecting the highest-yielding plant and self-pollinating it, you will generally find that your yield will deteriorate with each generation. You would be following the basic rule of saving seed from the best, but you would be destroying the variety instead of improving it.

We do not entirely understand inbreeding depression or hybrid vigor. Generally, though, they seem to be associated with the degree of genetic heterogeneity — that is, the proportion of loci that are heterozygous instead of homozygous. In an open-pollinated line, many loci are homozygous; there are a lot of *AA*'s, *bb*'s, and *CC*'s (instead of *Aa*'s, *Bb*'s, and *Cc*'s) within the genotype. When we cross two open-pollinated varieties, the F_1 hybrid will have many more heterozygous loci than either of the parents. The more distantly related the two varieties are, the more loci that will be heterozygous when we make the cross. Generally, crosses between distant relatives show the most hybrid vigor.

Inbreeding depression is serious for many, but not all, plants. The naturally inbreeding plants such as peas, beans, and lettuce show little or no inbreeding depression. It is easy to obtain vigorous varieties that are highly inbred — that, in fact, can be traced to a single foundation plant. The naturally outbreeding plants, however, often must be dealt with in such a way that a good deal of genetic heterogeneity at many loci is maintained. You usually would not want to try to start a variety of outbreeders from a single foundation plant.

Not all outbreeding plants display serious inbreeding depression. Most of the cucurbits, for example, show some hybrid vigor, but they don't necessarily suffer serious inbreeding depression. It is usually possible to derive good varieties of squash, muskmelons, cucumbers, and watermelons by inbreeding from single foundation plants. Appendix A and Table 1 include information about inbreeding depression for various vegetables.

How we go about developing crop varieties is affected profoundly by whether the crop displays inbreeding depression. With such crops we have to take the more complex approach of achieving uniformity for the most important characteristics, while simultaneously creating or maintaining general genetic heterogeneity and avoiding genetic bottlenecks. The approaches used with plants that don't show inbreeding depression often deliberately create genetic bottlenecks, because the easiest and fastest way to get a uniform pure breeding variety is to eliminate the genetic variability.

Outcrossing and Self-incompatibility

Many plants have mechanisms that help promote outbreeding and discourage inbreeding. The extreme case is dioecious plants such as spinach, in which the male flowers and female flowers are on separate plants. Monoecious plants, such as most squash and melons, have separate male and female flowers on each plant. This prevents the self-pollination of a flower, but not the pollination of a flower by pollen from another flower on the same plant. Many cucurbits have an additional trick. They tend to open only one new flower per vine per day, thus additionally discouraging pollination of a flower by pollen from another flower on the same plant. Flowers can, however, be pollinated by pollen from another flower on a different vine of the same plant.

Some plants with hermaphroditic flowers also are outbreeders. Onions, for example, are protandrous — that is, "early male." The stamens develop and release pollen before the style finishes developing and the stigma becomes receptive. However, the flower can be readily pollinated by pollen from other flowers on the same plant. Plants with early female parts also are common.

In yet other plants there is nothing anatomical to indicate that the plant is an outbreeder. Instead, there is a genetic incompatibility mechanism. The mechanism prevents self-pollination. It also prevents pollination by plants with an identical incompatibility genotype (such as plants developed by cloning). It also prevents crosses of some plants with entirely different plants if they happen to have the same incompatibility genotype.

Self-incompatibility can be very strong (there is nearly no selfing) or weak (there is some spontaneous selfing). It often varies from variety to variety within a crop. In addition, self-compatibility may be affected by the weather. A variety may be self-incompatible in cool weather and self-compatible, or more so, in warm weather, or vice versa. Some plants are more self-incompatible at one or the other end of the flowering cycle. They may be very self-incompatible early during their flowering but more self-compatible toward the end, for example.

Self-incompatibility can be an inconvenience to the breeder. Whenever we make an F_1 hybrid and intend to go on to the F_2 with an inbreeder, for example, we can grow just a few F_1 plants, because we can self-pollinate them. But if we're working with an outbreeder, we often grow a larger number of F_1 plants so that we are likely to have more than one incompatibility genotype. Then we make various crosses between them to get the F_2, or plant them in a patch away from others of their kind and let them cross-pollinate.

If an interesting plant turns up in an inbreeder, we can start to explore and work with it just by keeping its seed, which is inbreeding it. But if something interesting turns up in an outbreeder, the plant may resist our efforts to self-pollinate it. Various tricks for overcoming incompatibility barriers are listed in Appendixes A and B. In some cases, though, we may need to let other plants pollinate the special plant, then allow those progeny to cross with each other, to try to obtain what we want in a number of plants in a subsequent generation.

Saving Seed of Outbreeders

When we maintain seed of established varieties of outbreeders, we usually try to save seed from at least twenty or more plants each generation. And more is often better. Exactly how many plants we need to keep enough genetic heterogeneity to avoid inbreeding depression depends upon the specific crop. If inbreeding depression is a major problem, more is better. But, as with the cucurbits, where inbreeding depression isn't a major problem, we may be able to handle the plants virtually as if they were inbreeders, except we have to control pollination. Information on the extent of inbreeding depression is listed in Appendix A and Table 1.

Corn is one of the plants that is most severely and consistently affected by inbreeding depression. For this reason, it's best to maintain open-pollinated corn varieties by saving seed from at least a hundred good plants in each generation. With a hundred plants per generation, you can maintain most of the variability. With a thousand you can maintain nearly all of it.

Plants of outbreeding varieties are often perfectly willing to cross with plants of different varieties belonging to your neighbor down the street, or to wild relatives growing as weeds in or near your garden. To maintain outbreeding varieties, we often must isolate them from others of their kind in some fashion. Isolation distances considered suitable for each crop are given in Appendix A and Table 1. Technical aspects of maintaining outbreeding varieties, including how to cheat at the isolation distances, are covered in Part II.

Inbreeder or Outbreeder

As previously discussed, how we breed and maintain a variety depends on its basic breeding system — whether it's an inbreeder or an outbreeder, is self-incompatible or not, and so on. There are three ways to establish the basic breeding system: look it up; figure it out experimentally; guess. I use all three methods, depending on the situation.

Appendix A and Table 1 give basic breeding information for more than 800 plants. If I'm working with a rare or unusual plant, the information I need usually isn't known. Then I guess. In some cases there are anatomical differences that make the guess an excellent one. Most of the time, if I'm wrong, I just don't get the cross I want and have to try a new approach another year. Most of the obvious guesses work, though.

If the pollen is dustlike and blows around freely, the odds are that the plant outcrosses to at least some extent. Similarly, if bees or insects visit the flowers, there is usually some outcrossing — enough that isolation is required for strict pure production. But insects can facilitate either inbreeding or outbreeding, depending on the position, arrangement, and anatomy of the male and female parts of the flower.

Tomato flowers usually hang downward. The style and stigma are within a cone formed by the fused anthers, which release their pollen inward. Released pollen drops down over the stigma and fertilizes it. The opening into the fused anther cone is tiny, too tiny for wind access, as well as for all but the smallest insects. And it's rare to see an insect in or around a tomato flower. Tomato flowers also aren't especially noticeable, nor do they have a pronounced odor. All this makes it no surprise that tomatoes are strongly self-pollinating.

Large, bright flowers or flowers with odors suggest that the plant is trying to attract the attention of someone for some purpose. That someone is often a pollinator. In some varieties of tomatoes (including some of the heirlooms), the style sticks out beyond the rest of the flower. Insects can walk over it, and the wind can blow on it. These varieties exhibit a greater degree of crossing. The extent of outcrossing in tomatoes and many other crops often varies with the specific variety. For example, in some tropical areas, a specific insect visits tomato flowers; where the insect is present, the tomato is a partially outbreeding plant.

Even when there is information available on a plant's breeding system, keep your eyes open. You may have different weather and different insects than the person who established the information. Just because something is in a book doesn't mean that your plants and your insects will act accordingly.

You usually can't tell whether a plant is self-incompatible other than experimentally. With brassicas, whenever I want to self-pollinate an unfamiliar plant, I do at least some bud pollination so as to bypass the incompatibility mechanism if there is one. A good indication is what happens when a plant is grown by itself. If it sets seed, it is obviously self-fertile to at least some extent. If it doesn't, the evidence is less conclusive. Pollinators often don't find individual plants. You may be able to establish whether there is an incompatibility mechanism simply by doing controlled self-pollination and pollination with other pollen.

You can establish the extent of spontaneous inbreeding versus outbreeding by using different varieties whose hybrid is recognizable. Make the hybrid deliberately so that you can recognize it. Then plant a single plant of one variety in a patch of the other at your ordinary planting density. Note that the planting density often affects the extent of cross-pollination by determining whether most insect trips are between flowers on different plants or between flowers on the same plant.

Making and Breaking Hybrids

The commercial hybrid seed industry is a mixture of a biological and a proprietary situation. Hybrids are genuinely superior to open-pollinated varieties *for some crops in some cases for some purposes.* But commercial hybrid seed breeders have reason to develop and sell hybrids whether they are better for the customers or not. Often considerable effort is put into developing and promoting hybrids instead of open-pollinated varieties simply because it is more profitable for the seed company.

When a seed company develops and sells an open-pollinated variety, customers can save the seed if they want to, instead of returning and buying it from the seed company in future years. But when a company develops a hybrid, it keeps the parent lines secret so that only it can produce the seed. This means that customers have to buy the seed from the company every year or not grow the hybrid.

The problem is that many people garden partly for the independence of it and prefer not to be dependent on others for something so basic and essential as seed. If we give up the ability to produce our own seed, we are giving up something integral to the gardening experience. We are giving up our role as stewards — and breeders — of the germplasm we use. And we are giving up the pleasure of harvesting and cleaning the seed, holding it in our hands, and running our fingers through it — the essence of next year's crop.

In addition, where one or a few closely related hybrids replace all the heirloom and open-pollinated varieties in an area, the effect is a huge loss in overall biodiversity. The region comes close to being a monoculture of a single genotype. Epidemics of pests and diseases are far more frequent and severe in such situations than they are when many different varieties and crops are grown. Furthermore, when one of those more frequent epidemics occurs, if you are growing the standard hybrid, your crop is fully sensitive to it.

Fortunately, there is an option other than using a commercial hybrid. We can make our own hybrids, thereby retaining our independence and our control over our seed and deliberately maintaining and expanding agricultural biodiversity.

When I refer to a *hybrid,* I mean an F_1 of a cross between two different pure breeding varieties. This is the classic genetic meaning. In the seed industry the word is often used considerably more loosely. Sometimes F_2 material or even subsequent generations are referred to as hybrids. Sometimes even pure breeding open-pollinated varieties that were derived originally from a cross between two varieties are called hybrids. This latter use is incorrect or deceptive.

There are two major reasons for using a commercial hybrid instead of an open-pollinated variety — vigor and uniformity. Since the classic hybrid crop is corn, I'll use it to illustrate. First, with respect to vigor, the best commercial hybrid corn for your area will usually outperform every available open-pollinated corn you can find, and normally by a substantial margin. The highest commercial productivity for corn is almost always associated with the use of hybrids. This isn't necessarily true for other crops.

The crops for which hybrids show the greatest advantage over open-pollinated varieties are those that show the largest inbreeding depres-

sion. Corn and many of the brassicas display extreme inbreeding depression, so hybrids for them are especially advantageous.

For crops that are normally inbreeding or that are outbreeding but don't show much inbreeding depression, there is usually only a small advantage to hybrids, if any. The major reason we have so many tomato, squash, and melon hybrids is proprietary — it is simply better for the seed company (if it can fool us into using them). In most cases, however, it isn't necessarily better for us.

Second, with respect to uniformity, a good commercial corn hybrid will always be much more uniform than even the best open-pollinated variety. For machine tending and harvesting, this can be a huge advantage. Each hybrid plant is almost genetically identical to the others, they all grow to the same size and shape, and they mature simultaneously.

For machine harvest, any plants that are too different in size or shape don't get harvested properly. Any that are much earlier than the harvest date usually have deteriorated. Any that are later than the harvest date never mature at all. The more the plants are identical and mature identically, the larger the proportion of them that contribute to the harvest. Synchronous maturity may matter even for hand-harvested vegetables. I harvest my chickpeas by hand, but I would rather harvest all the plants in a patch at once instead of having to watch individuals in the patch and harvest them one at a time as they're ready.

The uniformity of hybrids is sometimes not desirable for home gardeners. A gardener who likes fresh broccoli might prefer her thirty broccoli plants to mature variably so that she has a head every few days, not thirty at once.

Territorial Seed Company has an innovative solution to the problem that the vigor of hybrids may be desirable for home gardeners in cases where the uniformity is not. For a number of crops, they mix seed of different hybrids that have different maturity dates. A customer can buy the pure varieties or the mix. The mix gives you the vigor of hybrids without the uniformity in genotype or maturity date. You can, of course, always buy a variety of hybrids with different maturity dates and plant them in separate patches, or mix the seed yourself.

The superior performance of commercial hybrids and open-pollinated varieties is virtually always associated with optimal (commercial) growing conditions — large amounts of chemical fertilizers and good weather. Under less than optimal conditions, good regionally adapted open-pollinated varieties may perform as well as or even better than the best hybrid. In good years those who plant hybrids usually have a higher yield. In bad years those who plant an open-pollinated variety may be the only ones with at least some yield. Normally the introduction of hybrids goes along with the use of high-nitrogen fertilizers, herbicides, and pesticides so as to bring out the full potential of the hybrid.

I don't know whether the tendency of modern hybrids to perform well under only optimal conditions is inherent in the nature of hybrids. I suspect it is not. I think we have that kind of hybrid simply because that is what the developers of the commercial hybrids have chosen. Their hybrids reflect their beliefs and values. If you want hybrids that perform optimally under low-input, sustainable, more organic conditions, you should be able to develop them. Just start with parental varieties that do well under those conditions and test your hybrids under those conditions. When you're done, your hybrids will reflect *your* beliefs and values.

An additional criticism of hybrids is that they are nutritionally inferior to open-pollinated varieties. I think this is often true, but I also think it has less to do with the inherent nature of hybrids than with the specific kinds of hybrids that are being produced. Little attention is given to nutritional value in the breeding of food crops. Until recently it was more or less invisible. Superior nutritional value was probably involved indirectly in the preservation of heirloom varieties, because people would keep what tasted good (which may or may not correlate with nutritional value) or because the livestock preferred a variety and performed better on it. I believe that it should be possible to develop hybrids of high nutritional value, too, if we select for it directly or indirectly.

Alan Kapuler, who is a geneticist, a biochemist, and a vegetarian, has a special interest in the nutritional value of vegetables. He's been testing heirloom and other varieties and building a collection of varieties that are especially nutritious. His research on the nutritional value of vegetable varieties and the varieties themselves are available through Peace Seeds (see Appendix D). His catalog is probably the best starting point for those interested in breeding more nutritious hybrids or open-pollinated varieties.

Peas, beans, and many other largely inbreeding crops don't lend themselves to the production of hybrids. In addition, there isn't much to be gained, because they don't display inbreeding depression. There are, however, many hybrid tomato varieties, and tomatoes also are inbreeders. A hybrid tomato industry is possible only because most tomatoes produce several hundred seeds per hand-pollination.

There are many open-pollinated lines of tomatoes that are as good as even the best of the hybrids. I believe the major reason that there are hybrid tomatoes is the proprietary one — seed companies would rather have us grow something only they can produce. The hybrids are generally not superior to all available open-pollinated varieties, and in many cases they may not be hybrids at all. They are open-pollinated lines that the seed companies would prefer to have us believe are hybrids.

In a number of cases, especially with tomatoes and various cucurbits, if you inbreed a so-called hybrid variety, you see no segregation in the next and all subsequent generations. That is, the variety behaves exactly as if it were a pure breeding variety. It's possible that the hybrid came from a cross between two varieties that were identical for all genes involving plant size, shape, and form; leaf size, shape, and form; and so on, but were different for something invisible, such as disease resistance. But I suspect that in most cases where no obvious segregation occurs when we inbreed a supposed F_1 hybrid, it is because it is *not* an F_1 hybrid; it is a pure breeding variety. It's easy to see why false claims of hybrid status might be advantageous and profitable.

I began to suspect the status of hybrids a number of years ago when I inbred a couple of corn varieties and saw no segregation for anything. A few years later I remarked to Alan Kapuler that I thought many "hybrids" might not be hybrids at all. His response? "Aha! So you've noticed that too!"

I suspect that seed companies sometimes list an open-pollinated variety as a hybrid in order to deter seed savers or competitors. Why not call the bluff? If you raise twenty or so putative F_2 progeny and see no segregation, you may already have your open-pollinated variety.

Alternatively, the commercial variety might be a true hybrid, but might represent a cross between two parent lines that were very similar in all the obvious visible characteristics but different in other critical ways — such as disease resistance. If so, your F_2 would actually be segregating for disease resistances. Just grow the plants for a few generations under your conditions, saving seed from the best of each generation. If the material contains genetic variability for disease resistance, you will automatically select for resistance to the diseases of your region and growing conditions.

When you derive an open-pollinated variety from a commercial hybrid, give yours a distinctive name. Many open-pollinated varieties can be derived from a single hybrid, and they need names to distinguish them from the parent as well as from each other.

In the next chapter, a number of examples involve creating open-pollinated varieties from hybrids, that is "dehybridizing the hybrids." In the last part of this chapter we'll cover why, when, and how to make our own hybrids.

There are six major reasons for making and using a hybrid instead of an open-pollinated variety. First, it may be a crop, such as corn, for which hybrids are usually much more vigorous and productive than inbred varieties. Second, you may have kids; making and testing hybrids is a good breeding project for them. Third, you may need the potential uniformity of hybrids. If so, you need to cross inbred lines of open-pollinated varieties. Fourth, it is easier and faster to make and use hybrids in certain situations than to develop open-pollinated varieties. Fifth, you may want proprietary control yourself. Finally, there may be no other option. Sometimes we can make a hybrid between two varieties that has all the characteristics we need, but it isn't possible to develop an equally good pure breeding variety from it.

If you cross almost any two open-pollinated corn varieties, you'll see a substantial increase in plant vigor and yield. If you have kids, making corn hybrids is an ideal project. It's an excellent illustration of reproductive biology, the hybrids so outperform the parent varieties, and the entire project only takes a year.

With kids, an enjoyable approach is to interplant two varieties that have different colors of kernels. When you interplant a black corn and a white corn, for example, the white ears will have some black kernels on them. Kids enjoy picking the black kernels out of the ears for planting. If you want to make corn hybrids more efficiently, you can use hand-pollination, or plant alternate rows of the two varieties and detassel one of them (see appendix A).

If you want the most productive hybrids possible, you'll make crosses between pairs of all your favorite varieties and test them under your conditions. You should be able to find hybrid combinations that yield as well as or even better than the commercial varieties, because you are testing in the context of your exact growing conditions and requirements, and because your hybrids do not need to have broad regional adaptability.

The hybrids you make by crossing two open-pollinated varieties will usually not be as uniform as commercial hybrids. If you are a home gardener, this may be preferable. If you are a market gardener or farmer, you may need the uniformity typical of commercial hybrids. To get it, you must cross two inbred lines. I suggest that you first test your open-pollinated varieties without inbreeding to identify which combination of two varieties produces the best hybrid. Then develop some inbred lines from each of those varieties.

To develop an inbred line, just self-pollinate a few plants for two to four generations. Then maintain the lines normally. One or more of the lines will probably be genetically uniform enough to generate uniform hybrids, but still heterogeneous enough to be reasonably vigorous. Inbred lines of corn do not necessarily look very uniform, incidentally. Their lack of vigor seems to cause them to be highly affected by minor differences in the environment. Your inbred lines will probably look inferior and variable and yield poorly compared to the lines from which they were derived.

With tomatoes, squash, and melons, we will sometimes make and use a hybrid simply because it is so fast. When Glenn Drowns crossed two different varieties of watermelon, the F_1 hybrid between them was early enough to ripen in his area, though neither of the parents or other varieties were. He could have used the hybrid for his crop instead of developing an open-pollinated variety. In some cases, the number of genetic markers involved, or their linkage, or sterility of the hybrid makes it difficult or impossible to develop an open-pollinated variety. In these cases, using the hybrid is our only option.

In some cases gardeners, especially market gardeners, will want to explore and develop their own hybrids and use the proprietary characteristics themselves. For the innovative vegetable gardener who earns a living by finding and introducing new crops and varieties, a major problem is that other growers immediately copy them whenever they find a winner, without spending any time, labor, and money on research. Part of the solution is to keep finding new crops so as to stay ahead of these copycats. Those who sell produce might want to consider paying some attention to developing their own exclusive, farm-brand hybrids that only they can grow. People who have read this book can, of course, use your hybrids to develop their own equivalent hybrids or open-pollinated varieties, but not without their own research and labor.

Plant Breeding Stories

I N PREVIOUS chapters, I've provided all the theoretical background you need to breed plants. In this chapter, I show how to put it all together and use it.

The basic genetic techniques are selection, inbreeding, making crosses of various kinds (including backcrossing), and creating and maintaining genetic heterogeneity. Selection is a powerful technique when there is genetic variability for the characteristic we're interested in. It's useless when there isn't. In this chapter you'll learn how to distinguish between genetic and environmental variability, to tell when to use selection, and to turn the basic technique of selection into *power selection*. You'll also learn how to use back-

crossing and recurrent backcrossing, and how to turn basic inbreeding into *power inbreeding*.

Crosses are the way to introduce genetic variability when your material doesn't have it already. Crosses can be used to combine characteristics from different varieties into one variety or to create entirely new phenotypes. In addition, a very powerful kind of crossing, *recurrent backcrossing,* can be used to transfer one or a few genes into some other variety. Many breeding projects start with a cross. The cross creates the needed genetic variability; the other methods manipulate the variability in various ways.

Inbreeding is used to reduce the genetic variability in the genetic material so that it becomes predictable and reliable for the characteristics we care about. It's a very fast, powerful way of creating new, stable varieties for crop plants that don't suffer from serious inbreeding depression.

Corn, most of the brassicas, and many other crop plants are outbreeders that suffer from inbreeding depression if they are made to be too genetically homogeneous. The obvious breeding method of just saving seed from the best plants can destroy the vigor and productivity of these kinds of varieties. Inbreeding depression affects every aspect of creating or maintaining these varieties. I dealt with this problem theoretically in the last chapter. In this chapter I explain how to take it into consideration practically and how to use methods that deliberately maintain adequate amounts of genetic heterogeneity.

Selection is the only method commonly used alone in a project. Most projects involve a number of methods used at various stages of the project. In this chapter I illustrate how to mix and match these methods by telling stories about my own breeding projects and those of my friends.

How we go about doing any specific breeding project often has a lot to do with the breeding system of the crop involved. Information about the breeding systems of various plants is given in Appendix A and Table 1. How to figure out the breeding system where no information is available is explained in chapter 9. Mechanical details about hand-pollination, making crosses, and tricks for overcoming incompatibility barriers are given in general in Appendix B and for specific vegetables in Appendix A.

Perennial Vegetable Buckwheat
(Selection.)

I recently obtained about a dozen seeds of a Himalayan perennial vegetable buckwheat, *Fagopyrum cymosum,* from J. L. Hudson. Alan Kapuler took a few of the seeds and planted them at his house. I planted a few at mine. I ended up with two plants, both with attractive heart-shaped leaves. Alan ended up with a single plant that was much bigger than mine, but he had planted earlier and in better soil.

The first of my plants to flower did not set any seed until the second began to flower. It appeared to be self-incompatible. Alan's plant set seed prolifically; it was apparently self-fertile. Of my two plants, one had much more intense purple stems and leaf veins. These kinds of differences between individual plants are common when you're working with rare vegetables or with relatively new material.

I didn't have any a priori ideas about which characteristics were most desirable, except that I am interested in perennial vegetables that are hardy in my climate. So I just waited to see whether the plants would survive the winter.

All three plants died down with the first frosts but then sprouted vigorously in the spring. The plants are very vigorous and produce copious young shoots, which are the edible parts. All are a mild-flavored cooking green and taste about the same.

Alan's plant and one of mine are very invasive. My invasive plant had shoots coming up from spreading roots all across the garden bed, up to about 3 feet away. That's pretty impressive for a plant started from seed just one season before — impressive and frightening. My other plant, in another bed, sent shoots up only right around the plant. This plant is the one with the most intense purple coloration and is quite attractive.

I have been ripping up shoots from the invasive plant, eating them, and feeding them to my guinea pigs. When I want more plants, I will make divisions of the better-behaved purple. Alan will probably eliminate his plant and clone from mine, too, if it remains well behaved.

What I've just described is selection at the simplest and most basic level. In many cases this is all you have to do. We grew only three plants, but each was different. All were very vigorous once established. My favorite plant has produced copious amounts of food. I chopped it off at a couple of feet high and have chopped off shoots frequently enough to keep it contained as a good-size bush. Now I need to learn more about how to use the plant. I'll also want to keep an eye on its invasiveness as its root system becomes larger and better established.

Many rare plants or collections of wild material are genetically very heterogeneous. All you have to do is look. It's important to look with an open mind — that is, grow some plants, watch them, live with them a bit, and then decide what you want. Consider both the virtues and disadvantages of every characteristic. A self-incompatible perennial buckwheat, if propagated vegetatively by itself away from others, cannot be propagated by seed, which can be considered

a disadvantage, for example. But it also will not be shedding seed all over the place to become weeds, which greatly facilitates caring for the perennial vegetable patch.

Orange Popbean
(Selection, determining whether variability is genetic or environmental, power selection.)

Some varieties of chickpeas are able to overwinter readily here in maritime Oregon; others are not. Most of the large cream-colored types that resemble the commercial chickpeas common in the United States don't overwinter here, but many of the colored types, as well as some of the tiny cream ones, do.

Last year I planted ten seeds of the orange popbean (PI374085) I described in Chapter 3 as part of my overwintering trials of 150 accessions. Only one plant survived, so I am saving seed from that plant. Perhaps the plant survived because it was genetically more cold resistant than the others. But maybe it survived only because it was lucky — it happened to be in a little more favorable spot and got better established before the bad weather hit, for example.

If the plant survived because it was genetically superior, I might want to use it as the foundation for its own variety. So how can I tell whether the variability I saw was genetic or environmental? Well, if the plant is genetically superior, it should be able to pass on that superiority to its offspring.

I've saved seed from that plant separate from other seed of the accession (that I planted later). I will treat that seed as if it were, indeed, a new variety. But next year I will test that seed against seed produced by the unselected plants. If some or all of the selected material performs much better than any of the plants in the unselected accession, I will know that the original plant was genetically superior. Likewise, if the average performance of the seed of the selected plant is superior to that of the unselected material, I will know that genetic differences are involved.

If, on the other hand, the average performance and the range of performance of the selected material aren't any different from those of the unselected material, I will conclude that the original plant wasn't genetically superior; it was just lucky. I won't try to develop a more hardy variety based on that plant as the foundation.

What I am doing physically involves only keeping seed from the best plant separate for at least one generation. But in actuality, it is quite sophisticated. When I compare offspring of the best plants with offspring of the average, I evaluate whether the differences I see in the overall population of plants are genetic or environmental. This evaluation allows me to tell whether the selection I did accomplished anything and whether further selection along the same lines will accomplish anything.

Let's suppose that I find that a number of plants from the selected plant's seed are much more vigorous and productive than those in the rest of the accession. I will then use the seed from that original plant as the basis for a new more cold-hardy variety.

If my new variety shows variability for cold hardiness, I will continue to select by saving seed from the best plants each year. I won't throw away the seed from the rest of the plants, though. I will plant some as a comparison. At some point I will have eliminated all or nearly all of the genetic variability for cold hardiness from my variety. At that point the seed from the best plants and that from the worst plants should perform about the same. When that happens, it means that the poor plants were poor for environmental reasons, that they were just as good at passing on cold hardiness to their offspring as the best plants were.

Whenever you see variability in a variety or a population of plants, you can test whether that variability is genetic or environmental by saving the seed from the best and the worst separately, then comparing the two the following year. This method works whether the characteristic is based on one gene or several. Comparing the best with the average is not quite as powerful, but usually it's less work and is good enough.

Whenever a variety or population contains genetic variability for a characteristic you're interested in, you can improve the variety just by simple selection. If the variability is not genetic, then no amount of selection will be effective in improving the variety. In order to test the population for the nature of its variability, you just proceed as if you were selecting, but you evaluate whether selection is actually happening. *Selection plus evaluation is power selection. Evaluation also tells you when you're done — when you've eliminated virtually all unwanted variability and can simply maintain the variety.*

Maintaining varieties also requires selection, however. It's usually referred to as *cull-*

ing or *roguing*. Whether we call it selection or culling/roguing depends on just how much we're eliminating. When we're creating a new variety, we save seed from only the best plants in each generation. Most plants aren't allowed to contribute to the next generation. We call that selection. Once we have our new variety, we keep seed from nearly all the plants. But there will still be occasional new mutations or undesirable combinations segregating out. So before the variety begins flowering, we examine the population and eliminate the occasional rogue so that it doesn't contribute to the next generation.

High-Yielding Chickpeas
(Selection.)

One of my plants of small, fast-cooking cream-colored chickpeas had more than eight hundred seeds on it. I kept the seed from that plant separate from the rest of the accession. If the yield of that lot is better than that of the rest of the accession, it will represent a first step in breeding for high yield. It is a common method for use with inbreeding crops and is called *single plant descent*. All it really amounts to is keeping your eyes open for really good plants and saving the seed from them separately. But remember to compare the plants grown from that seed with the rest of the population to see whether the seed really is different.

With inbreeding crop plants such as chickpeas, when you start with heterogeneous material such as a land race, you can do a large amount of breeding by simply observing the plants and noticing any that are different or superior. As I explained in

the last chapter, given the genetic nature of inbreeding plants, it is likely that in most cases the plants I pick out as superior are homozygous for whatever genes are involved in their superiority. When this is true, their offspring are a new variety. I may have a new, pure breeding variety of chickpeas simply by saving seed from a superior plant.

The Quest for 'Golden Snap' Peas
(Doing a cross and going to the F_2, followed by selection and inbreeding.)

Alan Kapuler wants a yellow-podded snap pea. There is an heirloom yellow-podded snow pea available — 'Golden Sweet' — that is one of the varieties Mendel worked with, Alan says. But a yellow snap pea would be a nice addition to the snap pea repertoire.

There aren't any yellow snap peas. Furthermore, there is no apparent variation in color within snap pea lines. If, when you grew a snap pea variety, you saw some plants with pods that were more yellowish, you might try to develop a yellow snap by simple selection. But you don't see that, so you can't start the project just by selecting. You first have to introduce the variability you need.

Alan started off by crossing 'Golden Sweet' with 'Sugar Snap', a green-podded snap pea. He wanted a new variety that would have the pod color of 'Golden Sweet' and the pod type of 'Sugar Snap'. To combine characteristics from two different varieties, you start by crossing them. The most common way of proceeding is to go to an F_2 and start selection with the F_2 or a later generation.

I've mentioned before that snow (edible-podded) peas differ from shelling peas by two recessive genes. Both are required for a snow pea that is edible when the pods are large and the peas have started to form. An additional gene, the N gene, is associated with pod thickness. Snap peas are *ppvvnn,* snow peas are *ppvvN_,* and shelling peas are usually *P_V_N_.* So a cross of a modern snow pea variety and a snap variety is likely to be *ppvvNN* × *ppvvnn.* In other words, it's a one-factor cross. Likewise, the difference between yellow and green pods is generally a one-factor cross.

I learned about the genetics of pod type and color by reading the chapter on peas in Mark Bassett's *Breeding Vegetable Crops* (see bibliography). It includes lists of the common genes involved in agriculturally important characteristics and their linkage groups. I give information on some of the most common genes in the most common vegetables in appendix A. Referrences to the professional literature on hundreds of vegetables are included in Table 1. You should be able to track back into the plant breeding literature and find out what is known about every common vegetable, as well as hundreds of less common ones.

It is not necessary to do that, however. I usually don't, even though I am a geneticist by training, have a full-scale agriculture library a few blocks from my home, and am already familiar with the specialized plant breeding literature. In many cases there is little or no information available on the vegetable I care about. Often, even if information is available, finding it would take more work than doing the cross and raising a few F_2s. Sometimes there is plenty of information, but not necessarily on the traits I am interested in (such as overwintering ability or slug resistance). Or there is some information, but it is contradictory.

In some cases, especially with the more common vegetables, you can find out enough relevant information to make better guesses about what results you are going to get when you do a cross. But you will usually do exactly the same thing whether you have this information or not: you do the cross and go to the F_2. When you get the F_2, you learn the essential genetics, whether you knew anything about it ahead of time or not. And what you find out tells you how to proceed.

Even if you read a lot of technical sources and make guesses about the genotypes of your varieties and the genetics involved, you often will have to revise your assessment based on your own results. When someone says that some trait involves one factor, that is only with respect to that person's two parental varieties. That tells you that the genetics of the trait in your cross may be simple; it doesn't guarantee it.

When Alan Kapuler started his 'Golden Snap' pea project, he did not know anything about the genetics of pod types. The snow versus snap characteristic could be a one-factor difference in his cross, or it could involve two factors or more. He suspected that the yellow versus green pod color was a one-factor difference, since this was one of the traits Mendel worked with, and he'd read about it. But just because yellow versus green was a one-factor difference in Mendel's crosses, there was no guarantee that it would be in Alan's, unless he was doing exactly the same cross. And he wasn't.

So Alan didn't know what specific genes

he was working with or what their patterns of inheritance would be. The simplest possibility would be that pod type would be associated with a single gene, that color also would be associated with a single gene, and that the two genes would be independent. When he crossed the green snap and the yellow snow pea, he got a green snow pea (F_1). If it was a two-factor cross, it was the double-recessive class he was trying to recover; the expected frequency of the class would be $\frac{1}{16}$.

Alan raised just a few F_1 plants. As we have seen, the F_1 plants from a cross between two pure breeding varieties are all the same, so we don't need many. Alan grew about twenty F_2 plants. More would have been better. As Appendix H shows, he should have had forty-six plants to be 95 percent sure of getting at least one of a class that has a probability of $\frac{1}{16}$. But Alan had room for only twenty plants.

Twenty is a useful number of F_2s in many cases. It may not be enough to give you what you want immediately, but it will generally give you a rough idea of the genetics involved so that you can figure out what to do next. Often it provides the material you need for the next stage. I often plant only twenty of the F_2 of a cross, even when I know more than two factors are involved.

It's usually easiest to decipher what is going on in a cross if you first consider each trait separately and then consider how the traits are assorting with respect to each other. Let's consider the two traits in Alan's cross — pod color and pod type — separately.

With respect to pod color, the cross was green × yellow and produced a green F_1. Of the twenty or so F_2 plants Alan raised, all had pods that were either green or yellow. There were no new colors or shades of colors. Most of the F_2 pods were green, but a few were yellow. This pattern suggested that pod color in the cross was associated with a one-factor difference between the two varieties and that there was simple dominance. More complicated explanations are possible, but in most cases where you see that pattern, one gene with simple dominance is what you're dealing with.

With respect to pod color, Alan's cross was GG (green) × gg (yellow), and the F_1 was Gg (green). I chose the letters G and g to use as the symbols; you can use anything, as long as the two symbols you choose have some sort of logical relationship that makes it easy to remember that they represent alleles of the same gene. Conventionally, capital letters are assigned to the dominant allele.

With respect to pod type, Alan also got results that were consistent with a single-factor cross. The cross was snow × snap. The F_1 was snow, and the F_2 had two discreet classes, snow and snap, with most of the plants being snow. We can portray the cross with respect to pod type as SS (snow) × ss (snap); the F_1 would be Ss (snow). (We wouldn't normally know what symbols the professionals have assigned to the genes, so we would choose working symbols of our own.)

By assigning a different set of symbols to represent pod type, we're assuming that differences in pod type and pod color are associated with different genes. This is a good guess, but it doesn't necessarily have to be true. Some genes are pleiotropic — that is, they have several different apparent phenotypic effects. For example, the genotype $A\_$ is necessary for formation of the purple pigment, anthocyanin, in peas. It can affect the color of the pods, the stems, the flowers, and other parts of the plant, depending on what other genes are present. So flower color and pod color can be associated with a single gene in some crosses. In other crosses flower color and pod color may involve two genes, with one determining whether there is any purple pigment in the plant and the other determining whether it shows up in the pods. It's common for a gene to have a phenotypic effect and an effect on growth rate, general vigor, or fertility as well.

Pod color and pod type segregated from each other in the F_2 plants of Alan's cross, so they must be associated with different genes. Alan's results were typical for a two-factor cross, with one factor associated with each trait and with simple dominance for both loci.

Having looked at each trait separately, we then look at how they behave together — that is, are they assorting independently, or are they showing evidence of linkage? In most cases we will find the two genes to be independent, because they will be on different chromosomes or will be far enough apart on the same chromosome so that crossing over recombines them freely.

The theoretical expectation in a two-factor cross is a 9:3:3:1 ratio if the two genes are assorting independently. If they are linked

we expect a preponderance of the parental types. But there isn't much use talking about the exact ratio when you only have twenty plants. What's more relevant is whether all the expected classes showed up, and how frequent they are, roughly speaking, relative to each other.

What an expectation of $\frac{9}{16}G\_S\_$ + $\frac{3}{16}G\_ss$ + $\frac{3}{16}ggS\_$ + $\frac{1}{16}ggss$ translates to if you only have twenty plants is that you expect more of the class with the two dominants than of anything else, and least of the class with the two recessives. With twenty plants, you would often not get any of the class with the two recessives, in fact. But twenty plants is enough to tell you that it's most likely that two factors are involved in the cross, one associated with each trait, and that they are inherited independently. That analysis forms the basis for your decision as to how to proceed.

Knowing that, you would realize that your cross was a completely reasonable approach to getting what you wanted, and should do the trick, but you just need a few more F_2 plants. And it might have been that what you needed was one of the more common classes, so twenty plants might have been enough. In addition, twenty plants is frequently enough to give you better material for the next step than the original cross.

As it turned out, Alan got all four classes in the F_2. He had more green snows than anything else, a number of green snaps and yellow snows, and only one yellow snap. The appearance of the F_1 and the classes in the F_2 indicated that the yellow snap must be genotype *ggss* — that is, it would be pure breeding for yellow snap peas. Alan could save the seed from it and start increasing it, with the expectation that he already had, in only two years, a pure breeding yellow snap pea variety.

But there was a catch. The yellow snap pea was setting only one pod per node. 'Sugar Snap' and 'Golden Sweet' both usually set double pods. The F_1 had probably set doubles. There was variability for pod number in the F_2, so the parental varieties must have had different genes for double podding. At least two genes must be involved.

In other words, when Alan did the cross to generate material that had the required variability for pod type and color, he introduced unwanted variability for pod number. This sort of thing happens all the time. Anytime you do a cross to introduce the variability you want, you usually also introduce variability you don't want and then have to eliminate it.

It wasn't possible to tell much about the inheritance of pod number in the cross. However, it seemed possible that the single podding might be associated with one or more recessives. That might mean that the yellow snap with single pods might give rise only to single-podded offspring, which could be increased only into a single-podded variety. No double-podded peas would segregate out. For this reason Alan did not want to depend on the offspring of the yellow snap for a new variety.

If single podding was associated with dominant genes, the single-podded yellow snap's progeny might include some double-podded types. In that case, Alan could derive a double-podded variety from the progeny of

the single-podded pea. So he didn't want to reject the single-podded yellow snap either.

He saved seed from the single-podded yellow snap. But he also saved seed from green snaps (G_ss). He knew that some of the green snaps would be heterozygous for yellow, that is, $Ggss$. In fact, two-thirds of them would be. The reason for the number *two-thirds*: The F_2 is (¼ homozygous green + ½ heterozygous green + ¼ yellow). There are twice as many heterozygous greens as homozygous greens. That is, *of the greens,* two-thirds are heterozygous for yellow. Their genotype is $Ggss$. And they will produce the desired yellow snaps as one-fourth of the next generation.

Alan also kept seed from the double-podded yellow snows ($ggS\_$). Two thirds of them would be $ggSs$, and would also produce the desired yellow snap class as one-fourth of the next generation.

When Alan raised only twenty F_2s from his cross, that turned out to be enough to get one of the double-recessive class — but it wasn't enough because of other things segregating in the cross. The twenty F_2s were enough, however, to give him essential information about the genetics involved, and he could use the information to make good choices about how to continue the project and get what he wanted with an additional step.

Normally, even if the desired class is obtained in the F_2 generation, you raise seed from a number of the plants and select further. Anytime you do a cross, other things will be segregating besides the genes you're primarily interested in. Some plants will be more vigorous than others, some will be earlier, and so on.

The basic pattern, though, is that you do the cross, raise a few F_1s, and then raise twenty or more F_2s. If you obtain the desired class in the F_2, you self-pollinate the plants (if it is an inbreeding crop). If you don't get what you want in the F_2, you use what you've learned about the genetics from the results of your cross to decide how best to proceed — whether to try a different cross, stick with the one you have and raise more F_2s, or use some of the F_2s to derive what you want in one or more additional steps. After you have some of the desired class, you increase your interim variety by inbreeding further and selecting the seed from the best plants in each generation.

If many genes are involved in the cross instead of a few, the process takes longer, but the same general approach applies. If some of the genes are linked instead of independent, that can either speed things up or slow them down, but in most (but not all) cases the basic approach is still effective.

You may or may not know anything about the specific genes involved in your cross, but in most cases that doesn't change what you actually do. You'll use your knowledge of the general principles as applied to the actual results from your own crosses to decipher the basic genetics involved. Then you'll be able to figure out how best to continue.

Purple-Podded Peas
(Crossing, selection, power inbreeding, backcrossing, recurrent backcrossing.)

I want a purple-podded snap pea. I don't know of any. The snap peas we have don't show any variability for purple color, so if I

want a purple snap, the best way to start would be to cross a purple snow pea and a green snap pea. But there aren't any purple snow peas either. Every purple pea I've seen has been a shelling pea.

The purple shellers are beautiful. The plants are intensely green, the stems are greenish purple, the flowers are two-tone maroon, and the purple pods are lovely against the background of the foliage. Even varieties described as being edible-podded have always turned out not to be when I checked them out.

I obviously need to start by crossing a purple nonsnap to a nonpurple snap and using genetic recombination as the basis for developing a variety that is both purple and a snap. But I also want the enation resistance and powdery mildew resistance that matter in my area. So I started by crossing a purple-podded shelling pea with 'Oregon Sugar Pod II' to develop a disease-resistant purple snow pea.

There are two main diseases involved, and the purple shelling pea isn't resistant to them. The enation probably involves one gene and the powdery mildew two. In addition, there are two genes involved in the difference between shelling and edible pods; the edible pods are the combination of the two recessives. That's five genes all together, and for at least four of them, I need the recessive combinations.

What I've read about pod color suggests that it takes three dominant genes to produce purple pods. One is involved in anthocyanin production in the entire plant; the others extend it to the pods. To use arbitrary letters, let's say $A\_B\_C\_$ is the genotype required for the purple pods, *ddeeff* is the genotype required for the resistance to the two diseases, and *ppvv* is the genotype associated with the edibility of the pods.

The cross would be *AABBCCDDEEFFPPVV* (purple, disease sensitive, shelling pea) × *aabbccddeeffppvv* (green, disease resistant, snow pea). I'm guessing about parts of the genotypes, assuming the worst. The purple variety might carry one of the disease-resistance-conferring genes, for example. And the green variety might have one of the dominants that helps produce purple pods. But I should assume the worst, since any cross can introduce additional factors that I don't know about. (Such as the single-pod-per-node characteristic I've already mentioned.)

This is an eight-factor cross. I want a new variety of genotype *AABBCCddeeffppvv*. I could derive it very easily if I could obtain an F_2 plant of the *A_B_C_ddeeffppvv* class. But how likely is that? Is it at all possible for a person whose total pea breeding patch is 18 feet of wide row?

I can calculate it, assuming all the genes are unlinked. With this many genes, that assumption is questionable, but it gives me a first guess. The probability of getting the desired phenotypic class is ¾ for each of the dominants and ¼ for each of the recessives. So the probability of getting an individual of phenotype *A_B_C_ddeeffppvv* in the F_2 is $(¾)^3(¼)^5$, which is $^{27}/_{65,536}$, or 0.0004. Four peas in ten thousand. That's totally outside the range of numbers I usually deal with. Even if I had the space for ten thousand peas, I wouldn't have the time to evaluate them to find the phenotype I want.

But there are other approaches and tricks.

First, I don't have to get the class I want all in one step in the F_2. I can do it in more than one step. Second, I can derive subvarieties that have some of the characteristics I need, then make a cross between two of those to obtain the final variety. Third, I can use a trick I refer to as *power inbreeding*. Fourth, I can use an especially powerful different kind of technique called *recurrent backcrossing*.

Let's say I am willing to devote 20 feet of wide row to the project, with the row five peas wide and thirty peas per foot. That's three hundred peas total. In all cases I would start the same way. I would cross the purple shelling pea to the green snow pea. Then I would grow a few F_2 plants; that would require just a few inches of space. Then I would grow three hundred F_2 plants.

For approaches 1 or 2 — using more than one step to obtain the desired class or creating subvarieties — I want the class *A_B_C_ddeeffppvv*, but I won't try to get it all at once. I realize that the purple part of the phenotype is associated with dominant genes. (The phenotype of the F_1s being purple tells me this.) I might, for example, save purple-podded snow peas and eliminate the others. That is, I would pay no attention to disease resistance. I would probably be planting very early so as to get the most and best-quality seed anyway (or overwintering the plants), and under these conditions, the diseases normally don't show up.

Purple-podded (*A_B_C_*) peas are expected at $(¾)^3$ of the F_2, or $^{27}/_{64}$. Snow peas (*ppvv*) are expected at $^1/_{16}$. So purple snow peas should be $(^{27}/_{64})(^1/_{16})$ of the total, or about 0.026 of 300, or about 8. (This assumes none of the genes are linked.)

I would then have a purple snow pea, and I would probably just start using it as such. The working purple snow line would be pure breeding for edible-podded pea type, but not necessarily for purple color. I would cull out any greens that appeared in subsequent generations. I could start growing and using the pea and select for resistance to the various diseases in whatever year they happened to occur in.

But what if none of the peas were resistant to all the diseases? Whenever a disease appeared, I would take advantage of its presence to isolate sublines of my variety that were resistant to it. I might not have a line resistant to both, but I would be very likely to have lines resistant to one or the other. I could then cross a line resistant to one disease with the line resistant to the other and recover a variety resistant to both in the F_2 or beyond.

Approach 3 is what I call *power inbreeding*. Start out as before by making and raising an F_1 and an F_2. Grow three hundred F_2 plants, as before. Save one seed from each of the three hundred plants and plant those seeds the following year. Then save one seed from each of those plants, and plant them the next year. And so on. In the purest form of this approach, you don't do any selection at all. In this case I would modify the design slightly and keep seed from just the purple plants each generation.

This approach makes use of the fact that inbreeding reduces the heterozygosity by one-half each generation at all loci except those involved in traits that we are selecting for (see Chapter 9). That is, in the F_2 the probability of each locus's being heterozygous is $\frac{1}{2}$. In the F_3 it is $\frac{1}{4}$ (assuming we aren't selecting deliberately for or against anything). In the F_4 only $\frac{1}{8}$ of the loci are heterozygous. And in the F_5 only $\frac{1}{16}$ are. This means that in an unselected F_5 generation, the genotypes in the population are $(\frac{15}{32}AA + \frac{1}{16}Aa + \frac{15}{32}aa)$. In other words, the probability of obtaining genotype *AABBCCddeeffppvv* is about $(\frac{1}{2})^8$, or $\frac{1}{256}$. And this is if we did no selection at all.

In other words, if we merely inbred each generation without any evaluation whatsoever or any selection at all, raising three hundred seeds each generation for five generations, we would have a 1 in 256 probability of obtaining the exact genotype we want, *AABBCCddeeffppvv*. That's within the realm of possibility.

The slight modification of keeping only the purple-podded plants would eliminate most of the genes associated with green pods by the F_5 so that nearly all the F_5 plants would be *AABBCC*. The probability of obtaining the rest of the desired genotype in the F_5 would be about $(\frac{1}{2})^5$, or $\frac{1}{32}$. With three hundred plants, we should get about nine with this genotype.

Practically speaking, we wouldn't be able to identify the disease-resistant part of the phenotype (assuming growing the peas very early so as to avoid disease and get the best seed). So we would just keep all the purple-podded snow peas and start growing them as a variety. Disease resistance could be selected later with material planted later in years when the various diseases materialized.

With naturally inbreeding crop plants, just inbreeding for several generations followed by evaluation at the F_4, F_5, or F_6 stage is one of the most powerful methods. With peas, where the crosses are difficult to do and a successful one produces only a few seeds, it is often necessary to go to an F_3 or an F_4 just to obtain enough seed so that rare classes are likely to be found. Inbreeding has the added advantage of allowing more of the classes to present themselves. Initially after the cross, the classes associated with the dominant genes predominate; those associated with the recessives are rare.

Whenever we do a cross, we never begin selection in the F_1. The earliest generation in which we begin selection is the F_2. But now you can see that there are many advantages to delaying selection to even later generations in many cases.

Delaying selection until late in the project also is often desirable for technical reasons. Some characteristics cannot be evaluated or selected for every year or under all conditions. They may depend on the weather or on what diseases matter that year. Other characteristics are simply difficult to evaluate — flavor, for example. I would much rather evaluate the flavor of peas in an F_5 generation where each pea is fairly likely to be pure breeding for nearly everything than in an early generation where the flavor might not yet be established.

Whenever you design a breeding project, figure out not just what class or classes you want but also how you are going to identify them and how difficult that is likely to be. It's much easier to evaluate green versus pur-ple pods than whether pods are edible types or shelling types, for example. And flavor is always difficult to evaluate because it is subjective and it usually varies with stage of maturity and time of day.

The final approach is backcrossing. To use this approach, I make and grow the F_1, as before. That is, I cross the purple sheller (*AABBCCddeeffPPVV*) with the disease-resistant snow pea (*aabbccddeeffppvv*) and obtain the F_1, which is a purple sheller (*AaBbCcDdEeFfPpVv*). Then I cross the F_1 plants to the snow pea parent. This is called a *backcross* — that is, a cross back to one of the parents.

Let's look at what happens at some specific locus, say *D*, when we do the given backcross. The F_1 is *Dd*. The backcross is *Dd* × *dd*. The *Dd* parent generates ($\frac{1}{2}D$ + $\frac{1}{2}d$) gametes. The *dd* parent contributes only *d* gametes, so the progeny of the cross will be ($\frac{1}{2}Dd$ + $\frac{1}{2}dd$). Ratios involving one-halves are also typical of all the other genes in the cross, whether dominant or recessive.

The class we want is *AaBbCcddeeffppvv*. The probability of getting each part of the genotype in the progeny of the backcross is $\frac{1}{2}$. The probability of getting the whole thing all in one step is $(\frac{1}{2})^8$, or $\frac{1}{256}$. In other words, if we grow three hundred plants from a backcross, we have a decent chance of getting the desired class in one step.

As before, though, we wouldn't be able to identify the disease resistance part of the genotype, so we would just take all the purple snow types and go from there. The purple part of the genotype is all heterozygous from this kind of cross — that is, *AaBbCc* — so we would have to select against greens in subsequent generations.

Let's look at another aspect of backcrossing. Let's suppose there is one dominant gene, *L*, that I would like in one of my pea varieties. I like my variety exactly the way it is; I just want *L* in it. I don't want other characteristics from the variety that carries *L*, just the one gene. First I would cross the two varieties. Then I would backcross to my standard variety. Half the offspring would show *L*. I would choose one with *L* and backcross again. Again, half would show *L*. I would choose one with *L* and backcross yet again.

Let's look at what is happening in the genome other than at the *L* locus. Consider some other gene that was different in the two varieties, say *M*. Suppose the *L* variety is *mm*, and my standard variety is *MM*, where *M* is dominant. The original cross is *LLmm* × *llMM*. The F_1 is *LlMm*. The progeny from the backcross are $(\frac{1}{2}Ll + \frac{1}{2}ll)(\frac{1}{2}MM + \frac{1}{2}Mm)$. We choose one of the *Ll*'s. Of those, half are *MM* and half are *Mm*. We backcross again. If the plant we chose is *MM*, all the progeny are *MM*; they are fixed for the genotype of our standard variety. If the plant we chose is *Mm*, the probability is 1 in 2 that the genotype will become *MM* among the progeny. Considering both possibilities, the probability is now ¾ that the genotype is that of our standard variety. With another generation of backcrossing, the probability will be ⅞ that the genotype is that of our standard. The same thing is happening at all the loci.

By backcrossing to a given variety, the recurrent variety, we can eliminate nearly all the genes of the other variety except the specific one we are selecting for in each generation. Recurrent backcrossing is useful only with dominant genes. With dominant genes, however, it allows us to transfer an individual gene from one variety into a desired variety. This is very important in plant breeding, because often we have a variety that already includes large numbers of genes that are necessary for the production of what we want in our area.

Wild relatives of crops often have dominant genes that confer resistance to insects or diseases. These are usually transferred into crop varieties by recurrent backcrossing. A cross with the wild relative often introduces so much undesirable genetic variability that good crop types cannot be recovered in an F_2, even if thousands are raised. Recurrent backcrossing allows the breeder to transfer just the desired gene (or as close as possible to just the desired gene) into a desired genetic background. Whenever you want just one characteristic from a given variety and want the rest of the characteristics of the new variety to be identical to those of an established variety, consider backcrossing.

Where the characteristic in question is associated with recessive genes, you can still use some backcrossing as part of your approach. You can alternate backcrossing and inbreeding, for example. The inbreeding is necessary to identify the desired gene.

You cannot always transfer just one desired gene. The genes that are tightly linked to it tend to come along. So after recurrent backcrossing starting with a wild plant, it may not always be possible to eliminate all the undesirable wild characteristics.

Let's go back to the purple-podded pea. If I use recurrent backcrossing, I don't have to evaluate for pod type or disease resistance. It allows me to fix a disease-resistant genotype, even if the diseases never appear at the appropriate times during the project. It also doesn't require that I grow three hundred plants. All I need to grow is enough plants to obtain just one of genotype $AaBbCc$ for the next backcross. The probability of that is $\frac{1}{8}$. By referring to Appendix H, we see that to be 95 percent sure of getting at least one plant of a type that has a probability of $\frac{1}{8}$, we should grow twenty-two plants. That will take me less than 1 foot of row. The entire project can be done in 1 foot of row or less per year until the backcrossing is over and it's time to inbreed. Even then, only a few feet of row will be required. With so little space required and no evaluation needed, the recurrent backcrossing approach can be done in a greenhouse so as to allow two or three generations per year instead of just one.

It's obvious that backcrossing has some powerful advantages. With peas that are naturally inbreeding, backcrossing is much more work than inbreeding, though. To obtain twenty-two seeds, I probably should have at least eight successful backcrosses. To get those, I would probably need to do about twenty. That's laborious. To get twenty-two F_2s, I would just have to let one plant set seed all by itself.

An added consideration is that inbreeding allows new phenotypes to show up that derive from genes from both parents — phenotypes you weren't thinking about or planning on. Backcrossing is much more controlled. You are much more likely to get just what you want and not to see all kinds of other possibilities.

With all these approaches possible, how am I going after purple-podded peas? Actually, I'm doing a little of everything. Last year I crossed a purple-podded sheller to one of my snap pea lines and obtained an F_1. This year I raised three F_1 plants and used some of their flowers for backcrosses to the snap pea line. I let the rest of the flowers produce F_2 seed. Interestingly, although one parent had ordinary green seed and the other had tan seed, the F_2 seed shows segregation for seed colors of various shades and with speckling of various shades and amounts. It's really beautiful. I have been spending more time than I will admit just pouring the seed back and forth in my hands and looking at all the lovely patterns.

I will raise most of the F_2 seed to get an F_3. And I will backcross the first backcross generation if any are purple-podded. (I have few enough that I might not have any.)

This year I also made a number of crosses with a different purple shelling pea and my snap pea material as well as also crossing it to a disease-resistant sugar pea. I'll use some power inbreeding and some backcrossing. You don't have to do just one or the other. You can, for example, backcross the F_1 once, then inbreed the progeny. This gives you a genome that is $\frac{3}{4}$ that of the variety you backcrossed to, on average, instead of $\frac{1}{2}$. Where most of what you want is associated with one variety, just a single round of back-crossing before you start inbreeding can improve the odds dramatically and be very useful.

Fixing Dominant and Recessive Genes
(Using progeny to identify genotypes.)

Fixing a gene or trait in a variety means to make the variety pure breeding for it. With respect to pod type in the purple-pod project, once I select a plant that has snow-type pods, it is genotype *ppvv,* and the pod type is fixed automatically with just the one act of selection. So fixing a recessive gene that is fully penetrant (that expresses itself every time) is very easy. I just choose plants that have the phenotype associated with the recessive gene.

With disease resistance, it may take more than one generation of selection, because not all sensitive plants necessarily show up. If no aphid that carries the disease bites the plant early enough to matter, the plant may be unaffected, even though it is genetically sensitive. Professional breeders frequently inoculate plants with virus or fungal spores to ascertain their genotypes. Most amateurs usually depend on just selecting in years when the disease is prevalent and not worrying about it otherwise.

Some genes express themselves only in combination with other genes. These genes also usually require more than one round of selection for fixing.

There are two basic approaches to fixing dominant genes. One is to eliminate the recessives by culling them. To take a specific example, in the purple-pod project we usually design the work to obtain individuals that are *A_B_C_.* But we want a variety to be pure breeding for purple pods. That is, we want *AABBCC.* The first approach is to mass-select — to save seed from all the purple-podded plants. In the next generation we save seed from purples again. Each generation the recessive genes associated with green color will become less and less frequent in the population. This is an easy method, but it doesn't totally eliminate all the undesired recessive genes.

The other approach is to find a plant that is *AABBCC* and save seed from only that one. But all three genes are dominant, so there is only one way to identify which purples are pure breeding for purple: save their seed and see which ones give rise only to purple offspring. In other words, we do exactly as we would otherwise, but we plant the seed from each *A_B_C_* plant separately. Some families will be all purple podded. Others will have both purple-podded plants and green-podded ones. We eliminate all the families that produce green-podded plants and save seed only from those that breed true for purple.

If a plant is an outbreeder with inbreeding depression, we normally would not want to derive a new variety by any means that causes a genetic bottleneck eliminating most of the needed genetic heterogeneity. For an outbreeder, we would not try to eliminate all the recessives by identifying a single homozygous plant and deriving the variety from it. We would probably use mass selection instead. This would allow us to maintain general heterogeneity; it would not allow us to eliminate all the unwanted recessives. We would probably feel that we could more readily deal with having to cull out occasional recessive phenotypes each generation than risk ending up with a weak, low-yielding, but totally pure breeding variety.

If we absolutely could not afford an occasional unwanted recessive phenotype in the variety, we might generate a number of families that were identified as homozygous by inbreeding. Then we might combine them and let them cross with each other to restore heterogeneity and vigor.

A More Winter-Hardy 'Green Wave' Mustard

(Crossing and mass selection.)

In the very severe winter of 1990, all my 'Green Wave' mustard died. I had planted several beds of it on the farm in Independence, Oregon, but not a single plant of thousands survived. I had never had 'Green Wave' die out completely like that, although there have been other winters in which the plants were damaged. I don't like either event. I want a more winter-hardy 'Green Wave'.

'Green in Snow' is generally considered to be the most winter-hardy mustard. I haven't grown it because it is listed in various seed catalogs as a mild type with no special virtues other than its cold resistance. This might be because we Americans don't know how to use it. I understand it is a Chinese variety and is one of their most popular types. It's salted there, and the salted vegetable is stir-fried. Both 'Green Wave' and 'Green in Snow' are *Brassica juncea* types. I have begun to accumulate these types to trial them and find out more about how to use them.

In the meantime, I have crossed 'Green Wave' and 'Green in Snow', with the idea of deriving a new variety that has the vigor, insect resistance, and flavor of 'Green Wave' but the cold hardiness of 'Green in Snow'. My hope is that I'll be able to develop a version of 'Green Wave' that will survive any winter here, that will be vigorous enough to be a good winter green-manure crop, and that will have the spectacular flavor of 'Green Wave'.

I made the cross in both directions — that is, I used some 'Green Wave' plants as the female as well as some 'Green in Snow'. Pods developed, and I obtained a few dozen good seeds from the crosses. I will plant most of the seed in the fall and let it overwinter to obtain big plants that produce a lot of seed. (I'll hold back some seed in case we have another severe winter; the F_1 hybrid might not be winter hardy.) Then I'll collect F_2 seed and plant a substantial plot of that somewhere to get F_3 seed.

Brassica juncea types are said to be mostly inbreeding, but they are insect-pollinated, so there will undoubtedly be both inbreeding and outbreeding going on in my F_2 and F_3 patches. I don't mind. I just want to create a population that contains plenty of cold-hardy plants. I won't start selecting until the F_3 and subsequent generations. If we have no severe winters before then, I'll do my initial selections based on plant vigor, form, flavor, and suitability as a winter green-manure plant for our area.

I also will release F_3 seed as breeding material through seed companies so that other gardeners can get involved. The F_3 of the cross between 'Green Wave' and 'Green in Snow' should contain all the variability needed to select mustards of many different types suitable for many different areas.

After the initial crosses to obtain the F_1, I won't do any controlled crosses at all. I'll just grow a mustard patch and allow the plants to cross- or self-pollinate as they wish. Brassicas frequently display inbreeding depression. I'll generate plenty of genetic heterogeneity from my initial cross. After that, I'll have hundreds of plants or more in each generation. In this way I should be able to maintain nearly all the heterogeneity created by the initial cross, except for genes specifically selected against in later stages.

Lettuce-Salsify

(Uncontrolled crosses to generate a genetically heterogeneous population, selection.)

Dan Borman (Meristem Plants) became interested in scorzonera a number of years ago. (*Scorzonera hispanica* is referred to as scorzonera, black salsify, or oyster plant.) Like me, Dan was interested in the possibility of using the leaves as a perennial lettuce instead of eating the roots.

I don't know of any information about the breeding system of scorzonera, but it has large, attractive yellow flowers that are visited by bees. A good guess is that scorzonera is at least mostly cross-pollinated. It may well be subject to inbreeding depression. If so, mass selection, not selection of individual

foundation plants, would probably be the best choice of methods. Dan started by doing uncontrolled mass crosses. (With inbreeders, if you want a cross, you have to make one. With outbreeders you have the option of letting them do it for you.)

Dan obtained all the varieties of scorzonera he could find from as many different sources as possible and interplanted them all in one patch. He would let them cross in all possible combinations so as to generate a population with as much genetic variability as possible. He planted the seed that resulted from his first patch of plants in another patch. Then he saved the best plants and culled out the rest, saved seed again, and planted in a third patch. By this time he was finding many plants with interesting new phenotypes. Some had leaves 4 inches across (instead of 1 or 2 inches), leaves with petioles (instead of sessile), and leaves that were much more tender than any of the material he'd started with.

With each generation Dan makes sure that he keeps at least a dozen plants, which he allows to cross-pollinate freely. In this way he hopes to maintain a good level of general genetic heterogeneity.

There has been one hitch in Dan's breeding program. Deer prefer his third-generation lettuce-salsify. They will walk right past the patch of his original material in order to eat his selected plants. Greater pest problems are often an inevitable problem where the variety is to be used as a raw vegetable.

The other side of the coin is that when you deliberately select for pest resistance, you need to check and make sure the plant is still edible. Many plants are pest-resistant because they are poisonous or unpalatable. A professional breeder recently developed a more pest-resistant potato — that turned out to make people sick. It had higher levels of the poisons that make wild potatoes resistant to pests — and less desirable to people. I sometimes identify milder, tastier wild plant individuals by checking those with the most insect damage. The most delicious daylily flower of the kinds I'm working with is the only one that shows marked pest damage. I am not the only one who thinks it is a superior raw vegetable, apparently.

Dan started his project by mixing a number of varieties and allowing them to cross. The purpose of the crosses was to create genetic heterogeneity. But there was no way of knowing which cross might generate the heterogeneity needed to select better leaf types. So Dan, in effect, did all the possible crosses.

A somewhat more rigorous way to generate heterogeneity would be to actually *do* all the possible crosses, pool the F_1 seed, grow the mixed seed without selection and allow it to cross however it wanted, and then start selecting the following year. This design does a better job of making sure that you get all combinations and possibilities, but it also takes much more work. Dan's approach generated many combinations, though probably not all, and it made optimal use of his time.

When you're working with cross-pollinating or partially cross-pollinating varieties, doing crosses by just mixing and interplanting the varieties is often the most practical method simply because it is so effortless. In many cases hybrids between the various varieties are so much more vigorous than the original material that they can be readily identified in the next generation.

The more breeding projects you have that take care of themselves, the more breeding projects you can have going at once. And with a lot of breeding projects, nearly every week presents at least some minor triumph.

Tomato, Corn, Squash, Melons
(Dehybridizing the hybrids.)

Many of the accomplishments of modern professional plant breeders are presented in the form of hybrids. If you know how to dehybridize the hybrids you can transform these accomplishments into seed that you control. Alan Kapuler is passionately committed to finding, preserving, and distributing *public-domain* open-pollinated varieties. When there is a particularly excellent hybrid, though, he doesn't do without it. Instead, he uses it to create an equivalent public-domain, open-pollinated variety. He dehybridizes the hybrids.

In a number of cases, Alan has run into what I referred to as *pseudohybrids*. That is, when he attempted to develop an open-pollinated version of the hybrid, the very first step he took — saving seed of the putative hybrid and growing out the F_2 — gave evidence that the "hybrid" probably wasn't a hybrid at all. It was apparently already an open-pollinated variety. In a number of other cases, the hybrids showed the expected segregation in the F_2, and Alan developed his open-pollinated version by selecting through subsequent generations.

Alan had a hybrid sweet tomato, for example, that was very sweet and very productive. He wanted an open-pollinated version. So he simply saved the seed and planted it. Tomatoes are highly self-pollinating. If the "hybrid" were really an F_1 hybrid, the plants resulting from its seed would be the F_2 generation after a cross. They would be expected to show segregation for all the characteristics that differed between the two parents that went into the original cross.

But Alan saw absolutely no segregation. The supposed F_2 plants all looked exactly the same. In other words, the "hybrid" variety didn't act like a hybrid genetically. It's offspring were all as similar as could be to each other and to the parent. That is, by definition, what we mean by a "pure-breeding" variety.

Alan selected a line of open-pollinated material that he offers through Peace Seeds as 'Peacevine Cherry'. He suspects that the original hybrid was actually an open-pollinated variety to start with, however, and that 'Peacevine Cherry' is identical or very similar to the original commercial variety.

I think seed companies may sometimes claim a purebreeding variety as a hybrid in order to deter people from saving its seeds. I have inbred a number of commercial hybrids and obtained no segregation. This means that either the "hybrids" were not hybrids, or that they represent crosses between parent lines that were different only in characteristics that are invisible (such as disease resistance, for example, when you don't have that disease in your field).

Most varieties listed as hybrids genuinely are hybrids, though. When you save the seed from them, they show segregation for various characteristics in the next generation.

Alan Kapuler was impressed with the virtues of a full-season hybrid white sweet corn, for example, that had nice big ears, good yield, excellent rich flavor, and attractive, vigorous pinkish-purple plants. He wanted an open-pollinated version. So he just grew the hybrid in a patch by itself and saved the seed. When he planted it the next year, he obtained vigorous plants with big, uniformly good-flavored white ears. But there was segregation for plant color. There were purples and pinks and greens. Alan liked the deep purple plants with their beautiful purple-husked ears.

So he detasseled the green plants and ate the ears, and saved seed from about two hundred to four hundred F_2 purple plants. He raised about a thousand F_3 plants. Only about 5 percent or so were green. He culled the greens again, always detasseling them before they flowered so that they could contribute neither as male nor female parents. His open-pollinated line began with the F_4 material, when he began offering it through Peace Seeds as 'True Platinum'.

In order to develop 'True Platinum', Alan practiced mass selection starting with the F_2 after a cross. The cross happened to be one someone else made, but that doesn't change anything. Notice the numbers of corn plants involved. Two to four hundred F_2 plants were the foundation of the line. Not one or two or twenty. These are the kinds of numbers you should be working with if you want to create or maintain a vigorous open-pollinated corn variety.

Alan developed his open-pollinated variety through mass selection instead of familial selection. This allowed him to keep the numbers high but the work low. Mass selection has the disadvantage that you never completely eliminate undesired recessive genes. This means that a few percent of green plants will always be characteristic of 'True Platinum'. With familial selection you could identify the genotypes of families and pool those that were homozygous for purple. But the work involved would be too much to do with more than small numbers. So the variety would end up being founded on a smaller number of plants.

With corn, it will usually be far more important to work with large enough numbers to maintain vigor than to eliminate every single unwanted recessive gene. Most vigorous open-pollinated varieties of outbreeding crop plants are developed via mass selection and contain a small frequency of various unwanted recessive genes. So they produce a low percentage of off-types every generation. This is part of the reason why roguing every generation is especially important with these crops.

Alan also liked a golden patty-pan–type summer squash that was a commercial hybrid. He wanted an open-pollinated line that was similar. So he hand-pollinated and selfed the variety, then saved the seeds. When he planted the seeds the next year, he got all kinds of things segregating out — squashes of green and gold, and of all shapes. It took him five years of inbreeding from plants of the desired type to select his open-pollinated version, which he offers as 'Summer Sun'.

Alan also dehybridized a popular hybrid mustard. He simply planted the mustard in a patch by itself and saved the seed. The F_2

showed lots of segregation for plant type and color. Alan mass-selected over a number of generations to develop his open-pollinated variety 'Prime Choi'.

'Rainbow Inca' Sweet Corn
(Mass Crosses and Mass Selection.)

The final story in this chapter is about Alan Kapuler's creation of 'Rainbow Inca' sweet corn. It was his first breeding project and took place before Peace Seeds was founded, a time when Alan was living on a commune in southern Oregon and growing vegetables. He grins when asked about it and says, "That's my hippie corn."

'Rainbow Inca' didn't start as a breeding project. It began as a spiritual act, a ceremony. Alan had grown a number of different varieties of corn the previous year. The ears were of all kinds and colors — flour corns, native Indian corns, heirloom sweet corns, and others. Some of the ears were especially beautiful.

Alan chose the twelve most beautiful ears and took them out to the field. Then he shelled each ear and planted it. He shelled the kernels off the ear in order, row by row, seed by seed, as he walked down the rows in the field — planting as he went — transferring the pattern of kernels in the ear onto the land. "It was like unrolling those ears of corn on my field," Alan said.

When he finished planting the kernels from one ear, wherever he was in the row, he started the next ear. So the corn from the different ears was all in one patch, in somewhat intermingled blocks.

Then came the moles. They were especially active that year, and they ate a lot of the corn. Alan had collected a number of heirloom sweet corns. He planted sweet corn into every spot in his field that was missing a corn plant, without paying any attention to varieties. This meant that all kinds of crosses could happen, even between very early and late types. In addition, it meant that he crossed various sweet corns with all the corns he had originally planted. He wasn't thinking about any of this at the time; it's just the way it happened.

One of the corns Alan planted was an Incan flour corn with huge, flat white seeds and plants about 12 feet high. When Alan harvested his corn, the ears on the Incan corn were especially beautiful, and there were one or two colored seeds on each ear that represented pollination by a colored variety. There were yellows, reds, purples, and blues; solid colors, stripes, blazes, and spots; clear colors and iridescent ones. There were 100 to 150 colored kernels altogether. Alan picked the colored kernels off the ears and saved them.

He planted about a hundred of the colored kernels the next year. When he harvested the patch, the ears showed an occasional crinkled seed, representing sweet types. The genes associated with sweet corn are recessive, so no crinkled kernels appeared in the original crosses involving the flour-type Incan mother plants. There were about forty crinkled kernels of all colors.

Alan kept and planted the crinkled kernels the following year. His harvest was about 5 pounds of kernels, all sweet and of all colors.

He selected for large, flat kernels and planted a couple of ounces of seed. In subsequent years he continued selecting for large flat kernels of all colors. He also selected for a plant height of about 8 feet instead of the 10 to 12 feet typical of the original Incan corn. (That's late enough to be marginal here. Eight-foot plants were enough earlier to be dependable.) He also selected for ears that were lower on the plant — "So I could reach them," he says. Lower ears also are larger, so selection for lower ears automatically selects for bigger ears and higher yield.

'Rainbow Inca' sweet corn preserves the cytoplasm of the original Incan flour corn and a large amount of genetic variability derived from many sources. The kernels are of all colors and patterns, huge compared to any other sweet corn, and broad and flat. The plants are about 8 feet tall. It's of late but reliable maturity here (meaning it would probably be considered midseason in most areas). It's undoubtedly been automatically selected for productivity in cool weather. The flavor is excellent.

Alan didn't realize that he had developed something special until he offered 'Rainbow Inca' through the Seed Savers Exchange and heard the reactions of those who grew it. It's excellent, unusual, unlike anything else, they said. 'Rainbow Inca' is currently offered through Peace Seeds and Seeds of Change and remains one of the most unusual corns available.

ELEVEN

Bigger, Brighter, and More Beautiful

UP TO this point I've been discussing the breeding of *diploid* plants, plants that have two copies of each kind of chromosome, and therefore two copies of each gene. Some important domesticated plants, however, are triploid, tetraploid, or hexaploid, and some have even higher degrees of ploidy. That is, they have three, four, six, or more copies of every chromosome. A tetraploid, for example, has four entire haploid genomes instead of the two a diploid has. Any plant with more than two complete haploid genomes is referred to as a *polyploid*.

Having higher than diploid levels of ploidy often causes a plant to have bigger flowers and fruit, more intensely colored flowers, bigger or thicker leaves, and bigger tubers. Potatoes and sweet potatoes are both polyploids, as are many fruits and ornamental plants.

A number of rare vegetables, as well as many wild edibles, could probably be turned into instant major vegetables if their edible parts could be made bigger. One approach to that is through standard breeding. An alternative is through spontaneous or deliberate chromosome doubling to create a polyploid.

Autopolyploids

In autopolyploids the haploid genome sets are all the same. The potato is an autotetraploid; the sweet potato is an autohexaploid. In allopolyploids, the haploid genomes are different. The standard commercial wheat is an allohexaploid. It has three complete genomes originally derived from three different species.

Autopolyploids presumably arise initially because cell division isn't perfect. Once in a while during mitosis or meiosis, chromosomes fail to segregate. If this occurs in a somatic cell, a tetraploid daughter cell results. This cell has four chromosomes of each kind instead of two. If properly positioned, this cell could divide and give rise to a tetraploid branch or section of plant.

Some species seem quite tolerant of changes in ploidy. In others any alteration in ploidy seems to be severely detrimental or lethal. We don't know why. Tetraploids of some species may have larger or thicker leaves and flowers, and the leaves or flowers may be more deeply colored. They also may grow more rapidly than diploids. (Higher levels of ploidy may or may not show such effects depending on the species.) In some species autopolyploids may be completely lethal. In others they may be obtainable, but they may grow more slowly than diploids, be smaller, or flower later. For most species there seems to be a ploidy level associated with the maximum size of flowers and other characteristics. Higher or lower levels are associated with smaller size.

Polyploids are not always vegetatively stable. Chromosomes may be lost during cell division so that the plant is a mosaic of tissues that differ genetically. In some cases, a triploid or other-ploid plant experiences chromosome loss accompanied by selection at the cellular level for the most vigorously growing combination, and ends up with an apex in which diploidy has been reestablished. When vegetative instability is occurring, you may see branches that are clearly different from neighboring branches, or sectors — patches of the plant that differ from the rest in color or morphology or flower characteristics or flower fertility. In these cases, two tubers or cuttings from the same plant may give rise to plants that are genetically different from each other and from the parent.

Many polyploids are vegetatively stable, however. Generally, the triploid, tetraploid, or other-ploid cultivars we use are vegetatively stable. They either are of species that are naturally stable when polyploid, or they have been allowed to grow and divide and "sort themselves out" until they have become vegetatively stable.

Autopolyploids behave differently than diploids when it comes to sexual reproduction, because they have more than two copies of every gene instead of two. All of Mendel's laws, predictions, and ratios depend on there being exactly two copies of each chromosome and each gene. We usually can obtain the classes we want, but the frequencies of various classes are unpredictable. In addition there may also be sterility problems.

Triploids

Triploids can be generated when an accidental unreduced diploid gamete pairs with a normal haploid one. They also may result from the cross of a diploid and a tetraploid. They are usually sterile. The chromosomes attempt to pair in meiosis, but they apparently can associate only two by two. All three chromosomes often associate, since chromosomes begin their pairing at numerous points along their lengths, so any given chromosome is often paired with one homolog in one region and the other in another region. When division occurs, there is no way for three chromosomes to sort themselves into two cells without creating massive genetic imbalances. The result is often complete pollen sterility and complete or nearly complete sterility as a female parent.

This doesn't mean that triploid plants are useless. It just means that they usually cannot be maintained through seed propagation. Often sterility is advantageous or even essential. The banana of commerce, for example, is a triploid. Diploid bananas produce hard seeds. The edible banana is a banana with aborted seeds; the same is true of the plantain. Many apples and pears are triploid, as is the seedless watermelon. The latter is produced by crossing a tetraploid variety with a diploid one.

Any perennial or plant that can be propagated from tubers, divisions, or root cuttings or by other vegetative means can be maintained and used even if it is a triploid. Triploid varieties of ornamentals may be especially useful. Flowers may be bigger and richer in color. And since they don't set seed, the individual flowers may last longer on the plant. The plant may have a prolonged blooming period — one that might be achievable with a diploid only by pinching off the flowers to prevent seed set.

Triploids and other plants that have odd numbers of haploid genomes can normally reproduce only vegetatively. But plants with even numbers of haploid genomes — tetra-

ploids, hexaploids, and octaploids — may be fertile.

Doubling Chromosome Numbers

Usually, the doubling of the chromosome number to produce an autotetraploid can be accomplished deliberately. It frequently occurs spontaneously, especially in regenerating tissue. Tetraploid suckers on tomato plants are common, for example. In some cases we can obtain polyploids simply by chopping off the top of the plant and watching for lateral shoots that look different. Chromosome numbers also can be doubled deliberately by using the poison colchicine.

Colchicine inhibits the formation of the spindle apparatus in cells by binding to the protein subunits from which the spindle is made. The spindle apparatus is part of the mechanical structure involved in moving chromosomes around during cell division. If colchicine is applied at the right concentration for the right time, some dividing cells will fail to have normal segregation of their chromosomes, but then will go on with normal growth and division. Some such cells will be tetraploid. The result, with somatic tissue, is a mosaic of diploid and tetraploid tissue. With a little luck, some new shoots arise from some of the tetraploid tissue and look different enough so you can identify and propagate from them.

Colchicine is usually applied to vegetative tissue as a cream. It must be applied to growing areas such as an apical tip or the shoots of a tuber. When used with seeds,

seeds are usually presoaked in water containing colchicine before planting.

Colchicine is a dangerous poison and is not readily available to amateurs. If you want to use it, and you haven't formal scientific background or affiliations, you may need to collaborate with a professional. Often, however, colchicine is not necessary to obtain what we want. Spontaneous chromosome doublings are common. If we have a triploid plant that is producing hundreds of sterile flowers, for example, often one branch or one flower has seeds. That branch or flower represents a spontaneous chromosome doubling, and the seeds are likely to be viable.

Allopolyploids

Allopolyploidy involves different genomes. Wheat and many grains as well as many brassicas are allopolyploids. Allopolyploidy arises naturally when a spontaneous wide cross between two different species is followed by spontaneous chromosome doubling. It is a common mechanism for evolution of new plant species. We can create allopolyploids deliberately by first doing a wide cross between two different species and then doubling the chromosome number. After chromosome doubling, the plant is functionally diploid, since it has two of each chromosome of each of the species from which it was derived. Such allopolyploids are called *amphidiploids*.

In the next chapter I explore the deliberate creation of allopolyploid vegetables further under the subject of wide crosses. To create an allopolyploid vegetable deliberately, we start by doing a wide cross.

TWELVE

Fun With Wide Crosses

W HEN MENDEL explored the genetics of diploids, the crosses he did were of varieties that were all close relatives. Often we want to make crosses that are of distant relatives or even entirely different species. These are called wide crosses. There are two kinds of wide crosses: crosses between plants that are distant relatives but of the same species and crosses between plants that are of entirely different species. Wide crosses between different species are rare in animals but common in plants. They are, in fact, a major mechanism of plant evolution.

Wide Crosses Within Species

When a wide cross is between two plants that are members of the same species,

obtaining the cross is usually easy, as is growing the F_1 plants. But when the plants are distant relatives, there is often chromosome nonhomology — that is, the two plants may have different arrangements of their genes into chromosomes. Such differences arise because breakage and reunion of chromosomes can occur, and chromosomes sometimes reunite with arrangements different from those they started with. Such chromosomal mutations can be preserved, just as single-gene mutations can, and are a part of the evolution of genomes of plants.

When distant relatives are crossed, a block of genes that is on one chromosome in one parent may be on a different one in the other. When the hybrid between them reproduces sexually, there may be aberrations in chromosome pairing and segregation. Some gametes can receive two copies of some genes and none of others. Some chromosomes get torn in two as the cell divides, having two of the region that attaches chromosomes to the division mechanism. Others lack that region entirely and get lost.

The wide-cross hybrid may produce many inviable gametes. Many zygotes also may die and lead to aborted seed or weak, aberrant plants. In short every nonhomologous chromosomal region in the hybrid causes problems during meiosis. In a hybrid between distant relatives, there can be dozens of nonhomologies, each creating all these problems with respect to segregation.

In such hybrids, low viability of pollen, eggs, or seed can be a problem. In addition, if we are following genes that are within or linked to a region of nonhomology, we may be unable to obtain the recombinants we want. We may even be unable to recover the genes that went into the cross because of linkage to or association with areas of the chromosome that are involved in the nonhomologies.

If the genes of interest to us aren't involved in the nonhomologies or linked to them, we usually can get what we want. But often we see all kinds of "junk" segregating out of the cross, there are problems with sterility, and normal Mendelian ratios don't apply. If we can recover the class we want at all, though, we can go ahead, even with the other problems. In future generations we tend to eliminate the nonhomologies and start seeing Mendelian genetic behavior. Many ordinary crosses have some degree of nonhomology and show "messy" ratios in the first generations after the cross.

Sometimes the problems with low pollen viability or general sterility are so great that it is difficult to do any breeding work with the hybrid. Often backcrossing is a useful way to deal with such hybrids. If we make the hybrid and then try to obtain an F_2, we're dealing with a situation in which the probability of obtaining both good eggs and good pollen is low, so the probability of a good egg's being fertilized by good pollen is even lower. If the probability of a good gamete of either kind were 1 in 1,000, for example, the probability of a good zygote would be 1 in 1 million. We might never get one.

But if we make the hybrid and then backcross it to one of the parents, one of the kinds of gametes is normal. In the preceding example, we would expect 1 in 1,000 good seeds instead of 1 in 1 million. Many plants produce several hundred or even several thousand seeds, so the odds of 1 in 1,000 can be quite workable. (With orchids, even odds of 1 in 1 million are workable, since orchids commonly produce millions of seeds per flower.)

Pollen seems more sensitive than eggs to chromosomal aberrations. It is more likely to work if we use good, normal pollen on questionable eggs than the reverse. In backcrossing we're most likely to succeed if we use the hybrid as female and fertilize it with pollen from one of the parents.

Wide Crosses Between Species

Many of the most interesting wide crosses are between entirely different species. These have the potential for giving rise to entirely new species of plants.

When plants from different species are crossed, the chromosomes are usually different enough that they don't pair at all. If a diploid plant with twelve chromosomes is crossed with a diploid plant from another species with sixteen chromosomes, for example, a gamete with a haploid set of six combines with a gamete with a haploid set of eight. The resulting hybrid has fourteen chromosomes, but none of the chromosomes in the haploid set of six is homologous with any in the haploid set of eight.

The result is a plant that, practically speaking, has fourteen chromosomes but is haploid. Such hybrids can usually grow vegetatively, but they can't reproduce sexually. As previously mentioned, sexual reproduction involves diploidy.

If the chromosomes of the wide-cross hybrid are doubled, however, it will have twenty-eight chromosomes in fourteen proper pairs. It will be a diploid, but one with more chromosomes or kinds of chromosomes than either of its parents. In some cases the plant remains sterile. But in many cases fertility is restored completely. The result is an entirely new species.

Where a variety propagates vegetatively from tubers or divisions, the chromosome doubling to recreate diploidy may not be necessary.

The results of wide crosses can be quite unpredictable. We don't know what the phenotype is going to be when we combine such different haploid genomes into one plant. In some cases there are partial chromosomal homologies instead of a complete lack of homology, so segregation may occur in subsequent generations, or it may not.

Somatic growth of hybrids may or may not be stable. If it isn't, selection for more rapidly growing and stable material happens as the plant grows, so that the genome from one part of the plant to another isn't the same. You may see obvious sectoring. Sometimes after enough growth and sorting out, when vegetative stability is restored, the plant may have the genome of just one of the parents again. Or it may have mostly the genome of one parent with a new chromosome or two from the other. Or it may have a genome that combines some chromosomes from one parent and some from the other, but not the complete genomes of both. In many cases, however, the hybrid does retain both entire parental genomes.

In most cases there is some difficulty in making wide crosses between plants of different species. This is virtually by definition. Different species are different because they don't ordinarily cross under normal conditions. Sometimes the difficulty is minor or technical. The plants normally grow on opposite sides of the world, for example. Or they flower at different times. In some such cases there may be no physical barriers to the cross when we hand-pollinate. In many cases, however, there are major incompatibility barriers that must be overcome in order to cross two different species. Various methods for overcoming incompatibility barriers in ordinary as well as wide crosses are given in the last half of Appendix B.

In general it is common to be able to obtain crosses of different species within the same genus, and occasionally it is possible to obtain crosses between plants in different genera in the same family. Sometimes a cross between two species can be made to work with a given individual or variety but not with others. It is very common for wide crosses to be possible in one direction only, so it is especially important to try crosses in both directions. Basically, we just have to try things.

Don't be discouraged from trying a wide cross because the experts say it's not possible. They are using different plants of different varieties and a different set of special tricks for overcoming incompatibility barriers, and they have different weather conditions. Failing to obtain a wide cross isn't significant; obtaining it is. And you need to obtain it only once.

*   *   *

I think the use of wide crosses, followed by chromosome doubling where necessary, has tremendous unexplored potential for creating new vegetables. The unpredictability of the results discourages much interest in the approach by professionals, but I think the very unpredictability makes it all the more fun for the amateur.

With much of standard breeding, we want a particular thing and we go about figuring out how to get it. But with wide crosses, I think the most productive approach is to do every cross that we think might be interesting. In cases where we are successful, we can then spend some serious time figuring out what the new plant is good for.

THIRTEEN

Happy Accidents

THERE ARE two basic approaches to plant breeding. First, there is the deliberate approach. We decide to try to get a specific something that is bigger, better, tastier, or more beautiful, and we design and conduct a project aimed at achieving our predetermined goal. But there is another approach. Something interesting turns up in our garden. We didn't plan it or ask for it, but there it is. Now that we have it, we decide we would like more of it.

Many important agricultural advances undoubtedly trace their origin to some genetic accident that happened in the garden of someone who was observant and curious. Knowing how to take advantage of the accidents can add a lot to our plant breeding repertoire. We can't plan to have an accident, but we can be ready for it when it occurs.

Genetic accidents are commonplace. I had five last year alone. Two were in my farm field vegetable patch that was about a third of an acre in extent. The other three were in my small home garden that is only about 200 square feet in size. The farm field probably had dozens of interesting accidents in it, but I wasn't there looking at every individual plant regularly. What's critical about genetic accidents isn't so much their occurrence as their being observed. A much-loved, much-watched home garden is especially fertile ground for the discovery and exploitation of happy accidents.

Three major classes of genetic accidents can provide interesting opportunities: somatic mutations, meiotic mutations, and accidental crosses. My accidents of last year were one obvious somatic mutation, one probable meiotic mutation, one accidental cross, and two events of less obvious origin. I discussed one of these — the appearance of a hairy, pest-proof, nonbolting mustard — in Chapter 5. In this chapter I illustrate how you can use happy accidents when they turn up in your garden.

The Seed Savers Exchange *Winter Yearbook* provides space where members can ask questions and invite interaction with others who are interested in similar things or have relevant information. In the 1991 edition, member Davis Huckabee asked about something that had turned up in his garden: "Do tomatoes produce bud sports as some tree fruits do? Last year one branch of a 'Porter' tomato produced fruit much larger. . . ."

The answer is yes. Tomato plants, like fruit trees and every other living thing on the planet, do experience mutation. If the mutation occurs in a somatic cell that gives rise to a branch, one entire branch will be genetically different from the rest of the plant.

Sport is one of those old, traditional words that dates back to the time when people didn't have any concept of genes or mutations. When one plant in the field was obviously different from its fellows, it was called a sport. When one branch on a tree was clearly different from the others, it was called a bud sport.

Somatic mutation has been especially important in developing new varieties of perennial plants such as fruit trees. Many traditional varieties were discovered because someone noticed a branch whose fruit was different from the other fruit on the tree. The person then made cuttings for rooting or grafting and established new trees with the new genotype. Vegetables also can have somatic mutations that, if properly positioned, can give rise to branches or parts of the plant that are genetically different from the rest of the plant.

Davis should be able to exploit his tomato sport quite easily. Since tomatoes are automatic inbreeders, saving the seed actually amounts to self-pollinating the flowers on the mutant branch — that is, inbreeding from them. That's an excellent approach. It's usually the best approach unless there is some reason to do otherwise.

When it comes to the next step, the important thing is to realize that the new tomato mutation is both heterozygous and dominant to its corresponding allele. Here's why: Tomatoes, being diploid, have two copies of every gene. It isn't likely that both copies will simultaneously mutate in the same plant. Instead, one gene presumably mutates, and its allele is still of the original form. That is, the mutant branch is heterozygous for the mutant allele. If the mutant allele were recessive, the branch would look the same as all the rest (by definition). But it doesn't. So the mutant allele must be dominant to (or codominant with) the original allele. In other words, if the genotype at some locus in the original variety was, say, *bb,* the unusual branch has a genotype of *Bb.*

Once we realize that the mutant tomato branch is heterozygous for the new mutation, we also realize that the seed from tomatoes on the mutant branch is going to show segregation for the new mutation. That is, the seed will not be pure breeding for the new mutation initially. So we grow a number of plants from the seed, because we know that only some of them are going to be of the mutant type. The expected genotypes in the next generation will be ($\frac{1}{4}BB$ + $\frac{1}{2}Bb$ + $\frac{1}{4}bb$). The genotypes Bb and bb correspond to large-fruited and normal-fruited phenotypes. Genotype BB is new; we can't predict what it will look like.

If there is simple dominance, BB will look the same as Bb. It might instead have even larger fruit than Bb. Or homozygosity for B might have detrimental effects. It's possible that BB plants will be lethal or that the fruit will be deformed and undesirable.

The most probable result is that BB and Bb will look the same, that there is simple dominance. In this case, we can derive a pure breeding variety with large fruits quite easily. We just inbreed and save seed from large-fruited types for a number of generations, as I've described previously.

Inbreeding and selecting large plants for a number of generations doesn't cost any work at all. Instead of growing just 'Porter', we grow some plants that are 'Porter' and more that are a larger 'Porter'. In addition, all we need is a few plants in each generation — no more than we would have been growing anyway.

There is a faster way to get the new pure breeding variety. It depends on deliberately identifying one or more plants that are homozygous for the mutation instead of on the general nature of inbreeding to generate homozygosity.

The first year we grow several (more than eight) plants from seed of the tomatoes on the mutant branch. Then we save seed from at least eight of the large-fruited plants. The second year we grow at least eleven plants from the seed of each of those eight, keeping track of which plants had which mothers. Most of the families have both large-fruited and normal-fruited types; their mothers were heterozygous for the mutation. But some families are all large-fruited; their mothers were homozygous. If we choose one of the plants from one of the latter families, we will have, as of year 2, a new pure breeding variety.

Why did we choose the numbers eight and eleven in the previous paragraph? Because we expect offspring of the mutant branch to be three-quarters large-fruited and one-quarter normal, and we realize that if we select large-fruited offspring, $\frac{2}{3}$ of the large-fruited offspring should be heterozygous (Bb) and $\frac{1}{3}$ homozygous (BB). Then we refer to Appendix H to figure out how many large plants we need to be at least 95 percent sure of having one homozygote. If something has a probability of $\frac{1}{3}$ we need eight plants to be 95 percent sure of getting one. Next we need to figure out which plant(s) is (are) homozygous.

We identify homozygous plants by growing a small batch of offspring from the seed of each. The ones capable of segregating out the recessive gene clearly aren't homozygous. We have to grow enough offspring of each to see the homozygous segregant if it is possible, but we don't want to grow more plants than we need to for that purpose. How many offspring do we need to grow to be 95 percent sure of finding at least one homozygote segregating from a heterozygote if that is what's going on? We look at our table again. If a mother plant is heterozygous, the probability of a homozygous offspring is ¼. How many plants do we need to be 95 percent sure of getting at least one of something that has a probability of ¼? The table tells us eleven.

So if we are happy with being 95 percent sure about things and we want to do the minimum possible work to generate a pure breeding variety in the least possible time, we grow eight plants from seed of tomatoes from the mutant branch, and we screen them by growing eleven plants from the seed of each. That's eighty-eight plants total in the second year. But then we're through; we have a pure breeding variety.

The family-line breeding program gives you a pure breeding variety faster than just saving seed from a single large-fruited plant or two each generation. But it is more work.

Davis Huckabee saved seed from his mutant tomato branch and planted it. Most of the plants had large tomatoes, but some had the standard 'Porter' size. He is now cheerfully growing and inbreeding his new larger 'Porter' to get a pure breeding variety.

You can often take cuttings from vegetables. It isn't normally done with annuals, because it is usually easier to start them from seed. My friend Miriam Ebert starts late plantings of tomatoes by simply cutting suckers off her early plants and sticking them in the soil. Brassicas can be propagated by shoot or root cuttings (see Appendix A). Little Brussels sprouts heads can be rooted and will flower and set seed as if they were baby cabbages.

If you have a small mutant branch of something, you may be able to increase it by such vegetative means. Or you might cut off the entire top of the plant above the mutant

branch so that it becomes the main leader. You can also increase material by grafting, or use grafting in emergencies.

My friend Dan Borman went out into his garden one day and found that a very special sunflower seedling had been nipped off by a slug. The top of the seedling was lying on the ground. He quickly sliced the base of the stem into a wedge. Then he cut the top off a seedling of a common variety and made a V-shaped slit into the region between the two seedling leaves. He slid the special seedling's top into the V. The graft took. The special plant was rescued. Neither he nor I have seen or read anything about either grafting sunflowers or grafting seedlings. What matters is that he tried it.

Not every different-looking branch represents a new mutation. Sometimes pests or diseases are responsible. In corn, for example, some fungal diseases produce hormones that alter the sexual expression of the plant so that there may be kernels on the tassels. In such cases the funny branch or tassel may not be genetically different at all.

My experience with somatic mutation was even simpler and more straightforward than the story of the 'Porter' tomato, because mine occurred in a vegetatively propagated perennial — the 'Egyptian' onion, also called the 'Walking' onion — a top-setting form of *Allium cepa*. 'Egyptian' onions make bulbs at the top of their flower scapes in the position where flowers would normally be. That is, they use part of the mechanism of sexual reproduction for vegetative reproduction and don't reproduce sexually at all.

The scapes bend over and plant the bulbs — thus the reference to "walking." Or you can plant the bulbs yourself, using them as sets, which is what I do. Some people use the top sets as little onions, but I have always found them too small to be worthwhile preparing that way. 'Egyptian' onions also divide at the base to form bunches, though not as prolifically as true bunching onions. I have about a dozen clumps.

Last year one section of one of my clumps produced scapes with two to four much larger top sets per scape instead of a lot of little ones. I think I might prefer having a few large top sets, as the bigger top sets make better planting stock. And these are so big that I might be able to plant them very deep and have an onion with a long, blanched stalk. That's possible with big top sets but not small ones.

Normally, to get blanched green-onion stalks, people grow seedlings, then transplant them into trenches and hill them up as they grow. But that's too much work for me. I've found that the biggest top sets of 'Egyptian' onions can be planted several inches deep and will come up and make a nice onion with a blanched stem. But only the biggest top sets have enough stored energy to do that, and my standard variety produces few big top sets. In addition, bigger top sets might be more useful to me in the kitchen.

I planted two top sets from the large-top-set-producing scapes in a second patch. One produced a plant with unusually large, thick scapes and large top sets — some up to an inch across. The leaves may or may not be bigger or thicker than usual; I can't tell. The

scapes are so thick and sturdy that they have remained erect, and it is now time to harvest the onions. In other words, the mutant walking onion doesn't walk anymore. If that's consistent, it could be handy.

I would much rather have scapes that stay upright until it's time to harvest the bulbs than ones that bend over and break and dump the bulbs in the mud before I get around to harvesting them. I would rather harvest the bulbs and use them or plant them where I want them than have them trying to plant themselves randomly in an already crowded garden. The plant produced twelve large top sets on four scapes. The largest was 1¼ inches across. Many of the top sets also developed scapes with batches of smaller top sets on them; there are thirty of these smaller, secondary top sets.

The plant that developed from the other large top set is a completely ordinary-looking plant with normal-size bulbs and scapes that bend over and dump the bulbs in the mud. Apparently, not all the tissue in the scape with the large bulbs was mutant. The original plant I noticed must have been a mosaic.

There are two obvious possible explanations for the new mutation. One is that it is a single-gene dominant mutation of some sort, as in the tomato example described previously. The other is that the mutation is a chromosomal one, most likely a spontaneous chromosome doubling, so that the mutant plant is tetraploid.

From a practical point of view, it doesn't make any difference whether the new mutation is of a single gene or a change in ploidy, since I'll be increasing and maintaining the material vegetatively. I'll just plant all the bulbs from the giant plant and keep any that also produce giants. All the scapes on the giant plant produced big bulbs, so I think this plant may be of purely mutant tissue, or very nearly so. If it is, all the bulbs should produce giant plants. If not, some of them may produce normal plants. I'll cull out any normals, of course.

As it stands, it looks as if I already have a new vegetable variety — and for no more work than just noticing something and picking two bulbs and shoving them into the ground. I wasn't especially thinking about how useful big top sets would be before the mutation happened. I wasn't thinking much about the 'Egyptian' onions at all. I was only eating them. But once the mutant appeared, I began thinking about its potential. With happy accidents, the question usually isn't "How can I get thus and so?" Instead it is "Here is something; what might it be good for?"

In fall of 1990 I planted about 800 seeds of 'Aprovecho Select' fava bean in a field in Independence. It's a large-seeded type with spectacular flavor as a green-shelled bean. It overwinters here in an ordinary year. But the winter of 1990–91 was unusually severe. All the fava bean plants died — except one. That one wasn't even damaged.

To me the situation represented not a setback, but a wondrous opportunity. There are no large-seeded favas that are fully as cold hardy as I would like. Perhaps the surviving plant could be used to breed one.

My tactic would be to inbreed the plant.

But favas can outcross to a significant extent, so I transplanted the plant to my home garden, where I could watch and pamper it, and where it would be isolated from other fava beans.

By mid-July, my special plant had produced nice large pods with big seeds. All the seeds had a pale, opalescent green skin and white hilums. 'Aprovecho Select' has tan seeds with black hilums. I don't know of any large-seeded fava with white hilums. And I had never seen that skin color before, either.

I showed the seed to Ianto Evans, who was the original source of the 'Aprovecho Select' seed I had planted. He said the green skin is the same color as that of certain Japanese varieties they have grown in past years — back before they realized how much isolation the favas require. The Japanese varieties don't have white hilums, however. A few of the 'Aprovecho Select' seeds I planted had white hilums, though. I think it likely that my special plant is a descendent of an accidental cross between a white-hilum 'Aprovecho Select' and a green-skinned Japanese variety.

I planted thirty seeds from the special plant in fall of 1991 and increased the seed stock to a few hundred. I've now passed most of the material over to Dan Borman. His ordinary winters are similar to my most severe ones, and he is passionate about fava beans. He'll do further seed increase and select for winter hardiness under his conditions.

By now you should see an obvious pattern to how to approach most happy accidents. Start by inbreeding, unless there is some reason to do otherwise — such as using a species with an incompatibility system or strong inbreeding depression. If the plant is a natural inbreeder, just save the seed. If it is enough of an outbreeder that isolation is usually required to maintain pure varieties, either hand-pollinate or isolate.

My happy accidents all involved plants that were inbreeders or that are vegetatively propagated. Happy accidents in outbreeders that don't suffer greatly from inbreeding depression are usually handled similarly. However, if the plant has an incompatibility system, you may not be able to self-pollinate it. Furthermore, if it is of a species that displays inbreeding depression, you cannot usually derive a vigorous variety on the genetic base of just a single plant. In these cases, you may need to cross the unusual plant to one or more others (or allow it to cross) to produce a population of plants from which you can obtain more of the unusual type in future generations.

If you find a single unusual corn plant, for example, you would realize that you cannot develop a vigorous variety through inbreeding from the one plant. Instead, you would use a careful combination of inbreeding and outbreeding. For example, you might cross the plant with a hundred ordinary plants, and save the seed. Then you plant that seed in a patch by itself. None of the plants might resemble the unusual one. But you wouldn't be discouraged. You would allow them to cross-pollinate each other. In the next generation, you will probably have many of the unusual plants, enough to form a solid genetic base for a vigorous variety.

When happy accidents appear in inbreeders, you usually develop their potential by inbreeding. When happy accidents appear in outbreeders, you often must do some outbreeding to generate an adequate population base. Then you inbreed and mass-select within the population.

Domesticating Wild Plants

EVERY CULTIVATED plant was once a wild plant somewhere. Every edible wild plant represents a possible new vegetable. Every niche in your yard that doesn't have edible things in it represents an area where an edible wild plant could be growing if you could figure out which one and find it. Our modern style of living in houses, for example, creates patches of partial or solid shade. So do the trees we like to live among. Most people try to grow sun-loving grass under their trees and north of their buildings. But the woods are full of edible plants, often perennials, that thrive on partial to full shade.

Generally, it seems to have been the sun-loving plants that were brought into cultivation. Now, with so many people living in homes with only small and usually partially shaded yards, might be a good time to look for edible plants to fill the shady niches.

There are said to be twenty thousand edible plant species. Of these, only a few hundred have received any breeding work from any member of humanity. Did our ancestors tally every edible species and choose exactly those with the most potential? Did they find and domesticate all the best possibilities? I'm sure that they did not. First, only some peoples were involved in domesticating plants; vast areas of the world were populated by people who weren't agriculturalists. None of the native edible plants of the northwestern United States were domesticated, for example, because the natives were migratory hunter-gatherers, not farmers. Our region is full of delicious edible wild plants that are already perfectly adapted to growing here.

Elsewhere in North America native agriculturalists concentrated on growing just a few staples. They collected most vegetables from the wild instead of domesticating them. All native North American edible wild vegetables represent essentially unexplored terrain for domestication.

Even if our ancestors had evaluated all possible wild edibles for their potential as domesticates, it wouldn't mean much. Even today no one can evaluate the potential of a plant before the breeding work with any degree of certainty. Even in areas of the world long occupied by domesticators and agriculturalists, there are undoubtedly many more wild plants that would make good domesticates.

Finally, what makes a good potential domesticated vegetable depends not just on the plant but on the cultural patterns of the potential users. There are new life-styles; new ways of handling homes, gardens, and land; new knowledge about and concern over nutrition; new food preparation methods. All create possible niches for new vegetables.

What about vegetables bred for microwaving, for example? Microwavable popbeans? Corn on the cob that tastes good when microwaved as whole ears without opening them up and removing the silks first? I recently read that waxy, moist potatoes such as 'Yellow Finn' are best for microwaving; that's the first reference I've seen to specific varieties and microwaves.

We have very efficient food dehydrators now, too. Some edible wild plants are not usually used because they have toxic substances that dissipate during drying. Such plants may now have potential for the casual gardener.

Until recently, gardens in America were planted in the spring and harvested in the fall, and then tilled. Perennial vegetables just didn't fit into that pattern. Now more and more people are using permanent beds for part of their vegetable gardening. Perennials fit into a permanent bed style of gardening extraordinarily well. Every alteration in our cultural patterns creates opportunities for creating new vegetables — new varieties of established vegetables as well as entirely new crop species.

Professional plant breeders are doing little work on domesticating new crop species. Professionals need to get grants for their work, and it's easier to get money for work on something that is already important. When Rich Hannan received phone calls from professional plant breeders who were interested in *Phaseolus vulgaris* popbeans (nuñas) and he told them about *Cicer arietinum* popbeans, they responded, "But what I have funding for is *Phaseolus*." I had funding for neither; I was free to be open-minded.

Very few professional plant breeders are working on domesticating new vegetable species, and those that are do so as only a small part of their efforts. I would guess that more than 99 percent of the professional vegetable breeding work is devoted to fewer than one hundred species.

Corn has been profoundly improved by modern professionals, but it was created by amateurs. The same is true of almost every food crop we have. The work of the modern professional plant breeder is directed almost exclusively at improving the crops that were created by the amateurs of yesterday. Who will create tomorrow's new crops and new crop species? Who else but today's amateurs.

How long does it take to domesticate a plant? It depends on what you mean by a *domesticated* plant. That's something that geneticists and anthropologists argue over. There is a tendency for them to claim that domesticating any plant takes hundreds of years but also to define domestication so that something isn't considered domesticated unless it has been cultivated for hundreds of years. That argument is somewhat circular. It's clear, though, that in some cases a single genetic mutation is enough to turn a wild plant into a much more useful domesticate. Such mutations often turn up in gardens by accident and are often present in wild populations. We can find them just by looking and collecting.

For my own purposes I think of domestication in the broadest and loosest way. Anytime we collect a wild, naturalized, or escaped plant and deliberately grow it near our domicile, we are engaged in domestication. Sometimes we have changed the plant genetically and sometimes not. Note, though, that when we choose a particular plant or population of plants in the wild to start our own population, we have done breeding in

the sense that our population differs genetically from the average for the wild population.

The best place to start in domesticating new vegetables is with the edible wild vegetables that are native to or have naturalized to your region. You can consult books on wild edibles or try to find out what plants the native people of your region used and how they used them. Then you go collecting, or refer to others who do or have. Note that many of the alternative and regional seed companies carry seed of native wild plants. In my own region, Abundant Life Seed Foundation, Meristem Plants, Peace Seeds, and Seeds Blüm all carry seed of native edible species.

My own preference is for plants that are more agroecologically adapted to my conditions. I already have plenty of vegetables, inherited from others elsewhere, that will grow here when watered out of season and pampered. I want new vegetables that produce tasty, different food, and for less work.

I also want vegetables with new, delicious flavors, not ones that are merely palatable. And I am not willing to do a lot of work to get a small edible part. Or to spend hours preparing anything. Or to boil anything in two changes of water and throw away most of the vitamins and minerals. Or to dry something, pound it into flour, and mix a small amount of that flour with wheat flour to make bread that would have tasted just as good without the addition. You will have your own preferences.

When you go collecting edible wild plants, keep the conditions in your yard or garden firmly in mind. Don't bother collecting from marshes unless you have a marsh. Nearly everyone has some areas of partial as well as deep shade where woodland plants might thrive. And most gardeners have at least some areas of full sun or nearly full sun that could be home to prairie or meadow wildlings.

The characteristics of the plant in the field usually aren't a very good indication of what it will be like under garden conditions, with richer soil and optimal spacing. It's a good idea to collect from a number of different areas whether the plants seem particularly impressive or not. They'll almost always be much bigger, more succulent, and more productive once they get their roots into good garden soil.

Individual wild plants and populations may vary a lot in ways that matter to you. They can be milder or stronger in flavor, be bigger or smaller, have leaves of various shapes, and so on. You'll probably start by collecting something haphazardly and growing it. If it's a winner in your garden, you will often go back to the wild and make additional collections looking for specific characteristics.

Whenever you collect, you will have to decide whether to lift plants from the wild or collect seed. If the plant is rare, or is rare in the area where you found it, you shouldn't collect whole plants. You should collect seeds. (Some areas even have laws against digging up wild plants.) An alternative might be to dig down and remove a section of the root for a root cutting or to remove a shoot or two to root. If the plant has a number of

crowns, you might be able to remove a small piece for yourself and leave the bulk of the plant where it is. The object is to obtain some germplasm while leaving the wild population unharmed.

Seed represents greater genetic diversity than a single plant or clones from one, and it can usually be transported and stored more easily. However, seed of wild plants frequently does have one or more special dormancy mechanisms, so you may have some work to do before you can figure out how to get your wild plant's seed to germinate. (The J. L. Hudson catalog is one of the best sources of information on how to germinate seed from rare or wild plant material.)

Some seed has a tough seed coat that must be penetrated in some fashion (nicked with a knife or file, for example, or abraded a bit with sandpaper) before it will germinate. Other seed needs a cold treatment to mimic winter. Seed that ripens in late summer or fall and germinates in spring is often most easily handled by planting in fall and waiting for spring.

Generally, edible wild vegetables are usable as soon as you get the plants established, which is usually a few months for most annuals, two or three years for perennials started from seed, and one or two years for perennials started from plants or cuttings.

Dan Borman (Meristem Plants) collected plants of *Lomatium nudicaule,* an edible in the family Apiaceae, from the wild and moved the plants to his garden. The greens have become one of his major vegetables and staples.

Alan Kapuler (Peace Seeds) is in the process of building a collection of all the wild Apiaceae of the Northwest. Many are edible; many were used by the Indians. Some are potherbs. Some produce large, succulent roots. Alan also has two different species of camas (*Camassia leichtlinii* and *Camassia quamash*). Camas was one of the Indians' staples. The plant was so important that harvesting rights to various camas fields were a major issue.

A number of professional plant explorers and breeders in Louisiana have been working extensively with the American groundnut (*Apios americana*), which was collected wild and used extensively by the Indians. (See the article by Blackmon and Reynolds listed in the bibliography and Janick and Simon's *Advances in New Crops*.) They have made extensive collections from the wild and are evaluating them. Among the characteristics they are selecting for are fused (bigger) tubers, plants that have tubers closely spaced underneath them instead of dispersed thinly over several square feet of ground, and plants that don't twine and are able to grow on the ground instead of needing support.

My major attempt at domestication so far involves the dandelion (*Taraxacum officinale*). Dandelion leaves can make a delicious dish of greens in the early spring, and the plants are perennial.

There are plenty of dandelions growing in fields around here. But the ones here in town are in lots frequented by dogs or in yards where people mow them down or spray them with herbicides. I can walk a few blocks to a park area and collect dandies, but that isn't very convenient. When I want to eat something, I want to be able to go to the garden and have it there. I need my dandelions handier.

I started by ordering seed of the variety 'French Broadleaf' and planting it. It grew slowly. That's not surprising for a young perennial. The plants looked and tasted exactly like the wild dandelions that were growing all over the place, though. There was nothing especially broad about the leaves. In the summer the plants were covered with powdery mildew and stopped growing entirely. They were still quite small; now they were both small and miserable looking. The wild dandelions growing in the lawn a few feet away looked happy and vigorous.

I was going about things backward, I decided. Why should I be trying to grow an imported dandelion when my region was full of naturalized dandelions — dandelions that have been selected over hundreds of generations to grow here without any special pampering? If the import tasted better or was more productive than the natives, that might be reason enough. But that didn't seem to be the case.

So I forgot about other people's dandelions and started paying attention to the native ones. They come in various sizes and shapes. Some are bitter even early in the spring, and some are mild flavored until quite late. Every time I went anywhere and saw a particularly large, luxurious dandelion in flower, I took away some of its seed. Dandelions along roadways, dandelions in clear-cuts, dandelions in the gardens of my friends — all were scrutinized.

One winter day when I was out walking my dog, I noticed a dandy that had produced large leaves in late winter before most plants did much. I carefully marked the plant. Later in the spring, after the dandelions began flowering, I went back and collected seed from the early one. Then, as I came back to the house with my thoughts still on dandelions, I happened to notice an especially large dandelion with very big leaves that was growing right in my driveway. I tasted it. It was mild — as mild or milder than any dandy I had ever tasted. And the leaves were huge. I had walked right by it on the way out to collect dandelions elsewhere. I named it 'Driveway Surprise'.

'Driveway Surprise' was in danger of getting mowed down. So I lifted it, separated it into a major plant and a number of root cuttings, and moved it to my garden.

But the clones of 'Driveway Surprise' produced plants with small, dissected leaves instead of big, broad leaves. And they weren't especially mild. They tasted pretty much like the rest of the dandies in my yard. In the summer they got powdery mildew, too, though not nearly as badly as the imported dandelion had. They never seemed to quit growing completely or to look quite as miserable as the import. It was time, however, to rethink my approach.

Maybe little plants produced small, dissected leaves. Many wild plants seemed to have both kinds of leaves, dissected ones and broad ones. Or maybe it had to do with the sun. As I thought about it, it seemed to me that the tall, luxurious, broad-leafed dandelions also were the mildest ones at any given time of year, and they were always growing in fairly heavy shade.

I was still going about things wrong, I decided. When I transplanted 'Driveway Surprise' to my garden, I had arranged for it to have adequate sun. Maybe that was counterproductive. It had been growing in a gravel driveway area north of the garage and under a tree. It must have had very close to full shade. The other thing I did when I transplanted the clones to my garden was that I put them in a year-round water situation. Wild dandies around here normally don't get any rain in the summer. That section of driveway was never watered except by natural rain, and the plant was big enough that it must have survived and thrived through at least a couple of years.

I left the dandy clones where they were in my garden, but I planted other plants south of them. This spring the dandelions were heavily shaded, and I got my first taste of large, mild, tender 'Driveway Surprise' dandelion greens. Interestingly enough, the flowers also are quite large — up to about 2 inches across and more a solid ball than flat — and the large, unopened flower buds make a nice vegetable. The flowers on the original 'Driveway Surprise' as it was growing in the driveway were nothing special. But the plant was growing in gravel and hard-packed clay, not good garden soil.

My 'Driveway Surprise' leaves stayed mild far into the summer, long after the wild dandelions — even those growing in full shade — had turned bitter. Throughout the spring and early summer, I could boil the leaves in water, drain them, and serve them with salt, pepper, and vinegar, just like any delicious green. (I didn't have to boil them in two changes of water, as is often done with wild dandelion greens.) I had only enough leaves to experiment with, not enough to satisfy. During midsummer, the leaves became inedible because they got powdery mildew, and ultimately they became bitter. The plants continued growing, however. I'll remove the old growth in the fall and let the plants start over making fresh, clean leaves.

I think I'm still doing things wrong, though. When I can get to it, I'll build a patch north of the garage in practically full shade for 'Driveway Surprise'. I won't water the patch at all. I'll subdivide and further clone the plants I have in late fall after the rains start. They will have all the rain they need to get established. They probably won't give me anything to eat the first spring, but the following year, in late winter and early spring, I should have plenty of greens. I should never have to water them. The plants will, presumably, go dormant in late summer when there is no water available. And with no watering, they probably won't get powdery mildew. Once established, the dandy patch should produce good spring greens year after year with no work at all. And the huge golden flowers will be attractive as well.

I don't yet know whether 'Driveway Surprise' is actually milder than the wild dandies around here. Part of the mildness seems to come from growing the plant in the shade. Another part of it, as with many plants, is keeping the leaves picked so that I am always harvesting young, fresh leaves. Leaves that have been around for a while are always bitter. Growing in good garden soil, the plant grows leaves much faster than the dandies in my lawn, so the fact that the leaves are milder doesn't necessarily mean there is a genetic difference.

It's also possible that other wild dandies would have flowers as big as 'Driveway Surprise' if they were equally pampered. I haven't done any side-by-side comparisons of 'Driveway Surprise' and other wild types. If I were going to offer it for sale, I would need to do that so I would know what special features my plant had. But I'm not. I need only to grow and eat dandelions. I don't need to sell them.

I also haven't grown out seed from 'Driveway Surprise'. I've just cloned it. My first priority is to establish a big enough patch of 'Driveway Surprise' so that I will have all the dandelion greens I want in the spring.

What about the seed of my dandy that might float all over and irritate my neighbors? Well, I usually eat the buds before they open. When I miss them, I leave the beautiful flowers but pick the seed heads before they open. I don't want a garden full of dandelions any more than my neighbors want them in their yards. And I want my dandies producing more leaves and buds, not seed.

When I first started playing with wild-collected dandelions, I treated them exactly as if they were a typical domesticated crop. Our normal garden crops are imported from elsewhere, and we usually grow them by altering our conditions to make them as much as possible like those where the vegetables originated. I did not realize the agroecological implications of working with a plant that was already perfectly adapted to growing in my area. I now view the agroecological implications as one of the most important reasons for collecting and cultivating wild edibles.

Dandelions growing wild in my region have been selected to be able to survive summer drought and to do their growing in the wet winter and spring. That means that they could make a nice permanent bed of edible and ornamental that would require no water and no care. Dandelions that grow wild in your area will probably have different characteristics. If you live in an area with severe winters, for example, your wild dandelions might make especially big roots and then die back and go dormant in winter. Such a dandelion might put on a heavy burst of leaves in early spring. Or it might have potential as a forcing crop — a vegetable that produces a root that is harvested and then put in a warm place to develop a crop of greens during the winter.

My dandelion might not be able to go dormant in winter and may not even survive in a harsher winter. Then again, the ability to go dormant whenever there is a reason to may be one of the basic characteristics of all dandelions. This seems to be the case with sea kale (*Crambe maritima*). It can go dormant in either winter or, because you forgot to water it, summer. It doesn't go dormant if you water it in summer. *Allium vineale,* a local naturalized wild onion or garlic, represents the other extreme. It goes dormant and dies back in summer whether it is watered or not.

Working with native edibles broadens your perspective on gardening. It invites you to reexamine all your gardening methods and how they fit into your life-style, your cultural patterns, your cooking and eating habits, and the ecology of your region. It helps you identify the underlying assumptions and imagine other possibilities.

We need native vegetables. But we need more than just the plants. We also need entirely different ways of growing and using them, ways that capitalize on their unique adaptation to our climate and their special properties.

When it comes to growing and using an unfamiliar plant, some precautions are advisable. First, make sure you've properly identified the plant. (I frequently consult the weed specialists at my local agricultural extension service; to them, this stuff is "weeds.") Second, read what is available about how your plant is processed or used as an edible. Some plants have edible parts and poisonous parts, for example. Some are edible only at certain times of the year or only with special processing.

Don't believe everything you see in print, though. Many authors copy from each other's writing without ever trying the plants themselves. Some of the plants the books call edible taste awful when prepared by the methods they describe. Don't believe anything until you've tried it yourself. And try things cautiously and carefully.

Realize that you may be unable to eat a specific plant even when others can. There are individual differences in ability to digest specific compounds; human biological variability frequently shows up when you experiment with an uncommon food plant. I know at least three people who suffer from severe digestive upset and abdominal pain if they try to eat sunroot (Jerusalem artichoke, *Helianthus tuberosum*). A good guess is that they can't digest inulin, the unusual starch characteristic of sunroots.

You also may be allergic to specific plants, even if they are a fine food for others. I am allergic to all plant pollen. Many people can use the flowers and buds of a number of species as edibles, but I have an instant allergic reaction to them. If I remove the stamens of daylilies before they shed pollen, I can use the flowers as a vegetable; otherwise I cannot.

Whenever you are trying a new food plant for the first time, whether it's a wild plant or not, always start with a single bite and then wait a day. The next time try a small portion and again wait a day. On the third try you can have a full serving. Once you've established that the plant is a good food for you, don't assume that it is a good food for a guest; he or she might be unable to digest it or be allergic to it.

The other major precaution with respect to experimenting with unusual or wild plants involves their potential for becoming pests. Is your plant invasive? Will it take over your entire yard and that of your neighbors? Will it shed floating seeds that spread all over? Neighbors tend to prefer that your plants stay in your yard. That nice edible-shoot bamboo can send out 40-foot underground runners and turn your entire neighborhood into one huge bamboo stand unless you confine it rigorously.

Some counties have laws against growing certain plants or allowing them to grow on your land. In some counties, for example, it's illegal to grow dandelions. Some edible wild plants are not allowed to be brought into certain states or grown there because of their known potential as agriculturally important weeds.

Many wild edibles are potentially major weeds even though they aren't illegal. You need to take care that your plant adventures don't create problems for yourself and others. Sometimes edible plants aren't cultivated for very good reasons. You should not limit your care to the requirements of the law. Follow the principle that your neighbors have a right to decide what to grow and shouldn't be growing something of yours unless they planted it themselves.

I grow unfamiliar species in small raised beds surrounded by regularly mowed grass. If they prove to be moderately invasive, I can keep them contained. If they seem dangerously invasive, I can cover the entire bed with plastic and destroy them.

Plant breeding starts as soon as you choose what germplasm to perpetuate, as soon as you choose to grow this plant or seed from this population and not that one — that is, as soon as you bring a plant home and start growing it. If you become enamored of a specific wildling, you may make many collections. And you may do crosses of this plant with that one to try to combine desirable characteristics of both. Everything you learned about how to breed the more common plants applies to the rare vegetable or wildling. And, of course, you will be practicing selection with every generation.

Expanding Horizons

The fact is there are 20,000 plants that have useful edible parts. I exclude species like pine trees, whose underbark is edible, and include only species that have sizeable seeds or roots or leaves or stems or flowers or pods that are eaten somewhere in the world. Out of those 20,000 only about a hundred have been brought to an advanced agronomic level. Indeed, it's been said that a mere 22 crops feed the world. Wheat, rice and maize by themselves are supposed to provide 50% of the weight of food for the world. This is a tiny, tiny larder from which to feed a planet. It is a dangerous vulnerability.

— Noel Vietmeyer, "The New Crop Era," in *Advances in New Crops*

THE IRISH potato famine occurred because Irish peasants depended almost exclusively on just one staple crop — potatoes — and on only a few closely related varieties. Growing just one staple worked well for years. That's why they did it. An individual farmer could support his family better with potatoes than anything else. But then suddenly, without warning, a year came when growing potatoes as the only staple didn't work; it didn't work at all.

Irish potatoes were sensitive to late blight. Because everyone grew potatoes and was growing blight-sensitive potatoes, once the blight broke out, it spread like wildfire. The entire crop was lost. Those who could emigrated. Thousands starved to death.

In the past few decades, agricultural patterns have changed so that this situation — fields planted in monoculture, entire regions dominated by single varieties of single crops — has ceased to be the exception and has become the rule. Such conditions are ideal breeding grounds for major epidemics of pests and diseases. From the Southern corn blight in the United States in the 1970s to the outbreak of whiteflies in California in the early 1990s, we are obtaining increasing evidence that our current agricultural patterns are seriously flawed.

We simply have too few food-crop species. And we are growing too few varieties of those we have. Combined with unsustainable and ecologically destructive agricultural practices, it is a recipe for disaster. We need to make some major changes in our food production patterns. And we need new crops to facilitate new patterns.

We must expand the food base. We need to be growing and using thousands of crop species, not a few dozen. We need hundreds of varieties of every major crop, not just a few. In addition, we need new crops and crop varieties that are fine-tuned to specific regions, crops that require minimal inputs and labor, and crops that help to preserve the land instead of destroying it. We also need to integrate the preservation of general biodiversity into agriculture. In short, we need to be practicing not agriculture but *conservation agriculture.* We need to expand the horizons of agriculture on this planet.

Many professional plant breeders, agronomists, horticulturists, and conservation biologists recognize the need for increased agricultural biodiversity as well as new, sustainable, environment-friendly methods. They are doing what they can. But there are too few professionals, and there is too much to do.

I believe that expanding the horizons of agriculture will depend predominantly on the contributions and involvement of amateurs. This book is an endeavor to enlist, empower, and encourage all gardeners and farmers everywhere to participate in this effort.

I began breeding plants just for fun. I thought I would just be applying what I already knew about genetics and gardening. But being involved in breeding my own varieties changed everything. I think about plants and gardening differently, see more possibilities, and am quicker to recognize and examine underlying assumptions. Breeding my own vegetable varieties has expanded my horizons.

Developing your own varieties is fun, and it adds new dimensions to your gardening. The value of breeding new, useful plants transcends the pleasure, though. A new food plant is a unique contribution to society. To be able to make that contribution is deeply rewarding.

When you start breeding your own plants, your contribution starts long before you develop anything, though. The minute you do a cross, bring home a wild plant, grow something unusual, or even just maintain and grow a rare variety, you achieve a small increase in agricultural biodiversity. The means are part of the end; you succeed in part just by beginning. As soon as you begin breeding your own vegetable varieties, you begin expanding horizons.

801 Interesting Plants

Here is information about more than eight hundred interesting plants: scientific names, common names, families, and life-styles; chromosome numbers, basic breeding systems, flowering patterns, flower types and modifications, average cross-pollination frequency, major pollen vectors, and incompatibility system information; recommended isolation distances, seed yields, location in the USDA-ARS National Plant Germplasm System, and references.

I've placed the greatest emphasis on cultivated vegetables, edible wild plants that can be used as vegetables, and relatives of both that might be used for breeding. A secondary emphasis is on cultivated and wild fruits, including tree fruits, because there is no clear distinction between fruits and vegetables. Most plants that are normally considered fruit crops can also be used as vegetables, and many are obviously both, depending on the variety, the part of the plant used, or what you happen to be doing with it. Many grains and nuts also are included.

All species are listed in alphabetical order by scientific name. If you don't know the scientific name, you can find it by looking up the common name in the index. The first species listed for each genus includes the family the genus is in.

Sometimes information about a single species is scattered among a number of entries because different sources use different scientific names. I cross-referenced where I could, but I usually did not try to correct the taxonomy; thus there are contradictions in this table because there are contradictions in the sources. In addition, information is sometimes listed under "spp." instead of under the exact species. Where contradictory information exists, I sometimes list a species twice to keep each set of information and its references intact. When you look up something, it's a good idea to scan through all the entries for the genus to check for contradictions or information listed under synonyms or "spp."

This table is, in part, a compilation of compilations. You can, however, find the original sources by using the references in each entry.

Plants are a mutable and variable lot. When I include a specific piece of information, all it means is that someone, somewhere, when looking at one particular cultivar or even one particular plant, thought he or she saw it that way. There could be differences from cultivar to cultivar or population to population. And people sometimes misidentify the plants they are describing.

No information on uses of the plants is included, as doing so would have made the table larger than the rest of the book. Almost all the entries are included in Stephen Facciola's book *Cornucopia,* which compiles all the available information about the uses of three thousand cultivated and edible wild species. *Cornucopia* gives references for everything and also includes sources for the plants. See the bibliography for all the references listed here.

Categories of Information and Key to Abbreviations

Listed below are the basic categories of information and the most commonly used terms and abbreviations, given in the order in which they appear in the entries.

l- Life-style. Whether *a*nnual, *b*iennial, or *p*erennial, and whether an *h*erb, a *s*hrub, a *t*ree, a *v*ine, *g*rass, or a *l*iana (woody vine). Examples: "l-a,v" refers to an annual vine; "l-p/a,h,s" is a perennial that is raised as an annual and may be an herb or a shrub (depending on cultivar, climate, growing methods, or growing conditions).

Much of the information about life-styles and the system of abbreviations describing it were taken from Duke, Hurst, and Terrell, *Economic Plants and Their Ecological Distribution.*

2n = Diploid chromosome number. Information is usually from Duke, Hurst, and Terrell, *Economic Plants;* Darlington and Wylie, *Chromosome Atlas of Flowering Plants;* or Zevan and Zhukovsky, *Dictionary of Cultivated Plants and Their Centres of Diversity.* These are compilations, but they give the original references.

gf- Genomic formula. Helps trace genetic and evolutionary relationship between species. See explanation in section on brassicas in Appendix B.

b- Basic breeding system. "b-o" and "b-i" refer to outbreeding and inbreeding, respectively. Note that, practically speaking, most plants can both inbreed and outbreed; what matters is the degree. Always look further in the listing for information on cross-pollination frequency and recommended isolation distances.

f- Flower types, flowering patterns, flower specializations and modifications. **f-m, f-d,** and **f-h** refer to species that are monoecious, dioecious, and hermaphroditic (perfect flowers), respectively. Flower modifications and specializations are given after the basic flowering-pattern description. Examples: "f-m plant protandrous" means the species is monoecious —that is, plants have separate male and female flowers, but the male flowers tend to develop before the female ones; "f-h flower protogynous" means the species has perfect flowers, but the female part of the flower develops and is receptive before the male part develops and releases pollen.

x- Average percent cross-pollination (contamination) frequency in inbreeding plants and major vectors. Example: "x-3% by insects." Note that "x" varies depending on region, weather, insects present, and other factors, so "x" figures are intended to give you just a rough idea. When the frequency matters, you can determine it yourself, as has been described in chapter 9.

s- Self-compatibility information. "s-i" and "s-c" are self-incompatible and self-compatible, respectively. A listing of "s-i" means that at least some cultivars are partially or totally self-incompatible. Other cultivars in the same species may be self-compatible. Likewise, where incompatibility exists, it may be slight, causing only a minor reduction in seed vigor or seed set, or it may be so major that no seed forms at all without cross-pollination.

v- Major pollen vector (in outbreeding plants). Example: "v-bees, flies."

d- Recommended isolation distances as given by various sources. Distances are given in meters unless otherwise specified. Note that there are many ways to get around or minimize the need for isolation, so don't get discouraged if you see a huge isolation distance. (See part II.) Since different sources have different levels of authoritativeness and usefulness (depending, for example, on your climate), recommendations of the various sources have been kept distinct by using a number following the "d." The d- figure is usually least authoritative, as it represents older information that has been superseded in many cases.

> **d-** Recommended isolation distances as given by Frankel and Galun, *Pollination Mechanisms, Reproduction, and Plant Breeding.*
> **d2-** Recommended isolation distances as given by George, *Vegetable Seed Production.* These distances are considered appropriate for producing foundation seed in Great Britain.
> **d3-** Isolation distances recommended by the curator in charge of the species for the USDA-ARS National Germplasm System collection or by some other appropriate professional breeder or curator of the group. Names and positions of sources are given.
> **d4-** Recommended isolation distances as given by Fehr and Hadley, *Hybridization of Crop Plants.*

y- and **y2-** Yield of seed when grown for seed:

> **y-** Information referring to large commercial plantings; mostly from George, *Vegetable Seed Production.* Usually in kilograms per hectare (kg/ha). Sometimes from Duke, *Handbook of Legumes of World Economic Importance.*
>
> **y2-** Information from Jeavons, *How to Grow More Vegetables,* 2nd ed. (There is a third edition now available.) Refers to small-scale, intensive planting in raised beds using organic/biodynamic methods. Given in lb/100 sq.ft.max., which is pounds per hundred square feet of raised bed and represents the maximum attained by Jeavons under excellent gardening conditions in California. These figures are usually more relevant to small-scale backyard growing.

USA- Location in USDA-ARS National Plant Germplasm System. Abbreviations following "USA-" are those used by the system. See Appendix C for meanings of abbreviations and addresses, phone numbers, and curators in charge of the various collections. Information is from an unpublished list. Example: "USA-CR-COR" means the species is in the clonal repository in Corvallis, Oregon. "CR-COR" is an abbreviation used by the National Plant Germplasm System. You can find it by looking in Appendix C.

r- References — sources of information for the entry. The major sources are indicated by the abbreviations that follow. Occasional sources are written out. Complete information is given in the bibliography. Sample reference: "r-d,f,g,Smith 1982" would mean the entry is a compilation of information from Duke, Hurst, and Terrell; Frankel and Galun; George; and Smith's book or article of 1982. Abbreviations used are as follows:

> **r-b** Bassett, *Breeding Vegetable Crops.*
>
> **r-d** Duke, Hurst, and Terrell, *Economic Plants and Their Ecological Distribution.* A list of one thousand important plants. This is where to go to find information about where things will grow.
>
> **r-d2** Duke, *Handbook of Legumes of World Economic Importance.* Best compilation of breeding, growing, and use information on all the legumes.
>
> **r-d3** Darlington and Wylie, *Chromosome Atlas of Flowering Plants.*
>
> **r-f** Frankel and Galun, *Pollination Mechanisms, Reproduction, and Plant Breeding.*
>
> **r-f2** Fehr and Hadley, *Hybridization of Crop Plants.*
>
> **r-f3** Facciola, *Cornucopia.* Compilation of all available information on uses of three thousand species, with references. Also gives sources for all the plants.
>
> **r-g** George, *Vegetable Seed Production.*
>
> **r-j** Jeavons, *How to Grow More Vegetables.*
>
> **r-l** Unpublished list. An internal document from the USDA-ARS National Plant Germplasm System that is not readily available.
>
> **r-pc** "Personal communication." Who told me is listed in parentheses after the reference. Example: "r-pc(Dan Borman)."
>
> **r-po** "Personal observation," which means I've worked with the species myself and noticed or established the fact.

r-r Ryder, *Leafy Salad Vegetables.*

r-s Shinohara, *Vegetable Seed Production Technology of Japan,* vol. 1.

r-s2 Simmonds, *Evolution of Crop Plants.*

r-y Yamaguchi, *World Vegetables.*

r-z Zevan and Zhukovsky, *Dictionary of Cultivated Plants and Their Centres of Diversity.* There is a more recent second edition that I didn't know about when I compiled this table.

General Abbreviations Used

cv., cvs.   cultivar(s)

kg/ha   kilograms per hectare

lb/100 sq.ft.max.   pounds per hundred square feet in raised bed; represents maximum yield under good conditions

m   meters ("miles" is spelled out)

max.   maximum

sp., spp.   species (sing.), species (pl.)

ssp.   subspecies

var.   variety

List of Plants

Abelmoschus esculentus (= *Hibiscus esculentus*) Malvaceae okra l-a,h 2n = 72,144 b-i f-h x-5–20% by insects, hummingbirds d-400m d2-500m y-1,500 kg/ha; 500 in the tropics y2-9.3 lb/100 sq.ft.max. USA-S-9 r-d,f,f3,g,j,l

Abelmoschus moschatus (= *Hibiscus moschatus*) ambrette, musk mallow l-a,b,p,h,s 2n = 72 USA-S-9 r-d,f3,l,z

Actinidia chinensis Actinidiaceae chinese gooseberry l-p,s,v 2n = about 116 + ,160 USA-CR-DAV r-d,f3,l

Aegle marmelos Rutaceae bael, bengal quince l-p,t 2n = 28,36 r-d,f3

Agave sisalana Agavaceae sisal l-p,h 2n = 138–150 b-o f-h protandrous v-probably wasps, bees, insects; possibly also bats, gravity r-d,f3,s2

Albizia julibrissin Mimosaceae silk tree l-p,t 2n = 26,52 r-d3,f3

Allium All alliums that flower are b-o and f-h and protandrous, many extremely protandrous. Usually s-c. Some form bulbs in flower heads and don't flower.

Allium allegheniense Liliaceae 2n = 14 r-d3

Allium ammophilum 2n = 16, 32 r-d3

Allium ampeloprasum (= *A. porrum* = *A. babingtonii*) leek, levant garlic, perennial sweet leek l-b/a,h 2n = 16,24,32,40,48(r-b) 2n = 32 (r-d3) b-o f-h s-c v-bees d2-1,000m y-500 kg/ha; 600 max. y2-9.8 lb/100 sq.ft.max. USA-W-6 r-b,d,d3,f,f3,g,j,l,s2,z

Allium ampeloprasum wild leek 2n = 32 r-d3

Allium ampeloprasum var. elephant garlic l-a,b/a,h r-f3

Allium amplectens 2n = (14),21,28 b-apomictic r-d3

Allium angulosum mouse garlic 2n = 16 100! USA-W-6 r-d3,l

Allium ardelii 2n = 14 r-d3

Allium artemisietorum 2n = 16

Allium aschersonianum 2n = 16 r-d3

Allium babingtonii see *A. ampeloprasum*

Allium bakeri 2n = 32 r-d3

Allium bidwelliae 2n = 28 r-d3

Allium canadense tree onion 2n = 16 r-d3,f3

Allium carinatum 2n = 16,24,25,26 r-d3

Allium carmeli 2n = 14 r-d3

Allium cepa onion l-b/a,h 2n = 16(r-b,s2) 2n = 16,32(r-d) 2n = 16,32(r-d3) b-o f-h protandrous Male sterility exists and is used to make hybrids. x-93% s-c Demonstrates some inbreeding depression. v-bees, flies, insects d-800m d2-1,000m y-500–1,000 kg/ha; 2,000 max. y2-(regular or tor-

pedo)10.3 lb/100 sq.ft.max.   USA-NE-9
r-b,d,f,f3,g,j,l,s2

Allium cepa var. ascalonicum (= *A. ascalonicum*)
shallot   2n = 16   r-b,d3,f2,g,j

Allium cepa   Aggregatum group   potato onion
2n = 16   r-b,f3

Allium cepa var. viviparum   USA-NE-9   r-l

Allium cepa × *Allium fistulosum*   Egyptian, tree,
top-setting, or walking onion   l-p,h   Propagated vegetatively from top sets or bulb divisions. Some varieties flower before making top
sets; others make top sets without flowering.
r-b,f3,po

Allium cernuum   wild onion, nodding onion,
lady's leek   l-p,h   2n = 14,(16!)   USA-W-6
r-d3,f3,l

Allium chinense   rakkyo   l-p/a,h   2n =
24,32(r-d)   2n = 16,32(r-z)   r-d,f3,z

Allium chloranthum   2n = 16   r-d3

Allium ciliare   2n = 32   r-d3

Allium condensatum   2n = 17   USA-W-6   r-d3,l

Allium coppoleri   2n = 16   r-d3

Allium cyaneum   2n = 32   r-d3

Allium deseglisei   2n = 32   r-d3

Allium desertorum   2n = 16   r-d3

Allium dumetorum   2n = 16   r-d3

Allium fistulosum   Welsh onion, Japanese bunching onion   l-p/a,p/b,h   2n = 16   b-o   f-h
s-c   y2-39.6 lb/100 sq.ft.max.   Yield listed
for "bunching onion," which I am assuming is
A. fistulosum, but it might not be. USA-NE-9
r-b,d,d3,f3,j,l,s2,z

Allium fuscoviolaceum   2n = 16   USA-W-6
r-d3,l

Allium fuscum   2n = 16   r-d3

Allium hirsutum   2n = 14   r-d3

Allium karataviense   2n = 18   USA-W-6   r-d3,l

Allium kunthianum   2n = 16   r-d3

Allium ledebourianum   2n = 16   USA-W-6
r-f3,l,z

Allium lusitanicum   perennial Welsh onion
2n = 16   r-d3

Allium macranthum   2n = 14,28   r-d3

Allium margaritaceum   2n = 16,32   r-d3

Allium materculae   2n = 16   r-d3

Allium meteoricum   2n = 16   r-d3

Allium modestum   2n = 16   r-d3

Allium moly   lily leek   2n = 14   b-apomictic
r-d3

Allium moschatum   2n = 16   r-d3

Allium neapolitanum   daffodil onion
2n = 14,21,35,28   r-d3

Allium nipponicum   2n = 16,32   r-d3,z

Allium nutans   2n = 16–108!   r-d3

Allium odorum   fragrant-flowered garlic, Chinese
leek flower   l-p,h   2n = 16,32   r-d3,f3

Allium oleraceum   field garlic   2n = 32,40
USA-W-6   r-d3,f3,l

Allium pallasii   (= *A. pallyssium*)   2n = 16

Allium pendulinum   2n = 14,(18!)   r-d3

Allium porrum   see *A. ampeloprasum*

Allium proliferum   2n = 16   r-d3

Allium pseudoflavum   2n = 18   r-d3

Allium pyrenaicum   2n = 16   USA-W-6   r-d3,l

Allium roseum   2n = 32   r-d3

Allium roseum var. bulbiferum   2n = 48   r-d3

Allium rotundum   2n = 32   r-d3

Allium sativum   garlic   l-p/a,h   2n = 16(r-b)
2n = 16,48(r-d)   2n = 16(r-z)   2n =
16(r-d3)   USA-W-6   r-b,d,d3,f3,j,l,s2,z

Allium sativum var. ophioscordon   (see also *A.
scorodoprasum*)   rocambole garlic   USA-W-6
r-f3

Allium schoenoprasum   chives   l-p,h
2n = 26,24,32(r-d)   2n = 16(24,32)(r-z)
2n = 16,24,32(r-d3)   b-o   f-h   s-i   v-bees
USA-W-6   r-d,d3,f,f3,l,z

Allium scorodoprasum   (see also *A. sativum* var.
ophioscorodon)   rocambole   2n = 16,24
USA-W-6   r-d3,l

Allium scorzoneraefolium   2n = 14   r-d3

Allium senescens   2n = 32,48   USA-W-6   r-d3,
f3,l

Allium sikkimense   2n = 32   r-d3

Allium splendens 2n = 32 l-h USA-W-6
  r-d3,f3,l
Allium spp. For uses and sources of these and
  additional wild and cultivated species, see r-f3.
Allium spp. See pp. 228–230.
Allium stellatum prairie onion 2n = 14
  r-d3,f3
Allium tartaricum tartar onion 2n = 28–32
  r-d3
Allium tel-avivense 2n = 16 r-d3
Allium thunbergii 2n = 16 r-d3,f3
Allium triquetrum 2n = 18 r-d3,f3
Allium tuberosum Chinese chives, leek, or leek
  flower; gynmigit l-p,h 2n = 16,24,32(r-d)
  2n = 32(r-d3) 2n = 16,32(r-z) b-o
  USA-W-6 r-d,d3,f3,l,z
Allium ursinum ramsons 2n = 14 r-d3,f3
Allium validum Pacific onion, swamp onion
  l-p,h 2n = 28 r-d3,f3
Allium victorialis l-h 2n = 16,32 USA-W-6
  r-d3,f3,l
Allium vineale crow garlic l-p,h 2n = 32,40
  r-d3,f3
Allium viviparum 2n = 16 r-d3
Allium yunnanense (= *A. mairei*) 2n = 16,32
  r-d3
Allium zebdanense 2n = 18 r-d3
Alocasia indica (= *A. macrorrhiza*?) Araceae
  2n = 28 r-s2
Alocasia macrorrhiza giant taro l-p/a,h
  2n = 26,28 r-d,l,s2
Althaea officinalis Malvaceae marsh mallow
  l-p,h 2n = 42(about 40–44) USA-NC-7
  r-d,f3,l,z
Amaranthus angustifolius Amaranthaceae l-a,h
  2n = 32(34) r-z
Amaranthus caudatus Inca wheat, quihuicha
  l-a,h 2n = 32,34 USA-NC-7 r-d,d2,
  d3,f3,l,z
Amaranthus caudatus var. love-lies-bleeding
  l-a,h y-0.5 lb/sq.yd. r-J. L. Hudson catalog
Amaranthus cruentus African, Chinese, Spanish,

or bush spinach or greens, amaranth l-a,h
  2n = 32,34 b-o v-wind d2-500m y-up
  to 2,000 kg/ha USA-NC-7 r-d,f3,g,l,z
Amaranthus dubius African, Chinese, Spanish, or
  bush spinach or greens, amaranth l-a,h
  2n = 64 (Might be a tetraploid form of *A.
  cruentus*.) b-o v-wind d2-500m y-up to
  2,000 kg/ha USA-NC-7 r-f3,g,l,z
Amaranthus gangeticus See *A. tricolor*
Amaranthus hybridus slim amaranth 2n = 32
  USA-NC-7 r-f3,z
Amaranthus hypochondriacus (= *A. leucocarpus*)
  grain amaranth, huazontle, princess feather
  l-a,h 2n = 32,34(r-d) 2n = 32(r-z)
  USA-NC-7 r-d,f3,l,z
Amaranthus lividus livid amaranth l-a,h
  2n = 34 USA-NC-7 r-d,f3,l,z
Amaranthus mangostanus See *A. tricolor*
Amaranthus mantegazzianus (= *A. edulis* = *A.
  caudata* ssp. mantegazzianus) 2n = 32
  r-d3,f3,z
Amaranthus paniculatus (= *A. cruentus*?)
  2n = 32 r-f3,z
Amaranthus spinosus thorny pigweed
  2n = 32(r-z) 2n = 34(r-d3) USA-NC-7
  r-d3,f3,l,z
Amaranthus spp. Most amaranths are b-i,o and
  v-wind, perhaps also bees with species with
  colorful flowers. Grain species are usually
  b-i,o f-m s-c. Arrangement of flowers and
  flowering pattern usually produces a mix of
  self- and cross-pollination. "Each of the many
  cymes of the inflorescence is initiated by a single staminate flower followed by an indefinite
  number of pistillate flowers, often over a
  hundred. Stigmas of the earliest pistillate flowers are receptive before the staminate flower
  opens; most of the later pistillate flowers
  develop after the staminate flower has ab-
  scissed. However, cymes of different ages are
  present on each indeterminate inflorescence
  and pollen transfer among them probably

makes selfing more common than crossing"
(r-s2).

Amaranthus spp.   grain types   y2-16 lb/100
sq.ft.max.   r-j

Amaranthus tricolor   (= *A. gangeticus* = *A.
melancholicus* = *A. tricolor* var. mangostanus)
African, Chinese, Spanish, or bush spinach or
greens, amaranth   l-a,h   2n = 32,34
2n = 32(r-z)   2n = 34(r-d3)   b-o   v-wind
d2-500m   y-up to 2,000 kg/ha   USA-NC-7
r-d,d3,f3,g,l,z

Amaranthus viridis   green amaranth   2n = 34
r-f3,d3

Amphicarpaea monoica   (= *Falcata comosa*)
Fabaceae   hog pea, ground peanut   2n = 20
r-f3,z

Anacardium occidentale   Anacardiaceae   cashew
l-p,t   2n = 40,42   USA-CR-MAY   r-d,f3,l

Ananas comosus   Bromeliaceae   pineapple   l-p,h
2n = 50,75   b-o   f-h   s-i   v-hummingbirds
USA-CR-MIA   r-d,f,f3,l

Anchusa officinalis   Boraginaceae   true bugloss
l-b,p,h   2n = 16   USA-NC-7   r-d3,f3,l

Anethum graveolens   Apiaceae   dill   l-a,b,h
2n = 22   USA-NC-7   r-d,f3,l

Angelica archangelica   Apiaceae   angelica
l-b,p,h   2n = 22   r-d,f3,l

Annona cherimola   Annonaceae   cherimoya
l-p,t   2n = 14,16   b-o   f-h   flowers
strongly protogynous   v-beetles   USA-CR-
MIA   r-d,f,f3,l

Annona muricata   soursop   l-p,s,t   2n = 14,16
r-d,f3,l

Annona reticulata   custard apple   l-p,s,t
2n = 14,16   r-d,f3,l

Annona squamosa   sugar apple   l-p,t
2n = 14,16,28   USA-CR-MIA   r-d,f3,l

Anthriscus cerefolium   Apiaceae   chervil   l-b,p/
a,h   2n = 18   r-d,d3,f3,l

Anthriscus sylvestris   woodland chervil, cow parsley   l-p,h   2n = 16,18   r-d3,f3

Apios americana   (= *A. tuberosa* = *Glycine apios*)
Fabaceae   American groundnut   l-p,v
2n = 22   b-o   s-i   v-insects   r-f3,z,
Blackmon & Reynolds 1986, Reynolds, et al.
in Janick & Simon, eds., *Advances in New
Crops* 1990

Apium carvi   see *Carum carvi*

Apium graveolens var. *dulce*   Apiaceae   celery
l-p/b?,a,b,h   2n = 22   b-o   f-h
flower protandrous   x-30% by insects
s-c   v-insects   d2-500m or more   y-500 kg/
ha   y2-9.9 lb/100 sq.ft.max.   USA-NE-9
r-d,f,f3,g,j,l,r

Apium graveolens var. *rapaceum*   celeriac   l-a,b,h
d2-500m or more   USA-NE-9   r-d,f3,g

Apium graveolens var. *secalinum*   leaf celery
r-f3, r

Arachis hypogaea   Fabaceae   groundnut, peanut
l-a,h   2n = 4x = 40   b-i   f-h   x-occasion-
ally by bees   d-0   y-1 tonnes/ha   y2-24 lb/
100 sq.ft.max.   USA-S-9   r-d,d2,f,f3,g,j,l

Aralia cordata   Araliaceae   udo   l-p,h
2n = 28   USA-S-9   r-d,f3,l,z

Arctium lappa   Asteraceae   great burdock, edible
burdock, gobo   l-p/a,b/a,h   2n = 32,36
r-d,f3,l

Armoracia rusticana   (= *Cochlearia armoracia*)
Brassicaceae   horseradish   l-p,h   2n =
32(r-s2)   2n = 28,32(r-d,z)   Propagated vegetatively.   Highly seed sterile.   r-d,f3,j,l,s2

Artemisia abrotanum   Asteraceae   southernwood
l-p,s   2n = 18   r-d,l

Artemisia dracunculus   tarragon   l-p,h
2n = 18,36,54   r-d,f3,l

Artemisia vulgaris   mugwort   l-p,b,h
2n = 18,16,36   USA-W-6   r-d,f3,l

Artocarpus altilis   Moraceae   breadfruit   l-p,t
2n = 56?,54?,81   r-d,l

Artocarpus heterophyllus   jackfruit   l-p,t
2n = 56?   USA-CR-MIA   r-d,l

Asclepias cornutii   see *A. syriaca*

Asclepias curassavica Asclepiadaceae blood-flower l-p 2n = 22(r-d3) r-d3,l

Asclepias incarnata milkweed 2n = 22 r-d3,f3

Asclepias spp. 2n = 22 for all eight species listed in d-3 USA-S-9 r-d3,l

Asclepias speciosa milkweed, showy milkweed l-p,h 2n = 22 r-d3,f3

Asclepias syriaca (= *A. cornutii*) common milk-weed, silkweed l-p,h 2n = (20),22,(24) (r-z) r-d3,f3,l,z

Asclepias tuberosa pleurisy root, butterfly weed l-p,h 2n = 22(r-d3) USA-W-6 r-d3,f3,l

Asimina triloba Annonaceae pawpaw l-p,t 2n = 18,16,2x? USA-NC-7 r-d,l

Asparagus acutifolius Liliaceae wild Mediterra-nean asparagus 2n = ? USA-NC-7 r-b,f3,l

Asparagus densiflorus sprenger asparagus l-p,h 2n = 60 USA-NC-7 r-d,l

Asparagus maritimus 2n = ? USA-NC-7 r-b,l

Asparagus officinalis asparagus l-p,h 2n = 20,40 b-o f-d However, male flow-ers have vestigal stigmas that sometimes set an occasional seed; female flowers have nonfunc-tional vestigal stamens. Also, andromonoecious plants (which have perfect and male flowers on the same plant) exist; their anatomy sug-gests that they are largely self-pollinating (r-b). v-insects Fresh pollen can be stored in vials in a home freezer for six months; longer with lower temperature and/or control of relative humidity (r-b). USA-NC-7 r-b,d,f,f3,l,z

Asparagus springeri 2n = ? Resistant to *Fusar-ium* but doesn't cross with *A. officinalis.* r-b

Asperula odorata Rubiaceae sweet woodruff l-p,h 2n = 44 r-d,f3,l

Atriplex canescens Chenopodiaceae fourwing saltbush l-p,s USA-W-6 r-d,f3,l

Atriplex hortensis orach, mountain or German spinach, butter leaves, sea purslane l-a,h 2n = 18 USA-W-6 r-d,f3,l

Atriplex spp. saltbush b-o f-d v-wind USA-W-6 r-f,l

Avena spp. Poaceae oat b-i f-h Largely cleistogamous; upper flower usually imperfect. x-up to 10% by wind d-0 y2-13.2+ lb/100 sq.ft.max. USA-NSGC r-f,f3,j,l

Averrhoa bilimbi Oxalidaceae bilimbi l-p,t 2n = 22,24 r-d,f3,l

Averrhoa carambola carambola l-p,t 2n = 22,24 USA-CR-MIA r-d,f3,l

Bactris gasipaes Arecaceae peach palm l-p,t 2n = 30 r-d,f3,l

Balanites aegyptiaca Zygophyllaceae desert date l-p,t 2n = 16,18 r-d,f3,l

Barbarea verna (= *B. praecox*) Brassicaceae creasy greens, upland or winter cress l-b,p,h 2n = 16 r-f3,z

Barbarea vulgaris yellow rocket, common winter cress, upland cress l-b,p/a,h 2n = 14,16 2n = 16(r-z) r-d,f3,l,z

Basella alba Bassellaceae Malabar spinach l-p/b,a,h 2n = 48,44 USA-S-9 r-d,f3,l

Bellis perennis Asteraceae daisy b-i f-h d-f,f3,l

Benincasa hispida Cucurbitaceae waxgourd, winter melon l-a,h,v 2n = 24 b-o f-m USA-NE-9 r-d,f3,l,s2(p306)

Beta vulgaris ssp. cycla Chenopodiaceae Swiss chard, chard, spinach beet l-a,b/a,h b-o f-h flower protandrous s-i v-wind d-2km; 5km from ssp. esculenta y-1.1 tonnes/ha; 2 max. y2-29 lb/100 sq.ft.max. USA-NC-7 r-d,f,f3,g,j,l

Beta vulgaris ssp. esculenta sugar, fodder, or red beet; beetroot; mangold; mangel l-a,b/a,h 2n = 18,27,36 b-o f-h flower protan-drous s-c,i v-wind d2-500–1,000m; 5km from ssp. cycla y-1,000 kg/ha; 2,000 max. USA-NC-7 r-d,f,f3,g,j,l

Bombax ceiba Bombacaceae red silk cotton l-p,t 2n = 46 r-d,f3,l

Borago officinalis Boraginaceae borage l-a,h
2n = 16 USA-W-6 r-d,f3,l

Borassus flabellifer Arecaceae palmyra palm
l-p,t 2n = 36 r-d,f3,l

Brassica campestris ssp. chinensis (= *B. chinensis*
= *B. rapa* ssp. chinensis) Brassicaceae
Chinese mustard, pak choy, pe-tsai, Chinese
leaf cabbage l-b/a,h 2n = 20 gf-AA
USA-NE-9 r-f2,l,r

Brassica campestris ssp. dichotoma (= *B. napus*
ssp. dichotoma) toria 2n = 20 gf-AA
USA-NC-7 r-f2,l

Brassica campestris ssp. eu-campestris bird rape,
wild turnip rape l-a,h 2n = 20 gf-AA
r-d,f2,l

Brassica campestris ssp. nipposinica (= *B. nipposinica*) curled mustard 2n = 20 gf-AA

Brassica campestris ssp. oleifera (= *B. rapa* ssp.
oleifera) turnip rape 2n = 20 gf-AA
USA-NC-7 r-d,f2,l

Brassica campestris ssp. pekinensis (= *B. pekinensis*) Chinese cabbage, Chinese head cabbage, celery cabbage l-b/a,h 2n = 20
gf-AA b-o f-h s-i v-bees y2-6.1
lb/100 sq.ft.max. USA-NE-9 r-d,f,f2,j,l

Brassica campestris ssp. rapifera (= *B. rapa*)
turnip l-b,h 2n = 20 gf-AA b-o f-h
s-i v-bees, insects, wind d2-1,000m y-
1,500 kg/ha; 1,500 max. y2-14.7 lb/100
sq.ft.max. USA-NC-7 r-d,f,f2,g,j,l

Brassica campestris ssp. ruvo? (= *B. rapa* ssp.
ruvo? = *B. ruvo?*) broccoli raab (sprouting
turnip) The Seed Savers Exchange lists broccoli raab as *B. rapa,* Ruvo group, and refers to
it as a sprouting turnip. However, the taxonomy
I am following here calls turnips *B. campestris.*
Most sources don't list broccoli raab at all.

Brassica campestris ssp. sarson sarson 2n = 20
gf-AA r-f2

Brassica carinata Abyssinian mustard 2n = 34
gf-BBCC USA-NC-7 r-f2,l

Brassica hirta (= *Sinapsis alba*) white mustard,
yellow mustard l-a,h 2n = 24 gf-DD
r-d,f2,l

Brassica juncea mustard, brown mustard, Oriental mustard, leaf mustard, Indian mustard
l-p/a,b,h 2n = 36 gf-AABB b-i f-h s-c
x-about 20% (r-f2) v-bees y2-5.7 lb/100
sq.ft.max. USA-NC-7 r-d,f,f2,j,l,r

Brassica kaber (= *Sinapsis arvensis*) charlock,
wild mustard 2n = 18 gf-SS r-f,f2,l

Brassica napus ssp. oleifera rape l-a,b,h
2n = 38 gf-AACC b-o f-h x-more than
10% s-c v-bees d-40m y2-5.4 lb/100
sq.ft.max. USA-NC-7 r-d,f,f2,j,l

Brassica napus ssp. pabularis rape-kale
2n = 38 gf-AACC USA-NC-7 r-f2,l

Brassica napus ssp. rapifera (= *B. napobrassica* = *B. napus* ssp. napobrassica) rutabaga,
swede l-b,a,h 2n = 38 gf-AACC b-i
f-h x-about 20% (r-f2) s-c v-bees, insects,
wind d2-1,000m y-1,500 kg/ha; 2,500
max. y2-5.4 lb/100 sq.ft.max. USA-NC-7
r-d,f,f2,g,j,l

Brassica nigra black mustard l-a,h 2n = 16
gf-BB USA-NC-7 r-d,f2,l

Brassica oleraceae ssp. acephala kale, collards
l-b/a,a,h 2n = 18 gf-CC b-o f-h s-i
v-bees, flies d2-1,000–1,500m from all
B. oleraceae varieties y2-(kale)3.8 lb/100
sq.ft.max. USA-NC-7 r-b,d,f,f2,g,j,l

Brassica oleraceae ssp. botrytis cauliflower, broccoli l-b,h 2n = 18 gf-CC b-o f-h s-i
However, summer cauliflowers self-pollinate
readily compared with other *B. oleraceae* subspecies. v-bees, flies d2-1,000–1,500m
from all *B. oleraceae* subspecies y-400
kg/ha y2(broccoli)-5.5 lb/100 sq.ft.max.
y2(cauliflower)-1 lb/100 sq.ft.max. USA-
NE-9 r-d,f,f2,g,j,l

Brassica oleraceae ssp. capitata cabbage l-b/a,h
2n = 18 gf-CC b-o f-h s-i v-bees, flies

d2-1,000–1,500m from all *B. oleraceae* subspecies   y-700 kg/ha   y2-3.6 lb/100 sq.ft.max.
USA-NE-9   r-b,d,f,f2,g,j,l

Brassica oleraceae var. gemmifera   Brussels sprouts   l-b/a,h   2n = 18   b-o   f-h   s-i v-bees, flies   d2-1,000–1,500m from all *B. oleraceae* subspecies   y-600 kg/ha   y2-2.8 lb/100 sq.ft.max.   USA-NE-9   r-d,f,g,j,l

Brassica oleraceae ssp. gongylodes   kohlrabi l-b,h   2n = 18   gf-CC   b-o   f-h   s-i v-bees, flies   d2-1,000–1,500m from all *B. oleraceae* subspecies   y-700 kg/ha   y2-20.1 lb/100 sq.ft.max.   USA-NE-9   r-d,f,f2,g,j,l

Brassica oleraceae ssp. italica   sprouting broccoli, asparagus broccoli   l-p,h   2n = 18   gf-CC v-bees, flies   d2-1,000–1,500m from all *B. oleraceae* subspecies   USA-NC-7   r-d,f2,g,l

Brassica oleraceae ssp. sylvestris   wild cabbage 2n = 18   r-f2

Brassica rapa   see *B. campestris*

Brassica spp.   See pages 230–234

Brassica spp.   See r-f3 for uses of brassicas.

Brassica tournefortii   wild turnip   2n = 20 gf-AA   r-f2

Brosimum alicastrum   Moraceae   ramon   l-p,t 2n = 26   r-d

Bunias orientalis   Apiaceae   Turkish rocket l-p,h   2n = 14,44?   r-d3,f3

Bunium bulbocastanum   Apiaceae   earth chestnut l-p,h   2n = ?   r-f3

Cajanus cajan   (= *C. indicus*)   Fabaceae   pigeon pea   l-a,p,p/a,s   2n = 22,44,66   b-o   f-h s-c   v-honeybees   x-13–65%   y-1.2 tonnes/ ha   y-600 kg/ha ave.; 2,500 max. (r-d2) y2-24 lb/100 sq.ft.max.   USA-S-9   r-d,d2, f,f3,g,j,l

Calendula officinalis   Asteraceae   pot marigold l-a,h   2n = 32,28   USA-NC-7   r-d,f3,l

Caltha palustris   Ranunculaceae   marsh marigold, American cowslip   l-p,h   2n = 28-60 USA-NC-7   r-d,f3,l

Camassia cusickii   Liliaceae   2n = 30   r-d3,f3

Camassia leichtlinii   (= *C. esculenta*)   camas, quamash   l-p,h   2n = 30   r-d3,f3,z

Camellia sinensis   Theaceae   tea   l-p,t 2n = 30,45,60   b-o   f-h   s-i   v-insects r-d,f,f3

Campanula rapunculus   Campanulaceae   rampion, ramps   l-b,p/a,h   2n = 68,102(r-d) 2n = 20,102(r-z)   r-d,f3,z

Canarium indicum   Burseraceae   java almond l-p,t   r-d,f3

Canarium ovatum   pili nut   l-p,t   b-o   f-d v-insects   r-f,f3

Canavalia ensiformis   Fabaceae   jackbean   l-a,p,p/a,h   2n = 22   b-i   x-20% by bees   s-c y2-700–5,400 kg/ha   USA-S-9   r-d,d2,f,f3,l

Canavalia gladiata   swordbean   l-a,p,p/a,h,l 2n = 22,44   x-more than 20%   s-c   v-bees y-1.3 tonnes/ha   y2-700–1,500 kg/ha USA-S-9   r-d,d2,f3,g,l

Canavalia plagiosperma   oblique jackbean   l-a,h 2n = 22   r-d,d2

Canna edulis   Cannaceae   edible canna, Australian arrowroot, achira   l-p,h   2n = 18,27 r-d,f3

Capparis spinosa   Capparaceae   caper   l-p,s 2n = 24,38   r-d,f3

Capsicum annum   Solanaceae   pepper   l-a,h b-i   f-h   Protogynous; anthers dehisce two to three days after flowers open.   x-5–10% by bees, thrips (r-f); up to 68% in India (r-g) Some will cross with *C. frutescens*.   d-30m y-100–200 kg/ha   USA-S-9   r-d,f,f3,g,l

Capsicum baccatum var. pendulum   pimentchien l-a,h   2n = 24   USA-S-9   r-d,f3,l

Capsicum chinense   aji   l-a,h   2n = 24   USA-S-9   r-d,f3,l

Capsicum frutescens   pepper, tabasco pepper   l-p/a,h,s   2n = 24   b-i   f-h   Protogynous; anthers dehisce two to three days after flowers open.   x-7–36% by bees, thrips (r-f); up to

68% in India (r-g) Some will cross with *C. annum.* USA-S-9 r-d,f,f3,g,l

Capsicum pubescens rocoto l-p,h,s 2n = 24 USA-S-9 r-d,f3,l

Capsicum sp. cayenne peppers y2-0.1 lb/100 sq.ft.max. r-j

Capsicum sp. sweet/green peppers y2-0.3 lb/100 sq.ft.max. r-j

Carduus marianum see *Silybum marianum*

Carica papaya Caricaceae papaya l-p,s,t 2n = 18,36 b-o f-d v-wind, insects USA-CR-HIL r-d,f,f3,l

Carica pubescens mountain papaya l-p,t 2n = 18 r-d,f3

Carissa carandas Apocynaceae karanda l-p,s,t 2n = 22 r-d,f3,l

Carissa edulis Egyptian carissa l-p,l,s 2n = 22 r-d,f3

Carissa macrocarpa natal plum l-p,s 2n = 22 r-d,f3

Carthamus tinctorius Asteraceae safflower l-a,h 2n = 24,32 b-o f-h s-c v-bees d-5–90m y2-17.4 lb/100 sq.ft.max. USA-W-6 r-d,f,f3,j,l

Carum carvi (= *Apium carvi* = *Seseli carvi*) Apiaceae caraway l-p/a,b,h 2n = 20,22 USA-NC-7 r-d,f3,l

Carya illinoensis Juglandaceae pecan l-p,t 2n = 32 b-o f-m Variable as to whether stigma becomes receptive before or after pollen is shed. s-c v-wind USA-PIO r-d,f,f3,l

Caryota urens Arecaceae fishtail palm l-p,t 2n = 32 r-d,f3

Casimiroa edulis Rutaceae Mexican apple l-p,t 2n = 36 r-d,f3

Castanea pumila Fagaceae Allegheny chinquapin l-p,s,t USA-CR-BRW r-d,f3,l

Castanea spp. American, Japanese, and Chinese chestnuts l-p,s,t 2n = 22,24 b-o f-m s-i v-insects USA-CR-BRW r-d,f3,l

Ceiba pentandra Bombacaceae kapok l-p,t 2n = 80,88,72 USA-CR-MIA r-d,f3,l

Celastrus scandens Alastraceae American bittersweet b-o f-d r-f,f3,l

Celosia argentea Amaranthaceae cock's comb, green soko, white soko v-insects y-up to 600 kg/ha USA-NC-7 r-f3,g,l

Centranthus angustifolius Valerianaceae 2n = 14 r-d3

Centranthus gilloti 2n = 14 r-d3

Centranthus ruber (= *Kentranthus ruber*) Jupiter's beard, red valerian l-p,h 2n = 14 USA-NC-7 r-d3,f3,l

Ceratonia siliqua Fabaceae carob l-p,t 2n = 24 b-o f-d v-wind, insects USA-S-9 r-d,d2,f,f3,l

Ceratotheca sesamoides Pedaliaceae bungu l-a,h 2n = 32 USA-S-9 r-d,f3,l

Chaerophyllum bulbosum Apiaceae turnip-rooted chervil l-b,p/a,h 2n = 22 r-d,f3,l

Chenopodium album Chenopodiaceae lamb's-quarters, mutton tops, fat hen l-a,h 2n = 18,36,54 USA-NC-7 r-d,f3,j,l,z

Chenopodium ambrosioides epazote, Mexican tea, American wormweed, Indian wormweed l-a,p,h 2n = 16,32,36,64 r-d,f3,z

Chenopodium bonus-henricus (= *C. esculentus*) Good King Henry, allgood, mercury, fat hen l-p,h 2n = 36 r-d3,f3,z

Chenopodium bonus-henricus b-o f-h Strongly protogynous. Pistils usually dry up before anthers emerge and dehisce. On smaller flowers on a scape, anthers may not emerge at all. A scape that has just started to flower has emerged pistils on dozens of the earliest flowers but no emerged stamens. v-wind r-po

Chenopodium botrys ambrosia, Jerusalem oak, feather geranium 2n = 18 r-f3,z

Chenopodium capitatum (= *C. foliosum*) strawberry blite, beetberry l-a,h 2n = 16,18 USA-NC-7 r-f3,l,z

Chenopodium esculentus   see *C. bonus-henricus*
Chenopodium foliosum   see *C. capitatum*
Chenopodium hybridum   2n = 18,36   r-d3
Chenopodium nuttalliae   (= *C. berlandieri?*)
huazontle   2n = 36   Seems to be just a late-
maturing *C. quinoa* with edible flower/seed
heads (a chenopodial broccoli). Breeding system presumably like that of quinoa.
r-f3,s2,z

Chenopodium pallidicaule   (= *C. cañihua*)   kan-
iwa,   l-a,h   2n = 18(r-s2 and others)
2n = 36(r-z)   b-i   f-h   cleistogamous   x-
rare; about 0.1% in greenhouse (r-s2)   y-up
to 2,400 kg/ha; 5,000 in experimental plots
USA-NC-7   r-f3,l,p,s2,z,*Lost Crops of the Incas*
1989(pp134–136)

Chenopodium petiolare   l-p,h   r-*Lost Crops of the
Incas* 1989(footnote, p137)   (If anyone has
any of this, please send me some. I want to try
it as another source for introducing perennial-
ism into chenopodial grains and vegetables.)

Chenopodium quinoa   quinoa, quinua, petty rice
l-a,h   2n = 36   b-i   Plants are gynomonoe-
cious — that is, they have a mixture of hermaphrodite and female (pistillate) flowers.
x-2–9%   Preceeding information from s2.
Simmonds has studied the breeding system of
chenopodiums. Contradictory breeding information exists.   r-f lists *C. quinoa* as b-o   f-h
v-wind.   J. Rea, in *Lost Crops of the Incas* 1989
(p159) is quoted in a one-line footnote as saying it is b-i + f-h   usually, but   f-m   in "a
few cases."   s2 notes that there is a report of
one population carrying a cytoplasmic/genetic
male sterile, so outbreeding must be common in at least that population.   USA-NC-7
r-d,f,f3,l,s2,z,*Lost Crops of the Incas* 1989
(p159)

Chrysanthemum coronarium   shungiku, garland
chrysanthemum   l-a,h   2n = 18,36   b-i
f-h   (except ray florets, which are pistillate)

x-about 10%   y-130 kg/10 a; 200 max. (r-s)
USA-NC-7   r-d,f3,l,s,z

Chrysophyllum cainito   Sapotaceae   star apple
l-p,t   2n = 24,26   r-d,f3

Cicer arietinum   Fabaceae   chickpea, garbanzo
l-a,h   2n = 16   b-i   f-h   Cleistogamous;
self-pollination occurs one to two days before
anthesis. "Flowers visited by bees, but cross-
pollination is rare." x-small amount by bees
d3-3 feet or whatever is needed to prevent
physical mixture of different varieties; isolation
distance needs established at Pullman, Washington; under those conditions, chickpea is
one of the most highly inbreeding species
(r-pc, Rich Hannan, curator of USDA *Cicer* collection, USA-W-6). Ashworth, in *Seed to Seed*
(1991), says that "several researchers" say that
cross-pollination occurs commonly and
chickpeas cross readily under organic
growing conditions (r-p0). One-half mile is
reasonable. y = 400–1,600 kg/ha ave.; can
exceed 2,000 kg; has reached 5,200 on
experimental plots (r-d2) y2-24 lb/100
sq.ft.max.   USA-W-6
r-d,d2,f,f2,f3,j,1,pc(Rich Hannan),s2

Cichorium endivia   Asteraceae   endive   l-a,b,h
2n = 18,36   b-i   f-h   x-15% by flies   d-
10m   Can cross with *C. intybus* (r-r).   y-500
kg/ha; 1,000 max.   USA-NC-7   r-d,f,f3,g,l,r

Cichorium intybus   chicory   l-p,p/b,h   2n = 18
b-o   f-h   s-i   d2-1,000m   Can cross with
C. endivia (r-r).   y-1 tonne/ha   USA-NC-7
r-d,f,f3,g,l,r

Citrullus colocynthis   Cucurbitaceae   colocynth
l-a,p,h,v   2n = 22,24   USA-S-9   r-d,f3,l

Citrullus lanatus   (= *C. vulgaris*)   watermelon
l-a,h,v   2n = 22   b-o   Most cvs. f-m; some,
especially older cvs., andromonoecious — that
is, have a combination of hermaphrodite and
male flowers on each plant.   s-c   v-bees   d-
800m   y-400 kg/ha for cvs. with normal seed

yield; 250 kg/ha for 'Charleston Grey' y2-2.6 lb/100 sq.ft.max.   USA-S-9   r-b,d,f,f3,g,j,l,s2

Citrus spp. Rutaceae citrus   b-i   f-h   Many cvs. apomictic. USA-CR-RIV or USA-PIO r-f,f3,l

Clausena dentata   Rutaceae   clausena   l-p,s   2n = 18   r-d

Clausena lansium   wampi   l-p,t   2n = 18   r-d,f3

Clitoria laurifolia   Fabaceae   laurel-leaved clitoria, butterfly pea   l-p,s   2n = 24   USA-S-9   r-d,d2,f3

Clitoria ternatea   butterfly pea   l-p,v   2n = 16   b-i   s-c   USA-S-9   r-d,d2,f3,l

Coccoloba uvifera   Polygonaceae   seagrape   l-p,s,t   2n = 80,132   r-d,f3

Cochlearia armoracia   see *Armoracia rusticana*

Cocos nucifera   Arecaceae   coconut   l-p,t   2n = 32   b-o   f-m   plant protandrous x-depends on cv.   v-wind, insects   r-d,f,f3

Coleus amboinicus   Lamiaceae   Indian borage   l-p,h   2n = 68,32   r-d

Coleus parviflorus   ratala   l-a,h   2n = 56,64   r-d,f3

Colocasia esculenta   Arecaceae   taro, cocoyam, l-p/a,h   2n = 28   3n = 42   USA-CR-MAY   r-d,f3,l,s2,y

Corchorus capsularis   Tiliaceae   white jute   l-a,h   2n = 14,28   USA-S-9   r-d,f3,l

Corchorus olitorius   tussa jute   l-a,h   2n = 14,28   USA-S-9   r-d,f3,l

Cordyline terminalis   Agavaceae   ti palm   l-p,s   2n = 152+   r-d,f3

Coriandrum sativum   Apiaceae   coriander   l-a,h   2n = 22   USA-NC-7   r-d,f3,l

Corydalis solida   (= *C. bulbosa*)   Fumariaceae   2n = 16,32,24   r-d3

Corylus spp.   Corylaceae   hazelnut, filbert   l-p,s,t   2n = 28,22   b-o   f-m   s-i   v-wind   USA-CR-COR   r-d,f,f3,l

Crambe abyssinica   Brassicaceae   crambe, Abyssinian kale   l-a,h   2n = 90   b-i mostly, but some outcrossing occurs   f-h   USA-NC-7   r-d,d3,f2,l

Crambe cordifolia   (= *C. tataria*)   colewort, Tatarian sea kale   l-p,h   2n = about 120   USA-NC-7   r-d3,f3(p50),l,z

Crambe filiformis   2n = 30   USA-NC-7   r-d3,l

Crambe fruticosa   2n = 30,60   r-d3,l

Crambe grandiflora   2n = about 120   USA-NC-7   r-d3,l

Crambe hispanica   Spanish colewort   l-a,h   2n = 60   r-d3,z

Crambe hispanica var. glabrata   2n = 30   USA-NC-7   r-f2,l

Crambe hispanica var. hispanica   2n = 60   USA-NC-7   r-f2,l

Crambe kralikii   2n = 30,60,90   USA-NC-7   r-f2,l

Crambe maritima   sea kale   l-p,h   2n = 30,60(r-d)   2n = 60(r-z)   2n = 30,60(r-d3)   b-o   f-h   s-i   v-bees   d-200m   USA-NC-7   r-d,f,f2,f3(p50),l,z,Péron in Janek & Simon, eds., *Advances in New Crops* 1990

Crambe orientalis var. juncea   2n = about 120   r-d3

Crambe orientalis var. koktebelica   2n = 30   r-d3,f3(p51)

Crambe tataria   see *C. cordifolia*

Crambe tatarica   Tartar bread plant   2n = 60,120   r-d3,f3(p51)

Crithmum maritimum   Apiaceae   samphire   2n = 20,22   r-d3,f3

Crocus sativus   Iridaceae   saffron crocus   l-p,h   2n = 14,15,16   USA-W-6   r-d,f3,l

Cryptotaenia japonica   Apiaceae   mitsuba, mitsube, Japanese honewort, celery, or parsley   l-p,h   2n = (18),20,22   r-f3,y,z

Cucumis anguria   Cucurbitaceae   West Indian gherkin   l-a,h   2n = 22,24   USA-NC-7   r-d,f3,l

Cucumis melo   cantaloupe, melon, muskmelon   l-a,v   2n = 24(r-s2)   2n = 20,22,24(r-d)

b-o   f-m or with perfect and male flowers on
one plant   plant protandrous   x-0–100%
s-c   v-bees   d2-500–1,000m   y-300 kg/ha;
600 max. y2-2.9 lb/100 sq.ft.max. Note that
'Armenian Cucumber' is a *C. melo,* not a
true cucumber (*C. sativus*).   USA-NC-7
r-d,f,f3,g,j,l,s2

Cucumis metuliferus   jelly melon, African
horned melon   l-a   2n = 24   USA-NC-7
r-d3,f3,l

Cucumis sativus   cucumber   l-a,h   2n = 14
b-o   f-m,h   plant protandrous   Some modern cvs. have only or almost only female flowers under normal growing conditions.   x-70%
s-c   v-bees   Does not cross with muskmelons, watermelons, squash, pumpkins, or
gourds. (That's a common myth.) However,
note that the 'Armenian Cucumber' is not a
cucumber but a *C. melo,* so it will cross with
muskmelons and other *C. melo* subspecies.
d2-1,000–1,500m   y-400 kg/ha; 700
max. y2-4.1 lb/100 sq.ft.max.   USA-NC-7
r-b,d,f,f3,g,j,l,s2

Cucurbita ficifolia   (= *C melanosperma*)   Cucurbitaceae   Malabar, buffalo, or figleaf gourd
l-p/a,v   2n = 40(r-s2)   2n = 40,42(r-d)
USA-S-9   r-d,d3,f3,l,s2

Cucurbita maxima   squash, marrow, pumpkin
l-a,b,h,v   2n = 40(r-s2)   2n = 40,44,48(r-d)
b-o   f-m   s-c   v-bees   d2-1,000–1,500m
y-500 kg/ha; 1,000 max.   y2-(winter
squash)5.7 lb/100 sq.ft.max.   USA-NE-9
r-b,d,f,f3,g,j,l,s2

Cucurbita melanosperma   see *C. ficifolia*

Cucurbita mixta   squash, cushaw, marrow, pumpkin   l-a,v   2n = 40   d2-1,000–1,500m
y-500 kg/ha; 1,000 max.   r-b,d,f3,g,l,s2

Cucurbita moschata   squash, marrow, pumpkin
l-a,v   2n = 40(r-s2)   2n = 24,40,48(r-d)
d2-1,000–1,500m   y-500 kg/ha; 1,000 max.
USA-S-9   r-b,d,f3,g,l,s2

Cucurbita pepo   squash, marrow, pumpkin

l-a,v,h   2n = 24,28,40   d2-1,000–1,500m
y-500 kg/ha; 1,000 max.   y2-(pumpkin)5.1
lb/100 sq.ft.max.   y2-(crookneck, patty pan,
or zucchini squash)6.1 lb/100 sq.ft.max.
USA-NC-7   r-b,d,f3,g,j,l

Cucurbita spp.   squash, marrow, pumpkin, gourd
b-o   f-m   s-c   v-bees   Do not usually display strong inbreeding depression.   r-f,
r-Allard 1960

Cucurbita spp.   See pp. 234–238

Curcuma domestica   Zingiberaceae   turmeric
l-p,h   2n = 32,64,63   r-d,f3

Curcuma zedoaria   zedoary   l-p,h   2n = 64,63
r-d,f3

Cyamopsis pedata   Fabaceae   archucha   l-p/a,v
2n = 32   r-d

Cyamopsis tetragonoloba   (= *C. psoralioides*)
guar, clusterbean   l-a,h   2n = 14   y-700
kg/ha; up to 1,750 irrigated (r-d2)   USA-S-9
r-d,d2,f3

Cyclanthera pedata   Cucurbitaceae   accocha,
archucha   2n = 32   USA-S-9   r-d,f3,l

Cydonia oblonga   Rosaceae   quince   l-p,t
2n = 34   USA-CR-COR   r-d,f3,l

Cymbopogon citratus   Poaceae   lemongrass   l-p,g
2n = 40,60   USA-S-9   r-d3,f3,l

Cynara cardunculus   Asteraceae   cardoon   l-p,h
2n = 34   USA-NE-9   r-d,f3,l

Cynara scolymus   globe artichoke, artichoke
l-p,p/a,h   2n = 34   USA-NE-9   r-d,f3,l,z

Cyperus esculentus   Cyperaceae   yellow nutsedge
l-p,h   2n = 18,108   r-d,f3

Cyperus rotundus   purple nutsedge   l-p,h
2n = 108   r-d,f3

Cyphomandra betacea   Solanaceae   tree tomato
l-p,s,t   USA-W-6   r-d,f3,l

Cyrtosperma chamissonis   (= *C. edule* = *C. merkusii*)   Araceae   giant swamp taro   r-f3,s2

Dahlia rosea   Asteraceae   dahlia   b-o   f-h   s-i
r-f

Daucus carota   Apiaceae   carrot   l-a,b,h
2n = 18,22   b-o   f-h   flower protandrous

Displays severe inbreeding depression (r-Allard 1960). s-c v-bees, flies, insects d-800–1,600m Crosses readily with the common wild carrot Queen Anne's lace. y-600 kg/ha; 1,000 max.; European types in the tropics 300 kg/ha; Asian types in the tropics 250 kg/ha y2-17.8 lb/100 sq.ft.max. USA-NC-7 r-b,d,f,f3,g,j,l

Dendrocalamus spp. Poaceae dendrocalamus l-p,g 2n = 72 r-d,f3

Desmanthus illinoensis Fabaceae Illinois bundle-flower l-p,h The Land Institute is working on domesticating this species. USA-S-9 r-f3,l,*The Land Report* Spring 1991

Dictamnus albus Rutaceae gas plant l-p,h 2n = 30,36 USA-W-6 r-d,f3,l

Dioscorea alata Dioscoreaceae winged yam l-p,v 2n = 20,30,40 r-d,f3

Dioscorea batatas see *D. opposita*

Dioscorea bulbifera air potato l-p,v 2n = 36,40,54 r-d,f3

Dioscorea composita composite yam l-p,v 2n = 36,54 r-d

Dioscorea convolvulacea convolvulus yam l-p,v 2n = 36 r-d

Dioscorea floribunda floribunda yam l-p,v 2n = 36,54,72 r-d

Dioscorea japonica glutinous yam 2n = 40 r-f3,z

Dioscorea macrostachya cuculmeca l-p,v r-d

Dioscorea opposita (= *D. batatas*) Chinese yam, cinnamon vine l-p,v 2n = about 140, about 144 r-d,f3,z

Dioscorea rotundata eboe yam 2n = 40 r-d

Dioscorea spp. yam (not related to American sweet potato, which is sometimes called a yam) b-o f-d Many have been maintained vegetatively, and in many seed set is rare; may be more likely if plants are allowed to mature beyond normal harvest time. USA-CR-MIA r-d,l,s2

Dioscoreophyllum cumminsii Menispermaceae serendipity berry l-p,v,s,t 2n = 24 r-d,f3

Diospyros digyna Ebanaceae black sapote l-p,t r-d,f3

Diospyros kaki Japanese persimmon l-p,t 2n = 54,56 r-d,f3

Diospyros spp. persimmon l-p,t b-o f-d v-insects USA-CR-DAV r-f,l

Diospyros virginiana common persimmon l-p,t 2n = 60,90 r-d,f3

Dipteryx odorata Polypodiaceae cumaru tonka bean l-p,t 2n = 32 b-o v-insects r-d,d2

Distichlis palmeri Poaceae l-p,g Wild halo-phyte used by Cucapa Indians to make bread and atole. Efforts by N. and S. B. Yensen in Tucson, Arizona, are under way to domesticate it into a salt-tolerant perennial grain. See Wagoner 1990 for review. r-f3,Wagoner 1990.

Dolichos biflorus see *Macrotyloma uniflorum*

Dolichos lablab see *Lablab purpureus*

Dolichos uniflorus see *Macrotyloma uniflorum*

Echinochloa crusgalli Poaceae barnyard grass l-p,a,g 2n = 36,54,56 USA-NC-7 r-d,f3,l

Echinochloa crusgalli var. *frumentacea* billion dollar grass l-a,g 2n = 36,54,56 USA-NC-7 r-d,f3,l

Echinochloa pyramidalis antelope grass l-p,g 2n = 36,54,72 USA-NC-7 r-d,l

Elaeis guineensis African oil palm l-p,t 2n = 32,36 b-o f-m r-d,f

Elaeis oleifera Arecaceae American oil palm l-p,s,l 2n = 32 r-d

Eleocharis dulcis (= *E. tuberosa*) Cyperaceae waternut, water chestnut l-p,h 2n = 30? r-d,d3,f3

Elettaria cardamomum Zingiberaceae carda-mom, cardamum l-p,b,h 2n = 48,52 USA-S-9 r-d,f3,l

Eleusine coracana Poaceae finger millet l-a,g 2n = 36 USA-S-9 r-d,f3,l

Eleusine indica   goosegrass   l-a,g   2n = 18,36
USA-S-9   r-d,f3,l

Elymus canadensis   Poaceae   Canada wild rye
l-p,g   2n = 28,42   USA-W-6   r-d,f3,l

Elymus cinereus   basin wild rye   l-p,g
2n = 28,56   r-d

Elymus condensatus   giant wild rye   l-p,g
2n = 28,56   r-d

Elymus glaucus   blue wild rye   l-p,g   2n = 28
USA-W-6   r-d,l

Elymus junceus   Russian wild rye   l-p,g
2n = 14   r-d

Elymus spp.   wild rye   b-i   f-h   d-o   USA-W-
6   r-f,l

Eragrostis chloromelas   Poaceae   Boer lovegrass
l-p,g   2n = 40,60,63   USA-W-6   r-d,l

Eragrostis curvula   weeping lovegrass   l-p,g
2n = 20,40,50   USA-W-6   r-d,l

Eragrostis lehmanniana   lehmann lovegrass
l-p,a,g   2n = 40,60   USA-W-6   r-d,l

Eragrostis tef   teff   l-a,g   2n = 40   USA-W-6
r-d,f3,l

Eragrostis trichodes   sand lovegrass   l-p,g   b-i
f-h   d-0   USA-W-6   r-d,f,l

Eriobotrya japonica   Rosaceae   loquat   l-p,s,t
2n = 32,34   USA-CR-COR   r-d,f3,l

Eruca vesicaria ssp. sativa   (= *E. sativa*) Brassicaceae   arugula, rocket, garden rocket, rocket
salad, roquette   l-a,h   2n = 22   gf-EE
USA-NC-7   r-d,f2,f3,l,z

Eryngium foetidun   Apiacea   false coriander
l-b,h   2n = 16   r-d,f3

Eucalyptus spp. Myrtaceae   eucalyptus   l-p,t
2n = 22   b-o   f-h   plant protandrous   s-i
v-bees   r-d,f,f3

Eugenia uniflora   Myrtaceae   Surinam cherry
l-p,s,t   2n = 22   r-d,f3

Euryale ferox   Euryalaceae   Gorgon water lily
2n = 58   r-d3,f3

Eutrema wasabi   see *Wasabia japonica*

Fagopyrum cymosum   Polygonaceae   perennial
buckwheat, shakuchiri-soba   l-p,h   2n =
16(r-s2)   b-unknown, but "it may reasonably
be assumed to share with *F. esculentum* a het-
eromorphic incompatibility system" (r-s2)
r-f2,f3,s2

Fagopyrum esculentum   buckwheat   l-a,h
2n = 16,32(r-d)   2n = 16(r-s2)   b-o   f-h
s-i   Strong but not absolute incompatibility.
Incompatibility is associated with two anatomically different flower types. Plants have either
pin or thrum flowers. Pin flowers have short
stamens and long pistils. Thrum flowers have
long stamens and short pistils. Usually, thrums
can be pollinated successfully only by pins
and pins by thrums, so an individual plant
doesn't normally self-pollinate, even though it
serves as both male and female. Whenever
self-pollination occurs, severe inbreeding
depression usually results.   v-insects
d-hasn't been established (r-f2)   y2-30 lb/100
sq.ft.max.   r-d,d3,f2,f3,j,l,s2,z

Fagopyrum tataricum   Tatarian or Medawaska
buckwheat; duck or India wheat   l-a,h
2n = 16   b-?   f-h   s-c   There is just one
anatomical type of flower. Some seed dormancy?   USA-NE-9   r-d,f2,f3,l,p,s2,z

Feijoa sellowiana   Myrtaceae   feijoa   l-p,s
2n = 22   r-d,f3

Feijoa sp.   feijoa   b-o   f-h   s-i,c   v-bees   r-f

Ferula communis   (see also *Foeniculum vulgare*)
Apiaceae   giant anise fennel   l-p,h   2n = 22
r-d3,f3

Ficus carica   Moraceae   fig   l-p,s,t   2n = 26
b-o   f-d   v-wasps   USA-CR-DAV   r-d,f,f3,l

Ficus elastica   rubber plant   l-p,t   2n = 26,39
r-d

Ficus vogelii   vogel fig   l-p,t   2n = 26   r-d

Flacourtia indica   Flacourtiaceae   l-p,s,t
2n = 22,18   r-d,f3

Foeniculum vulgare   Apiaceae   fennel   l-p,b/a,h
2n = 22   Bronze, wild, bitter, sweet, Flor-

ence, and salad fennels are all *F. vulgare*. I've seen giant anise fennel listed as both *Foeniculum vulgare* and *Ferula communis*. I don't know whether these are the same. USA-NC-7 r-d,f3,l

Fortunella spp. Rutaceae kumquat l-p,s,t 2n = 18,36 USA-PIO r-d,l

Fragaria chiloensis Rosaceae Chilean strawberry l-p,h 2n = 56 USA-CR-COR r-d,l

Fragaria spp. strawberry b-o f-h, h + female, female (male sterile) flowers protogynous s-c v-bees USA-CR-COR r-f,l

Fragaria vesca European strawberry l-p,h 2n = 14,35 USA-CR-COR r-d,l

Fragaria virginiana Virginia strawberry l-p,h 2n = 56 USA-CR-COR r-d,l

Fragaria × ananassa garden strawberry l-p,h 2n = 56 USA-CR-COR r-d,l

Garcinia mangostana Clusiaceae mangosteen l-p,t 2n = about 76,96 r-d,f3

Garcinia tinctoria gamboge tree l-p,t 2n = 80 r-d

Gaultheria procumbens Ericaceae wintergreen l-p,s 2n = 24 USA-CR-COR r-d,f3,l

Gaylussacia baccata Ephedraceae black huckleberry l-p,s 2n = 24 USA-CR-COR r-d,f3,l

Gladiolus spp. Iridaceae gladiolus b-o f-h s-c r-f,f3

Glycine apios see *Apios americana*

Glycine max Fabaceae soybean l-a,h 2n = 40 b-i f-h d-0 y-1.3 tonnes/ha y2-14.4+ lb/100 sq.ft.max USA-SOY-N r-d,d2,f,f3,g,j,l

Glycine wightii (= *Notonia wightii*) perennial soybean l-p,v,l 2n = 22,44 most b-i; tetraploids may be b-o (r-d2) r-d,d2,f3,l

Glycyrrhiza glabra Fabaceae common licorice l-p,h 2n = 16 USA-W-6 r-d,d2,f3,l

Glycyrrhiza lepidota American licorice l-p,h 2n = 16 USA-W-6 r-d,d2,f3,l

Grewia asiatica Tiliaceae phalsa l-p,t 2n = 36 r-d,f3,l

Helianthus annuus Asteraceae sunflower l-a,h 2n = 14,34(r-d) 2n = 34(r-d3) b-o f-h flower protandrous x-20–75% s-variable,i Does not display strong inbreeding depression (r-Allard 1960). v-bees d-800m USA-NC-7 r-d,d3,f3,j,l,Allard 1960

Helianthus maximilianii l-p,h 2n = 34 USA-NC-7 r-d,d3,f3,l

Helianthus tuberosus Jerusalem artichoke l-p,p/a,h 2n = 102 hexaploid b-o f-h probably s-i Normally propagated vegetatively. Seed may be obtained when more than one variety is grown if it is not too cold during the period of flowering and seed development (r-po). USA-NC-7 r-d,d3,l,po,z

Hemerocallis spp. Liliaceae daylily l-p,h 2n = 22,33,44 b-o USA-NC-7 r-d3,f3,l

Heracleum lanatum (= *H. maximum*) Ariaceae American cow parsnip l-p,h 2n = ? USA-W-6 r-f3,l

Heracleum sphondylium branca-ursina, European cow parsnip l-p,h 2n = 22 r-d3,f3

Hesperis matronalis Brassicaceae sweet rocket l-p 2n = 24 USA-NC-7 r-f3,l,z

Hibiscus cannabinus Malvaceae kenaf l-a,h 2n = 36,72 b-i f-h x-2–45% by bees USA-S-9 r-d,f,f3,l

Hibiscus esculentus see *Abelmoschus esculentus*

Hibiscus moschatus see *Abelmoschus moschatus*

Hibiscus sabdariffa roselle l-a,s 2n = 36,72 b-i f-h USA-S-9 r-d,f,f3,l

Hippophae rhamnoides Elaeagnaceae sea buckthorn l-p,t b-o f-d v-wind USA-NC-7 r-d,f,f3,l

Hordeum bulbosum Poaceae bulbous barley l-p,g 2n = 14,28 USA-NSGC r-d,f3,l

Hordeum jubatum foxtail barley l-g b-i f-h d-0 USA-NSGC r-f,f3,l

Hordeum sp. y2-(barley, beardless)24 lb/100 sq.ft.max r-j

Hordeum spp.   "*Hordeum* species are classified into four sections . . . Grain-producing forms are in the section Cerealia. The basic chromosome number is seven with species in cerealia being diploid, and those in other sections being either diploid, tetraploid, or hexaploid." Cerealia includes three cultivated species and two wild species   (*H. vulgare, H. distichum,* and *H. irregulare* and *H. spontaneum* and *H. agriocrithon,* respectively). Crosses can be made among all of them with almost complete fertility, so some consider all five as variants of a single species. r-f2

Hordeum spp.   l-p,g   Wild perennial relative(s) exist that can be crossed with barley (*H. vulgare*). r-Wagoner 1990

Hordeum vulgare   barley   l-a,g   2n = 14,28 b-i   f-h   Many cvs. are cleistogamous; can depend on environment and weather. x-up to 10% by wind (r-f); "generally less than 0.2%" (r-f2) "Six-rowed types tend to have more natural crossing than two-rowed types, and natural crossing tends to be higher in awnless, awnleted, and naked types" (r-f2).   d-0   USA-NSGC   r-d,f, f2,f3,l

Hordeum vulgare   hybrids   barley hybrids   b-i f-h + f-male-sterile   d-200m   d-f

Houttuynia cordata   Saururaceae   tsi   l-p,h 2n = 56,96   r-d,f3

Humulus lupulus   Cannabaceae   hops   l-p,v 2n = 20   b-o   f-d   v-wind   USA-CR-COR r-d,f3,l

Hyscyamus niger   Solanaceae   black henbane l-a,b,p,h   2n = 34   USA-NC-7   r-d,l

Hyssopus officinalis   Lamiaceae   hyssop   l-p,h 2n = 12,13   r-d,f3,l

Inga edulis   Fabaceae   ice cream bean   l-p,t 2n = 26   r-d,d2,f3

Inula helenium   Asteraceae   elecampane   l-b,p,h 2n = 20   r-d3,f3

Ipomoea aquatica   Convolvulaceae   swamp morning glory, water spinach   l-p,v,h 2n = 30   USA-S-9   r-d,f3,l,z

Ipomoea batatas   sweet potato (unrelated to the yam, *Dioscorea* spp., although the orange varieties are called yams in the United States) l-p,p/a,v   2n = 6x = 90(r-b,f2,s2) 2n = 84,90(r-d)   b-o   f-h   flower protogynous   s-i   v-bees, bumblebees, insects   Varieties differ widely in ability to flower and set seed. Some require a shock, such as grafting or cutting the vine one-third of the way through and then propping the cut open with a matchstick. Seed may need scarification to germinate.   USA-S-9   r-b,d,f2,f3,j,l,s2

Iris spp.   Iridaceae   iris   b-o   f-h   s-c USA-W-6   r-f,f3,l

Jubaea chilensis   Arecaceae   chile coco palm l-p,t   2n = 32   r-d,f3

Juglans cinerea   Juglandaceae   butternut   l-p,t 2n = 32   r-d,l

Juglans nigra   black walnut   l-p,t   2n = 32 USA-CR-DAV   r-d,l

Juglans regia   English walnut   l-p,t 2n = 32,36   b-o   f-m   plant protandrous s-c   v-wind   USA-CR-DAV   r-d,f,l

Kaempferia galanga   Zingiberaceae   kentjoer l-p,h   2n = 22,54   r-d,f3

Kerstingiella geocarpa   see *Macrotyloma geocarpum*

Kochia scoparia   Chenopodiaceae   kochia, summer cypress   l-p/a,b,h   2n = 18   USA-W-6 r-d,z

Lablab purpureus   (= *Dolichos lablab* = *Lablab niger*)   Fabaceae   hyacinth bean, lablab l-a,p,h,v   2n = 22,24   y-450 kg/ha intercropped; 1,460 kg/ha in monoculture   y-0.9 tonnes/ha (r-g)   USA-S-9   r-d,d2,f3,g,l

Lactuca denticulata   Asteraceae   2n = 10   r-z

Lactuca indica   Indian lettuce   2n = 18 USA-W-6   r-f3,l,z

Lactuca perennis   perennial lettuce   l-p,h 2n = 18(r-d3)   USA-W-6   r-d3,f3,l

Lactuca quercina   2n = 18   r-l,z

Lactuca sativa lettuce l-a,h 2n = 18(r-r) 2n = 18,36(r-d) b-i f-h partly cleistogamous x-1–6% by flies, insects d-10m d2-2m Isolate from *L. serriola*. y-0.5 tonnes/ha (r-g) y2-(head)1.2 lb/100 sq.ft.max. y2-(leaf) 2 lb/100 sq.ft.max. USA-W-6 r-b,d,f3,g,j,l,r Also see pp. 213–216

Lactuca sativa Angustan group (= var. asparagina) asparagus lettuce, celtuce, stem lettuce r-f3,g,z

Lactuca serriola wild lettuce, prickly lettuce f-h b-i Crosses freely with *L. sativa*.

Lactuca spp. Species with 2n = 16: *aurea, bracteata, bourgaei, macrantha,* and others. Species with 2n = 18: *chondrillaeflora, perennis, raddiana, saligna, sativa, serriola, squarrosa, tatarica, viminea, virosa,* and others. Species with 2n = 34: *campestris, canadensis, ludoviciana, spicata,* and others. *L. orientalis* is listed with 2n = 36. r-d3

Lactuca virosa bitter lettuce, lettuce opium l-a,b,h 2n = 16,18(r-d) 2n = 18(r-d3) USA-W-6 r-d,d3,f3,l

Lagenaria siceraria Cucurbitaceae clabash, edible gourd, white-flowered gourd l-a,v 2n = 22 USA-S-9 r-f3,l

Lallemantia iberica Lamiaceae dragon's head l-a,h USA-W-6 r-d,f3,l

Lansium domesticum Meliaceae langsat l-p,t 2n = 72 r-d,f3

Lathyrus cicera Fabaceae garousse 2n = 14 b-i f-h USA-W-6 r-d3,f,l

Lathyrus hirsutus rough pea l-a,h,v 2n = 14 USA-W-6 r-d,d2,l

Lathyrus sativus grass pea l-a,h,v 2n = 14 b-i f-h USA-W-6 r-d,d2,f3,l

Lathyrus tingitanus Tangier pea b-i f-h USA-W-6 r-d,f,l

Lathyrus tuberosus earth chestnut l-p 2n = 14 r-d3,f3

Lavandula angustifolia Lamiaceae lavender l-p,s 2n = 54,50,36 USA-W-6 r-d,f3,l

Lavandula latifolia broad-leaved lavender l-p,s 2n = 54 r-d,f3

Lavandula stoechas French lavender l-p,s 2n = 30 r-d,f3,l

Lens culinaris Fabaceae lentil l-a,h 2n = 14 b-i f-h x-rarely (r-d2) d3-3 feet or whatever is needed to prevent physical mixture of different varieties; one of the most highly inbreeding species; isolation distances tested and established in Pullman, Washington (r-pc from Rich Hannan, curator of U.S. *Lens* collection, USA-W-6). y-450–675 kg/ha dry cultivated; 1,688 irrigated; 3,000 max. (r-d2) y2-8.0+ lb/100 sq.ft.max. USA-W-6 r-d,d2,f,f3,pc(Rich Hannan),j,l

Lepidium latifolium Brassicaceae l-p,h 2n = 24 r-f3,l,z

Lepidium meyenii maca 2n = ? r-f3,z

Lepidium sativum garden cress, common cress l-a,b,h 2n = 16,32(r-z) 2n = 16,24,32 (r-d) USA-NC-7 r-d,f3,l,z

Lepidium virginicum Virginia pepperweed l-a,b,h 2n = 32 r-f3,d

Leucaena leucocephala (= *L. latisiliqua* = *L. glauca*) Mimosaceae leucaena l-p,s,t 2n = 104,36 USA-S-9 r-d,d2,f3,l

Levisticum officinale Apiaceae lovage l-p,h 2n = 22 USA-NC-7 r-d,f3,l

Leymus arenarius Poaceae strand wheat (wild perennial grain cultivated by the Vikings) l-p,g r-f3,Wagoner 1990

Leymus racemosus Volga wild rye (Cool-season perennial grass with large, edible seeds; relative of rye. Efforts are under way at The Land Institute to develop perennial grains from this species.) l-p,g USA-W-6 r-f3,l,*The Land Report* Spring 1991,Wagoner 1990

Lilium spp. Liliaceae lily l-p,h 2n = 24(r-d3) b-o f-h s-i USA-NC-7 r-d3, f,f3,l

Linum usitatissimum Linaceae flax l-a,h

2n = 30,32   b-i   f-h   x-3% by bees   d-0
USA-NC-7   r-d,f,f3,l

Litchi chinensis   Sapindaceae   litchi   l-p,t
2n = 28,30   USA-CR-MIA   r-d,f3,l

Lomatium californicum   Apiaceae   wild celery
parsley   l-p,h   2n = ?   r-f3

Lomatium dissectum   fern-leaved biscuit-root
l-p,h   2n = ?   r-f3

Lomatium macrocarpum   large-fruited biscuit-root
r-f3

Lomatium nudicale   lomatium, cow parsley, smyrnium   l-p,h   2n = ?   r-f3

Lomatium utriculatum   desert gold, pomo celery
r-f3

Luffa acutangula   Cucurbitaceae   angled luffa,
Chinese okra   2n = 26   b-o   f-m   USA-S-9
r-d,f3,l,s2(p306)

Luffa aegyptiaca   (= *L. cylindrica?*) smooth luffa,
spongegourd   l-a,v   2n = 26   USA-S-9
r-d,f3,l

Luffa cylindrica   (= *L. aegyptiaca?*)   2n = 26
b-o   f-m   r-f3,l,s2(p306)

Lupinus albus   Fabaceae   white lupine   l-a,h
2n = 26,22   b-i but bee-pollinated   f-h   x-
10% by bees   v-bees   USA-W-6   r-d,d2,f,f3,l

Lupinus angustifolium   blue lupine   l-a,h
2n = 40,48   b-i bees not required   USA-W-
6   r-d,d2,l

Lupinus luteus   yellow lupine   l-a,h
2n = 52,104   b-i mainly bee-pollinated   f-h
x-10-25% by bees   USA-W-6   r-d,d2,f,f3,l

Lupinus perennis   lupine   2n = 48,96   b-i   f-h
USA-W-6   r-f,f3,l,z

Lycium chinense   Solanaceae   Chinese boxthorn
l-p,s   2n = 24   r-d,f3

Lycopersicon esculentum   Solanaceae   tomato
l-a,b,p,h   2n = 24   b-i   f-h   Protogynous,
but flower and anther cone are pendant, thus
facilitating self-pollination.   x-less than 2% by
solitary bees, thrips; may be much higher in
the tropics or in long-styled cvs.   Note: Some
hierlooms and modern cvs. are long-styled, so

you should examine the flowers.   d-30m
d2-30–200m, largely to prevent physical mixing   y-250–500 kg/ha in the U.S.; 10–50 kg/
ha in Africa   y2-5.5 lb/100 sq.ft.max.   USA-
NE-9   r-b,d,f,f2,f3,g,j,l   Also see pp. 206–213

Lycopersicon pimpinellifolium   currant tomato
l-p,a,h   2n = 24,48   b-i,   s-c   USA-NE-9
r-b,d,f3,l

Macadamia spp.   Proteaceae   macadamia nut
l-p,t   2n = 28,56   USA-PIO   r-d,f3,l

Macadamia ternifolia   Macadamia nut   b-o
s-partially c   USA-PIO   d-f,l

Macroptilium atropurpureum   (= *Phaseolus atropurpureus*) Fabaceae   siratro   l-p,h,v
2n = 22   r-d2

Macrotyloma geocarpum   (= *Kerstingiella geocarpa* = *Voandezia poissonii*) Fabaceae   Kersting's groundnut, groundbean, potato bean
l-a,h   2n = 20,22   r-d,d2

Macrotyloma uniflorum   (= *Dolichos uniflorus* = *Dolichos biflorus*)   horsegram, Madras
gram   l-a,h   2n = 20,22,24(r-d2)   USA-S-9
r-d2,f3,l

Malpighia punicifolia   Malphighiaceae   acerola
l-p,s,t   2n = 20   USA-CR-HIL   r-d,f3,l

Malus spp.   Rosaceae   apple   l-p,t   USA-CR-
GEN   r-d,f3,l

Malva alcea   Malvaceae   2n = about 84
USA-NC-7   r-d3,l

Malva brasiliensis   2n = about 112   r-d3

Malva crispa   curled leaf salad mallow   l-a,h
2n = about 112   r-d3,f3

Malva moschata   musk mallow   l-p,h   2n = 42
b-o   f-h   v-bees   USA-NC-7   r-d3,l,pc
(Heine Refsing),po

Malva neglecta   (= *M. rotundifolia*)   2n = 42
r-d3

Malva nicaeensis   2n = 42   r-d3

Malva parviflora   2n = 42   USA-NC-7
r-d3,f3,l

Malva pusilla   (= *M. borealis*)   2n = 42,76
r-d3,l

Malva sylvestris   common mallow   2n = 42
r-d3,f3

Malva sylvestris var. mauritiana   l-p,b,h   r-pc
(Heine Refsing)

Malva verticillata   2n = about 84   USA-NC-7
r-d3,f3,l

Mangifera indica   Anacardiaceae   mango   l-p,t
2n = 40   b-o   f-m; some apomixis   s-c,i
v-flies   USA-CR-MIA   r-d,f,f3,l,s2

Manihot esculenta   Euphorbiaceae   cassava, man-
ioc, tapioca   l-p,s,t   2n = 36   allotetraploid
b-o   f-m   plant protogynous   x-0–100%
v-wasps, bees, insects   d4-30m; 500m for
genetic studies   Seed tends to be dormant for
three to twelve months after harvest.   r-d,f,f2,
f3,s2

Manihot glaziovii   ceara rubber   l-p,t   2n = 36
USA-CR-MIA   r-d,f3,l

Manilkara bidentata   Sapotaceae   balata   l-p,t
r-d,f3

Manilkara zapota   sapodilla   l-p,t   2n = 26
USA-CR-MIA   r-d,f3,l

Maranta arundinacea   Arantaceae   arrowroot
l-p,h   2n = 18,48   r-d,f3

Marrubium vulgare   Lamiaceae   horehound
l-p,h   2n = 34,36   USA-W-6   r-d,f3,l

Matricaria chamomilla   Asteraceae   wild cha-
momile   l-a,h   2n = 18   r-d,f3,l

Medicago sativa   Fabaceae   alfalfa, lucerne   l-
p,h   2n = 16,32,64   b-o   f-h   Must be
tripped by insects.   x-more than 80%
s-i,c   Displays extreme inbreeding depression
(r-Allard 1960).   v-honeybees, bumblebees, insects   d-90m   y2-1.1 lb/100
sq.ft.max.   USA-W-6   r-d,d2,f2,j,l, Allard
1960

Medicago spp.   "Essentially all annual species are
cleistogamous and are exclusively self-pollinated. Generally, the perennial species require
tripping, and will set seed from either self- or
cross-pollination" (r-f2). For breeding and life-

style information on other species, see r-d2;
for use as edibles see r-f3.

Melicoccus bijugatus   Sapindaceae   Spanish lime
l-p,t   2n = 32   r-d,f3

Melilotus alba   Fabaceae   white sweetclover   l-
b,a,h   2n = 16,24,32   v-honeybees   Insects
required for pollination.   USA-NC-7   r-d2,l

Melilotus spp.   See r-d2,f3.

Melissa officinalis   Lamiaceae   lemon balm
l-p,h   2n = 32,64   r-d,f3

Mentha piperita   Lamiaceae   peppermint   l-p,h
2n = 36,48,64   b-o   f-h   v-flies   USA-CR-
COR   r-f,l

Mentha spp. and crosses include field mint, red
mint, apple mint, spearmint, and European
pennyroyal.   All l-p,h.   USA-CR-COR   r-d,f,
f3,l

Michelia champaca   Magnoliaceae   champac
l-p,t   2n = 38   r-d,f3

Momordica balsamina   Cucurbitaceae   balsam
apple, bitter melon   l-a,v   2n = 22
USA-S-9   r-d,f3,l

Momordica charantia   balsam pear, bitter melon,
bitter gourd   l-a,v   2n = 22   b-o   f-m
USA-S-9   r-d,f3,l,s2(p306)

Monstera deliciosa   Araceae   ceriman   l-p,l,v
2n = 24,56,60   r-d,f3

Montia perfoliata   Portulacaceae   miner's lettuce
l-a,h   2n = 12,24,36   r-d,f3

Morinda citrifolia   Rubiaceae   Indian mulberry
l-p,t   r-d,f3

Moringa oleifera   Moringaceae   horseradish tree
l-p,t   2n = 28   r-d,f3

Morus alba   Moraceae   white mulberry   l-p,s,t
2n = 28   b-o   f-d   v-wind   USA-CR-DAV
r-d,f,f3,l

Morus nigra   black mulberry   l-p,t   2n =
89–308   b-o   f-d   v-wind   USA-CR-DAV
r-d,f,f3,l

Morus rubra   red mulberry   l-p,t   2n = 28
USA-CR-DAV   r-d,f3,l

Mucuna spp. Fabaceae velvetbean l-a,h,v
2n = 22 b-i x-rare USA-S-9 r-d,d2,f3

Murraya koenigii Rutaceae curryleaf tree
l-p,s,t 2n = 18 r-d,f3

Musa acuminata Musaceae dwarf banana
l-p,h,s 2n = 22,23 r-d

Musa × *paradisiaca* banana l-p,h,s
2n = 22,32–35 r-d,f3

Myrciaria cauliflora Myrtaceae Brazilian grape
tree l-p,t r-d,f3

Myristica fragrans Myristicaceae nutmeg l-p,t
2n = 42,44 r-d,f3

Myrrhis odorata Apiaceae myrrh, sweet cicely
l-p,h 2n = 22 r-d,f3

Myrtus communis Myrtaceae myrtle l-p,s
2n = 22 r-d,f3

Nasturtium microphyllum see *Rorippa microphylla*

Nasturtium officinale see *Rorippa nasturtium-*
aquaticum

Nelumbo lutea Nelumbonaceae American lotus
2n = 16 r-d3,f3

Nelumbo nucifera (= *N. speciosum*) lotus
l-p,h 2n = 16 r-d,d3,f3

Nephelium lappaceum Sapindaceae rambutan
l-p,t 2n = 22 r-d,f3

Nephelium mutabile pulasan l p,t r d,f3

Nicotiana rustica Solanaceae b-i f-h USA-
TOBAC r-f,l

Nicotiana tabacum tobacco l-p/a,h
2n = 24,48,72 b-i x-2–3% by hummingbirds, bees d-50m USA-TOBAC
r-d,f,f3,l

Nigella sativa Ranunculaceae black cumin
l-a,h 2n = 12 r-d,f3

Notonia wightii see *Glycine wightii*

Nuphar lutea Nymphaeaceae yellow water lily
2n = 34 r-d3,f3

Nymphaea capensis Nymphaeaceae Cape water
lily 2n = 28 r-d3

Nymphaea flava (= *N. mexicana*) 2n = 56
r-d3

Nymphaea gigantea Australian water lily
2n = 224 r-d3

Nymphaea lotus Egyptian water lily 2n = 56
r-d3,f3

Nymphaea odorata American water lily
2n = 84 r-d3,f3

Nymphaea rubra red water lily 2n = 56 r-d3

Nymphaea stellata 2n = 28 r-d3,f3

Nymphaea tetragona pygmy water lily
2n = 112 r-d3

Nymphaea tuberosa magnolia water lily
2n = 84 r-d3,f3

Ocimum basilicum Lamiaceae sweet basil l-a,h
2n = 48 USA-NC-7 r-d,f3,l

Ocimum kilimandscharicum hoary basil l-a,h,s
2n = 76 USA-NC-7 r-d,f3,l

Ocimum sanctum holy basil l-a,p,h 2n = 64
USA-NC-7 r-d,f3,l

Oenanthe javanica Apiaceae water dropwort
l-p,h 2n = 20 r-d,f3

Olea europaea Oleaceae olive l-p,t 2n = 46
USA-CR-DAV r-d,f3,l

Opuntia ficus-indica Cactaceae prickly pear
l-p,s 2n = 22,88 r-d,f3

Origanum vulgare Lamiaccac wild marjoram
l-p,h 2n = 30,32 USA-NC-7 r-d,f3,l

Origanum vulgare marjoram b-i f-h d-90m
USA-W-6 r-d,f,f3,l

Ornithopus sativus Fabaceae serradella l-a,h
2n = 14,16 b-i r-d,d2,l

Oryza glaberrima Poaceae African rice l-a,h
2n = 24 USA-NSGC r-d,l

Oryza longistaminata African wild rice (wild
perennial relative that can be crossed with
rice) l-p,g r-f3,Wagoner 1990

Oryza rufipogon wild perennial relative that can
be crossed with rice l-p,g r-Wagoner 1990

Oryza sativa rice l-a,g 2n = 24 b-i f-h
Some cvs. are cleistogamous; varies with
climate. d-3m y2-24 lb/100 sq.ft.max.
USA-NSGC r-d,f,f3,j,l

Oryzopsis hymenoides Poaceae Indian rice grass, rice grass, bunchgrass l-p,g 2n = 48 USA-W-6 r-d,f3,l

Oryzopsis miliacea smilograss l-p,g 2n = 24 USA-W-6 r-d,l

Oxalis tuberosa Oxalidaceae oca l-p/a,h 2n = 14,60–70 USA-NC-7 r-d,f3,l

Pachyrhizus erosus (= *P. tuberosus*) Fabaceae jicama, yambean l-p/a,v 2n = 22 y-0.6 tonne/ha USA-S-9 r-d,d2,f3,g,l

Panicum miliaceum Poaceae proso millet l-a,g 2n = 36,54,72 b-i f-h x-more than 10% by wind d-400m y2-(millet, "regular")30 lb/100 sq.ft.max. USA-NC-7 r-d,f,f3,j,l

Panicum obtusum vine mesquite l-p,g 2n = 20,36,40 USA-S-9 r-d,f3,l

Panicum virgatum switchgrass l-p,g 2n = 18–108 b-o f-h may be apomictic v-wind d-90m USA-NC-7 r-d,f,l

Papaver bracteatum Papaveraceae scarlet poppy l-p,h 2n = 14 USA-W-6 r-d,l

Papaver somniferum opium or breadseed poppy l-a,h 2n = 22,20 USA-W-6 r-d,f3,l

Passiflora spp. Passifloraceae passionfruit, granadilla, waterlemon, barbadine l-p,v 2n = 18 USA-CR-MIA r-d,f3,l

Pastinaca sativa Apiaceae parsnip l-b/a,h 2n = 22 b-o f-h flower protandrous v-30% by bees, flies, insects (Listed as b-i with x-30% by r-f.) d-500m Must be isolated from commercial crops because a small fraction will bolt in the first year. y-1,000 kg/ha; 2,000 max. USA-NC-7 r-d,f,f3,g,l

Pennisetum americanum (= *P. glaucum*) Poaceae pearl millet l-a,g 2n = 14 b-o f-h s-i,c y2-18.3 lb/100 sq.ft.max USA-S-9 r-d,f,f3,j,l

Pennisetum clandestinum kikuyugrass l-p,g 2n = 36 USA-S-9 r-d,l

Pennisetum purpureum elephant grass l-p,g

2n = 28,27,56 USA-S-9 r-d,l,Wagoner 1990

Perilla frutescens Lamiaceae perilla l-a,h 2n = 38,40 USA-NC-7 r-d,f3,l

Persea americana Lauraceae avocado l-p,t 2n = 24 USA-NC-7 r-d,f3,l

Persea spp. avocado b-o f-h plant protogynous s-i v-bees r-f,f3

Petasites japonicus Asteraceae fuki, huki, sweet coltsfoot l-p,h 2n = about 87(r-z) 2n = 87(r-d3) r-d3,f3,z

Petroselinum crispum Apiaceae Hortense group = leaf parsley; Radicosum group = root parsley (Hamburg)

Petroselinum crispum var. Neapolitanum (leaf) parsley l-b,p,h 2n = 22 b-o f-h v-honeybees, flies d-500–1,000m y-800 kg/ha USA-NC-7 r-f,f3,g,j,l,r

Phaseolus acutifolius Fabaceae tepary bean l-a,h 2n = 22 b-i presumably (r-d2) USA-W-6 r-d,d2,f3,l

Phaseolus atropurpureus see *Macroptilium atropurpureum*

Phaseolus aureus see *Vigna radiata*

Phaseolus coccineus (= *P. multiflorus*) (scarlet) runner bean l-p/a,h 2n = 22 b-o (r-f); i (r-g) f-h x-more than 30% (r-f); less than 40% (r-g) s-c v-honeybees, bumblebees (required for self- or cross-pollination) d-100m; more for stock seed (r-g) y-1,000 kg/ha USA-W-6 r-d,d2,f,f3,g,l

Phaseolus lunatus lima bean l-p,p/a,b/a,h,v 2n = 22 b-i f-h x-0–80% by bees (r-f); less than 18% (r-d2) d-45m y-1 tonne/ha y2-(bush)17.8 lb/100 sq.ft.max. y2-(pole)22.3 lb/100 sq.ft.max. USA-W-6 r-d,d2,f,f3,g,l

Phaseolus multiflorus see *P. coccineus*

Phaseolus mungo see *Vigna mungo*

Phaseolus vulgaris green, snap, string, common, or dry bean l-a,h (some wild forms l-p)

$2n = 22$ b-i f-h protogynous x-0–1% in many locations; 6–10% in some places; 15–20% in Puerto Rico; often high outcrossing in the tropics. Thrips cause little outcrossing. Carpenter bees are responsible for high outcrossing in Puerto Rico. d-45m (r-f); 50–150m (r-g); see Appendix B. y-1,500 kg/ha; 2,000 max. y2-(dry beans: kidney, pinto, red, or white)24 lb/100 sq.ft.max. y2-(bush green beans)17 lb/100 sq.ft.max. y2-(pole green beans) 29.7 lb/100 sq.ft.max. USA-W-6 r-b,d,d2,f,f2,f3,j,l,s2 Also see pp. 221–228

Phoenix dactylifera Arecaceae date palm l-p,t $2n = 28,36$ b-o f-d v-wind r-d,f,f3

Phragmites australis Poaceae reed l-p,g $2n = 48,36,54$ USA-S-9 r-d,f3,l

Phyllostachys dulcis Poaceae sweetshoot or vegetable bamboo l-p,g r-f3,Northern Groves catalog

Phyllostachys spp. phyllostachys l-p,g $2n = 48,54$ USA-PIO r-d,f3,l

Physalis ixocarpa Solanaceae tomatillo l-p/a,h $2n = 24$ USA-NE-9 r-d,f3,l

Physalis peruviana Peruvian ground cherry, husk tomato l-p/a,s $2n = 24,48$ USA-NE-9 r-d,f3

Pimpinella anisum Apiaceae anise l-a,h $2n = 18,20$ USA-NC-7 r-d,f3,l

Pinus edulis Pinaceae silver pine l-p,t $2n = 24$ r-d,f3

Pinus quadrifolia parry pinyon l-p,t r-d,f3

Pinus spp. pine b-o f-m v-wind USA-S-9 r-f,l

Pistacia vera Anacardiaceae pistachio l-p,t $2n = 30$ b-o f-d v-wind USA-CR-DAV r-d,f,f3,l

Pisum arvense see *P. sativum*

Pisum sativum Fabaceae pea l-a,h $2n = 14$ b-i f-h x-usually less than 1% in the U.S., (r-b) but some cvs. up to 25%; depends on what insects are around d-0 d2-20–100m,

but very short in some countries and mainly to prevent physical mixing y-2,000 kg/ha y2-(bush)21.6 lb/100 sq.ft.max. y2-(pole)12.1 lb/100 sq.ft.max. USA-NE-9 r-b,d,f,f2,f3,g,j,l Also see pp. 216–221

Plantago coronopus Plantaginaceae buckhorn plantain l-p,h $2n = 10,11,12$ r-d3,f3

Plantago major $2n = 12,24$ r-d,d3,f3

Plantago maritima sea plantain $2n = 12,(18),24$ r-d3,f3

Plantago ovata large round-leaf plantain l-p,h $2n = 8$ USA-W-6 r-d3,f3

Plantago psyllium psyllium seed $2n = 12$ r-d3,f3

Plectranthus esculentus Lamiaceae kaffir potato l-p,h $2n = 24$ r-d,f3

Polianthes tuberosa Agavaceae tuberose l-p,h $2n = 60,50$ r-d,f3

Polymnia sonchifolia Asteraceae yacon, llacon l-p,h $2n = 60$ USA-S-9 r-f3,l,z, *Lost Crops of the Incas* 1989

Populus spp. Salicaceae poplar l-p,t b-o f-d v-wind USA-NC-7 r-d,f,f3,l

Portulaca oleracea Portulacaceae purslane l-a,h $2n = 54$ USA-S-9 r-d,f3,j,l

Prunus spp. Rosaceae almond, apricot, cherry, nectarine, peach, plum b-o f-h s-c,i v-bees USA-CR-DAV r-d,f,f3,l

Psidium guajava Myrtaceae guava l-p,s,t $2n = 22,33,44$ b-o f-h x-35% s-c v-bees USA-CR-MAY r-d,f,f3,l

Psophocarpus tetragonolobus Fabaceae winged bean, asparagus pea l-p,p/a,h,v $2n = 26$ y-1.1 tonnes/ha USA-S-9 r-d,d2,f3,g,l

Pueraria lobata Fabaceae kudzu l-p,h $2n = 24$ b-o v-bees USA-S-9 r-d,d2,f3,j,l

Pueraria phaseoloides tropical kudzu l-p,v $2n = 22$ r-d,f3

Punica granatum Punicaceae pomegranate l-p,s,t $2n = 16,18,19$ USA-CR-DAV r-d,f3,l

Pyrus spp. Rosaceae pear l-p,t b-o f-h
s-c,i v-bees USA-CR-COR r-d,f,f3,l

Quercus spp. Fagaceae oak l-p,t b-o f-m
protandrous or protogynous v-wind USA-
S-9 r-d,f,f3,l

Raphanus sativus Brassicaceae garden radish
l-a,b,h 2n = 18 gf-RR b-o f-h x-more
than 85% s-c,i v-bees, insects d-200–
1,000m y-1,000 kg/ha; 2,000 max.
y2-20.6 lb/100 sq.ft.max. USA-NE-9 r-f,f2,
f3,g,j,l

Rheum spp. Polygonaceae rhubarb l-p,h
2n = 22,44 USA-W-6 r-d,f3,j,l,z

Ribes hirtellum Grossulariaceae hairy gooseberry l-p,s 2n = 30 USA-CR-COR
r-d,f3,l

Ribes nigrum black currant l-p,s 2n = 16
USA-CR-COR r-d,f3,l

Ribes rubrum red currant l-p,s 2n = 16
USA-CR-COR r-d,l

Ribes sativum common red ribes l-p,s
2n = 16 r-d,f3,l

Ribes uva-crispa European gooseberry l-p,s
2n = 16 USA-CR-COR r-d,l

Rorippa microphylla (= *Naturtium microphyllum*)
watercress 2n = 64 s-c r-r

Rorippa nasturtium-aquaticum (= *Nasturtium
officinale*) Brassicaceae watercress, green or
summer watercress l-p,h 2n = 32 Can
cross- or self-pollinate. s-c v-flies, insects
Produces fertile seed. r-d,g,r

Rorippa nasturtium-aquaticum × *microphylla*
brown watercress 2n = 48 Sterile triploid
resulting from a cross of the diploid and tetraploid species. r-r

Rosa spp. Rosaceae rose l-p,s b-o f-h s-i
USA-NC-7 r-d,f,l

Rosmarinus officinalis Lamiaceae rosemary
l-p,s 2n = 24 b-o v-bees r-d,f,f3

Rubus idaeus var. strigosus Rosaceae American
raspberry l-p,s 2n = 14,21 USA-CR-COR
r-d,l

Rubus occidentalis black raspberry l-p,s
2n = 14 USA-CR-COR r-d,l

Rubus occidentalis × *idaeus* purple raspberry
l-p,s

Rubus spp. blackberry l-p,h b-o f-m s-i
v-bees USA-CR-COR r-d,f,l

Rubus spp. dewberry l-p,h USA-CR-COR
r-d,l

Rubus spp. raspberry l-p,s b-o f-h s-c
v-bees USA-CR-COR r-d,f,l

Rumex acetosa Polygonaceae sorrel, French
sorrel l-p,h 2n = 12 + XX, females;
12 + XY1Y2, males f-d r-d3,f3

Ruta graveolens Rutaceae common rue l-p,h,s
2n = 72,81 r-d

Saccharum officinarum Poaceae sugarcane l-
p,g 2n = 80,60,90 USA-CR-MIA r-d,f3,l

Salix spp. Salicaceae willow l-p,t b-o f-d
v-insects USA-NC-7 r-d,f,f3,l

Salvia hispanica Lamiaceae chia l-a,h
USA-W-6 r-d,f3,l

Salvia officinalis sage l-p,h,s 2n = 14,16
USA-W-6 r-d,f3,l

Salvia sclarea clary l-b,p,h 2n = 22
USA-W-6 r-d,f3,l

Sambucus canadensis Caprifoliaceae American
elder, Canadian elderberry l-p,s
2n = 36,37,38 USA-CR-COR r-d,f3,l

Sambucus glauca blueberry, elderberry l-p,s,t
2n = 38 r-d,f3

Sanguisorba minor Rosaceae burnet, salad burnet l-p,h 2n = 28,54,56 USA-W-6
r-d,f3,l

Sansevieria trifasciata Agavaceae snake plant
l-p,h 2n = 36,40,42 r-d,f3

Satureja hortensis Euphorbiaceae summer
savory l-a,h 2n = 45–48 USA-W-6
r-d,f3,l

Satureja montana winter savory l-p,s
2n = 12,30 r-d,f3

Schinus molle Anacardiaceae California pepper-
tree l-p,s,t 2n = 28,30 r-d,f3

Sclerocarya caffra Anacardiaceae marula nut
l-p,t r-d,f3

Scolymus hispanicus Asteraceae golden thistle,
Spanish oyster plant 2n = 20 r-f3,z

Scorzonera hispanica Asteraceae scorzonera,
black salsify l-p/a,h 2n = 14 r-d,
f3,z

Secale cereale Poaceae rye l-a,g 2n = 14–
29 b-o f-h s-i Does not display strong
inbreeding depression (r-Allard 1960). v-
wind d-200m y2-24 lb/100 sq.ft.max.
USA-NSGC r-d,f,f3,j,l,Allard 1960

Secale montanum wild perennial relative that can
be crossed with rye (*S. cereale*) l-p,g r-Wagoner 1990

Sechium edule Cucurbitaceae chayote, vegetable pear l-p,v 2n = 28(r-s2) 2n =
24(r-d) b-o f-m r-d,f3,s2(p306)

Sesamum alatum Pedaliaceae wing-sesame
2n = 26 USA-S-9 r-d,f3,l

Sesamum indicum sesame l-a,h
2n = 26,52,58 b-i f-h protandrous
x-5–65% depending on cv., by bees d-180–
360m y2-5.6+ lb/100 sq.ft.max. USA-S-9
r-d,f3,j,l

Sesamum radiatum wild sesame l-p,h
2n = 64 USA-S-9 r-d,f3,l

Sesbania bispinosa Fabaceae agati, canicha
l-a,s 2n = 12,24 b-i d-0 r-d,d2,f3,l

Sesbania exaltata hemp, sesbania, Colorado river
hemp l-p,a,h 2n = 12 r-d,d2

Seseli carvi see Carum carvi

Sicana odorifera Cucurbitaceae casabanana
l-p,v USA-S-9 r-d,f3,l

Silene cucubalus Caryophyllaceae bladder cam-
pion l-p,h USA-W-6 r-f3,l

Silybum marianum (= *Carduus marianum*)
Asteraceae St. Mary's thistle, milk thistle
r-f3,l

Simarouba glauca Simmondsiaceae aceituna
l-p,s,t r-d,f,f3

Simmondsia chinensis (= *S. californica*) Simmondsiaceae jojoba l-p,s,t 2n = 56,about
100 b-o f-d USA-W-6 r-d,f3,l

Sinapsis alba see Brassica hirta

Sinapsis arvensis see Brassica kaber

Sium sisarum Apiaceae skirret l-p,h
2n = 22 USA-NC-7 r-d3,f3,l

Smyrnium olustrum Apiaceae Alexander's salad
greens, black lovage l-b,p,h 2n = 22
r-f3,l

Solanum aethiopicum Solanaceae mock tomato
l-p,s 2n = 24 USA-S-9 r-d,f3,l

Solanum aviculare Australian nightshade l-p,s,t
2n = 46,48,92 r-d,f3,l

Solanum burbankii Mrs. B's nonbitter garden
huckleberry l-a,h USA-S-9 r-f3,l

Solanum fendleri Navajo, fendler, or wild potato
2n = 48 USA-IR-1 r-d,f3,l

Solanum ferox ram-begun l-p,h 2n = 24
USA-S-9 r-d,l

Solanum gilo gilo l-p,s 2n = 24 USA-S-9
r-d,l

Solanum hyporhodium cocona l-p,s USA-S-9
r-d,l

Solanum incanum Sodom apple 2n = 24
USA-S-9 r-d,f3,l

Solanum indicum Indian nightshade 2n = 24
r-d,f3

Solanum khasianum 2n = 24 USA-S-9 r-d,l

Solanum laciniatum kangaroo apple l-p/a,s,t
2n = 48,92 USA-S-9 r-d,f3,l

Solanum macrocarpon native eggplant l-h,s
2n = 36 USA-S-9 r-d,f3,l

Solanum melanocerasum garden huckleberry,
wonderberry, sunberry l-a,h USA-S-9
r-f3,l

Solanum melongena eggplant l-p/a,h
2n = 24,36,48 b-i f-h x-7% by bees,
insects; 6–20% in India at distances
less than 50m; no cross-pollination at
greater distances y-150 kg/ha; 200 max.
y2-0.6 lb/100 sq.ft.max. USA-S-9 r-d,f,
f3,g,j,l

Solanum muricatum   melon-pear   l-p,s
  2n = 24   USA-S-9   r-d,f3,l

Solanum nigrum   black nightshade   l-a,h
  2n = 24,36,48   USA-S-9   r-d,f3,l

Solanum quitoense   naranjilla, lulo   l-p,h,s
  2n = 24   USA-S-9   r-d,f3,l

Solanum torvum   terongan   2n = 24   r-d

Solanum tuberosum   potato   l-a,h
  2n = 24,36,48   b-i   f-h   Many cvs. produce
  nonfunctional pollen.   USA-IR-1   r-d,f,f3,j,l

Solenostemon rotundifolius   Lamiaceae   Hausa
  potato   l-p,h   2n = about 84   r-d,f3

Sorghum bicolor   Poaceae   sorghum   l-a,g
  2n = 20   USA-S-9   r-d,l

Sorghum halepense   johnsongrass   l-p,g
  2n = 20,40   USA-S-9   The Land Institute
  has a program of crossing and breeding with
  S. halepense and *S. bicolor* in an effort to
  develop perennial, winter-hardy sorghum.
  r-d,l,*The Land Report* Spring 1991,Wagoner
  1990

Sorghum sudanense   Sudan grass   l-a,g
  2n = 20   r-d,l

Sorghum vulgare   sorghum   b-i   f-h   d-300m;
  400m from *S. sudanense*   y2-25 lb/100
  sq.ft.max.   r-f,j

Sorghum vulgare var. sudanensis   Sudan grass
  l-g   b-i   f-h   d-0   r-d,f

Sorghum × almum   almum sorghum   l-p,g
  2n = 40   USA-S-9   r-d,l

Sphenostylis stenocarpa   Fabaceae   yam bean,
  yam pea   l-p,p/a,h   2n = 18   USA-S-9
  r-d,d2,f3,l

Spilanthes acmella   Asteraceae   toothache plant,
  para cress   l-a,h   2n = 14,24,52   r-d,f3

Spinacia oleracea   Chenopodiaceae   spinach
  l-a,h   2n = 12   b-o   f-d   Populations
  include male plants, female plants, and monoecious hermaphrodite plants.   v-wind
  d2-500–1,000m   y-800 kg/ha; 2,000 max.
  y2-10.8 lb/100 sq.ft.max.   USA-NC-7
  r-d,f,f3,g,j,l,r

Spondias mombin   Anacardiaceae   hog plum   l-
  p,t   2n = 32   r-d,f3,l

Spondias purpurea   red mombin   l-p,t   r-d,f3,l

Sporobolus airoides   Poaceae   alkali sacaton
  l-p,g   2n = 84-126   USA-W-6   r-d,f3,l

Sporobolus cryptandrus   sand dropseed   l-p,g
  2n = 18,36,72   r-d,f3,l

Stachys affinis   (= *S. sieboldii*)   Labiatae
  Chinese artichoke   l-p,h   2n = ?   USA-NE-
  9   r-d,f3,l,z

Stevia rebaudiana   Asteraceae   kaa he'e   l-p,s
  2n = 22   USA-CR-MIA   r-d,f3,l

Syagrus coronata   Erecaceae   ouricury palm
  l-p,t   2n = 32   r-d,f3

Symphytum asperum   (= *S. asperrimum*)   Borigi-
  naceae   prickly comfrey   2n = 40   r-d,z

Symphytum officinale   common comfrey
  2n = 26,about 36,40,about 40,48   USA-NC-
  7   r-d,f3,l,z

Symphytum peregrinum   see *Symphytum × uplandicum*

Symphytum × uplandicum   Quaker comfrey
  l-p,h,s   2n = 36   USA-NC-7   r-d,
  f3,l,z

Synsepalum dulcificum   Sapotaceae   miracle fruit,
  miraculous berry   l-p,s,t   r-d,f3,l

Tamarindus indica   Fabaceae   tamarind   l-p,t
  2n = 24   r-d,f3,l

Tanacetum vulgare   Asteraceae   tansy   l-p,h
  2n = 18   r-d,f3,l

Taraxacum kok-saghyz   Asteraceae   Russian dandelion   l-p,h   2n = 16   r-d

Taraxacum officinale   dandelion   l-p,h
  2n = 8,24,36(r-d)   2n = 16,24,46(r-r)   May
  be apomictic.   r-d,f3,j,l,r

Telfairia occidentalis   Cucurbitaceae   oyster nut
  tree   l-p,l   r-d,f3

Tephrosia candida   Fabaceae   white tephrosia
  l-p,s   2n = 22   USA-S-9   r-d,l

Tephrosia vogelii   vogel tephrosia   l-p,s
  2n = 22   USA-S-9   r-d,l

Terminalia catappa   Combretaceae   tropical

almond l-p,t 2n = 24 USA-CR-MIA
r-d,f3,l

Tetragonia tetragonioides Aizoaceae New Zealand spinach l-a,h 2n = 32 y2-17.2 lb/100 sq.ft.max. USA-S-9 r-d,f3,j,l

Tetragonolobus purpureus Fabaceae asparagus pea, winged pea l-a,h 2n = 14 USA-NC-7 r-d,d2,f3,l

Thaumatococcus daniellii Marantaceae miracle fruit l-h 2n = 20 r-d

Theobroma bicolor Sterculiaceae bacao l-p,s,t 2n = 20 r-d,f3

Theobroma cacao cacao l-p,t 2n = 16,20,26 b-o f-h x-more than 30% s-i,c v-midges, ants r-d,f,f3

Theobroma grandiflorum cupuacu l-p,t r-d,f3

Thymus serphyllum Lamiaceae creeping thyme l-p,s 2n = 20,24 r-d,f3

Thymus sp. USA-NC-7 r-l

Thymus vulgaris common thyme l-p,h 2n = 30 r-d,f3

Tragopogon porrifolius Asteraceae salsify, oyster plant l-b/a,h 2n = 12 y2-27.7 lb/100 sq.ft.max. USA-W-6 r-d,f3,j,l

Trapa natans Trapaceae caltrop, water chestnut l-a,p,h 2n = 36,40,48 r-d,f3,l

Treculia africana Moraceae muzinda l-p,t r-d,f3

Trichosanthes cucumerina (= *T anquina*) Cucurbitaceae snake gourd, club gourd l-a,v 2n = 22 b-o f-m USA-NE-9 r-d,f3,l,s2(p306)

Trifolium hybridum Fabaceae alsike clover l-p,p/b,h 2n = 16 b-o f-h x-more than 90% s-mostly i v-bees USA-NE-9 r-d,d2,f3,j,l

Trifolium incarnatum crimson clover l-a,h 2n = 14,16 v-bees necessary for self- or cross-pollination USA-S-9 r-d,d2, f3,j

Trifolium pratense red clover l-b,p,h

2n = 14,28 b-o f-h x-more than 90% s-mostly i v-bees USA-NE-9 r-d, f,f3,j,l

Trifolium repens white clover l-p,h 2n = 32,48,64 b-o f-h x-more than 90% s-mostly i v-bees USA-NE-9 r-d,f3,j,l

Trifolium spp. b-i,o f-h About 30% of species are s-i and v-bees, and some are cleistogamous. About 70% are b-i. Small-flowered species are often b-i; large-flowered species, b-o and v-bees. r-f2

Trifolium spp. y2-(sweet, hubam clover)2.2 + lb/100 sq.ft.max. y2-(timothy clover)0.46 + lb/100 sq.ft.max. r-j

Trifolium spp. See r-f3 for uses of additional species. See r-d2 for breeding systems of additional species.

Trigonella foenum-graecum Fabaceae fenugreek l-a,h 2n = 16 USA-W-6 r-d,d2,f3,l

Triphasia trifolia Rutaceae limeberry l-p,s 2n = 18 r-d,f3

Tripsacum dactyloides Poaceae eastern gamagrass (warm-season bunchgrass) l-p,g 2n − 36,72 Wild perennial grain with large, edible seeds. Efforts to domesticate it into a perennial grain crop are under way at The Land Institute. For review, see Wagoner 1990. USA-NC-7 r-d3,l,*The Land Report* Spring 1991,Wagoner 1990

Triticosecale spp. (*Triticum x secale*?) Poaceae triticale l-a,g 2n = 42,56 r-d

Triticum aestivum Poaceae wheat l-a,g 2n = 42 USA-NSGC r-d,l

Triticum carthlicum Persian wheat l-a,g 2n = 28 USA-NSGC r-d,l

Triticum compactum club wheat l-a,g 2n = 42 USA-NSGC r-d,l

Triticum dicoccon emmer l-a,g 2n = 28 USA-NSGC r-d,l

Triticum durum durum wheat l-a,g 2n = 28 USA-NSGC r-d,l

Triticum monococcum einkorn l-a,g 2n = 14 USA-NSGC r-d,l

Triticum polonicum Polish wheat l-a,g 2n = 28 USA-NSGC r-d,l

Triticum spelta spelt l-a,g 2n = 42 USA-NSGC r-d,l

Triticum spp. triticale and wheat b-i f-h largely cleistogamous x-less than 6% by wind d-0 USA-NSGC d-f,l

Triticum spp. wheat y2-(durum wheat)26 lb/100 sq.ft.max. y2-(Early Stone Age wheat)17+ lb/100 sq.ft.max. y2-(hard red spring, red winter, or white wheat)26 lb/100 sq.ft.max. USA-NSGC r-j,l

Triticum spp. wheat hybrids f-h + female (male sterile) d-200m USA-NSGC d-f,l

Triticum spp. See r-f3 for additional information.

Triticum timopheevii timopheev wheat l-a,g 2n = 28 USA-NSGC r-d,l

Triticum turanicum Oriental wheat l-a,g 2n = 28 USA-NSGC r-d,l

Triticum turgidum poulard wheat l-a,g 2n = 28,42 USA-NSGC r-d,l

Tropaeolum majus Tropaeolaceae nasturtium l-a,h 2n = 28 USA-S-9 r-d3,f3,l

Tropaeolum minus climbing or tall nasturtium l-a,p?,h 2n = 28 r-d,d3,f3

Tropaeolum tuberosum anu, mashua, isano l-p,h 2n = 42 r-d,d3,f3,*Lost Crops of the Incas* 1989

Urena lobata Malvaceae ullucu l-p,h 2n = 24,36 USA-S-9 r-d,f3,l

Urtica dioica Urticaceae stinging nettle l-p,h 2n = 32,48,52 r-d,f3

Vaccinium angustifolium Ericaceae lowbush blueberry l-p,s 2n = 24,48 USA-CR-COR r-d,f3,l

Vaccinium ashei rabbiteye blueberry l-p,s 2n = 72 USA-CR-COR r-d,f3,l

Vaccinium corymbosum highbush blueberry l-p,s 2n = 48,72 USA-CR-COR r-d,f3,l

Vaccinium macrocarpon cranberry l-p,l 2n = 24 USA-CR-COR r-d,f3,l

Vaccinium vitis-idaea mountain cranberry l-p,s,h,v 2n = 24 USA-CR-COR r-d,f3,l

Valerianella locusta (= *V. olitoria*) Valerianaceae cornsalad, mache l-a,h 2n = 14 r-d,d3,f3

Vicia faba (= *V. fava*) Fabaceae fava or faba bean; broad, bell, horse, tick, or field bean l-a,h 2n = 12 b-i,o f-h x-25–50% by bees (r-f2) d2-most people say 300m, but more recent information says 1,000m. y-1,500–2,000 kg/ha y2-("bell bean") 4.5 lb/100 sq.ft.max. USA-W-6 r-d,d2,f,f2,f3,g,l,s2

Vicia monantha bard vetch l-a,h 2n = 14 USA-S-9 r-d,d2,l

Vicia pannonica Hungarian vetch l-a,h 2n = 12 b-i bees required f-h d-3m USA-S-9 r-d,d2,f,l

Vicia sativa vetch l-a,h,v 2n = 10 b-i bees required f-h d-3m USA-W-6 r-d,d2,f,l

Vicia sativa ssp. *nigra* blackpod vetch l-a,v 2n = 12 USA-W-6 r-d,l

Vicia spp. For uses see r-d2,f3.

Vicia villosa winter/hairy vetch l-a,b,p,h,v 2n = 14 v-bees y2-1.1+ lb/100 sq.ft.max. USA-S-9 r-d,d2,j,l

Vigna aconitifolia Fabaceae moth bean l-a,h 2n = 22 s-c USA-S-9 r-d,d2,f3,l

Vigna angularis adzuki bean l-a,h 2n = 22 x-common s-c USA-S-9 r-d,d2,f3,l

Vigna mungo (= *Phaseolus mungo*) urd bean, black gram l-a,h 2n = 22,24 b-i f-h up to 42% cleistogamous USA-S-9 r-d,d2,f,f3,l

Vigna radiata (= *Phaseolus aureus* = *Phaseolus mungo*) mung bean, green or golden gram l-a,h 2n = 22 b-i f-h d-0 y-0.5 tonne/ha y2-24 lb/100 sq.ft.max. USA-S-9 r-d,d2,f,f3,j,l

Vigna umbellata   rice bean   l-a,h   2n = 22   s-c
USA-S-9   r-d,d2,f3,l

Vigna unguiculata   cowpea, yard-long bean   l-a,h
2n = 22   y-0.7 tonne/ha   y2-24 lb/100
sq.ft.max.   USA-S-9   r-d,d2,f3,g,j,l

Vigna unguiculata ssp. cylindrica   catjang, cowpea   l-a,h   2n = 22   USA-S-9   r-d,d2,f3,l

Vigna unguiculata ssp. sesquipedalis   yard-long
bean, asparagus bean   l-a,p,h   2n = 22   b-i
x-10–15% by ants, flies, bees   v-heavy insects
USA-S-9   r-d,d2,f3,l

Vigna vexillata   aka sasage, zombi pea   l-p,h,v
2n = 22   USA-S-9   r-d,d2,f3,l

Vitis labrusca   Vitaceae   fox grape   l-p,l,v
2n = 38   USA-GD   r-d,f3,l

Vitis rotundifolia   muscadine grape   l-p,l,v
2n = 40   b-o   f-d   v-insects   USA-CR-DAV
r-d,f,f3,l

Vitis vinifera   wine grape   l-p,l,v
2n = 38,40,57   b-i   f-h   s-some cvs. partly
or fully s-i   USA-GD   r-d,f,f3,l

Voandezia poissonii   see *Macrotyloma geocarpum*

Voandzeia subterranea   Fabaceae   bambara
groundnut   l-a,h   2n = 22   b-i,o   v-ants
y-0.6 tonne/ha   r-d,d2,f3,g,l

Wasabia japonica   (= *W. pungens* = *Eutrema
wasabi*)   Brassicaceae   wasabi, Japanese
horseradish   l-p,h   2n = 28   b-o   f-h   v-
flies?   r-f3,l,pc(Dan Borman),z

Xanthosoma sagittifolium   Araceae   yautia   l-p,h
2n = 24,26(r-d)   2n = 26(r-s2)   r-d,f3,l,s2

Xanthosoma spp.   zanier, yautia, cocoyam
2n = 26   r-s2

Yucca elephantipes   Agavaceae   ozote   l-p,t
r-d,f3,l

Zea diploperennis   Poaceae   diploid perennial
teosinte   l-p,g   2n = 20   Can be crossed
with *Z. mays* to produce fertile hybrids.
USA-NC-7   r-f3,l,Wagoner 1990

Zea mays   Will cross with wild perennials such
as *Z. perennis*, *Z. diploperennis*, and *Tripsacum*
spp.   r-Wagoner 1990

Zea mays ssp. mays   corn, maize   l-a,g
2n = 20   b-o   f-m   Often somewhat protandrous.   Displays severe inbreeding depression (r-Allard 1960).   v-wind   d-200m   d2-
1 km   y-1,500 kg/ha; 2,500 max.   y2-(sweet
corn)10.3 lb/100 sq.ft.max.   y2-(fodder
corn)22.6 lb/100 sq.ft.max.   USA-NC-7
r-b,d,f3,g,j,l,Allard 1960

Zea mays ssp. mexicana   teosinte   l-a,g
2n = 20   USA-NC-7   r-j,l

Zea perennis   perennial teosinte   2n = 40
USA-NC-7   r-f3,l,Wagoner 1990

Zingiber mioga   Zingiberaceae   temperate ginger,
Japanese wild ginger   l-p,h   2n = 55
r-d3,pc(Alan Kapuler)

Zingiber officinale   ginger   l-b/a,p/a,h
2n = 22,24   r-d,f3,l

Zizania aquatica   Poaceae   Indian, Tuscarora, or
wild rice   l-g   2n = 30   r-f3,l

Zizania palustris   northern wild rice   l-a,g
2n = 30   r-d,l

Ziziphus jujuba   Rhamnaceae   common jujube,
Chinese date   l-p,s,t   2n = 48   USA-CR-
DAV   r-d,f3,l

Ziziphus mauritiana   Indian jujube   USA-CR-
MIA   r-d,f3,l

PART II
Seed Saving
Practice

SIXTEEN

An Introduction to Seed Saving

ONE WALL of my living room is lined with shelves full of jars of my own seed. No wall decoration could provide more beauty, comfort, and security. Nothing else is such a good conversation piece. Here is a touchstone, a shrine to what is essential. Here are memories of past seasons and accomplishments, and dreams for those to come. Here is a symbol that this is, indeed, hearth and home and homestead. Here is hope for the future.

Why Save Seeds?

Saving seeds is fun. Cleaning the seed, holding the clean seed in your hands, is magical. Gaze at the seed, run your fingers through it, play with it, and you can feel the connections.

You're like a child with a gallon bucket of marbles, or a squirrel sitting on a hollow log full of acorns. Unquenchable joy arises. It is so intense it puzzles you initially. Then you recognize it. It is the joy that comes from being who you are supposed to be and doing what you are meant to do.

Seed saving is practical. If you know how to save your own seeds you can grow rare varieties. Many of the most spectacularly flavorful, unique varieties are not readily available commercially, either as fruits or seed. One of my favorite winter squash is 'Blue Banana', for example. This squash has a flavor that is superb, intense, and so different from all other squash that it is like an entirely different vegetable. But the seed is not available commercially. To grow rare varieties, you often have to get the seed when and where it is available, then maintain the variety yourself.

Many superb varieties are not readily available commercially because they have narrow adaptations to particular regions. 'Narragansett Indian Flint', for example, is said to be the corn that was given to the Pilgrims by the Indians, the corn that made the jonnycakes that were eaten at the first Thanksgiving. However, 'Narragansett Indian Flint' (also known as 'Rhode Island White Cap') has a narrow ecological adaptation. It likes the quasi-Mediterranean climate of Southern New England. It is too long-season for Northern New England, and too heat-intolerant for the Southeast or Midwest. It is not widely enough adapted nor popular enough to warrant its production as a seed crop via normal channels.

'Narragansett' makes wonderful jonnycakes — the very best, I'm told. I don't know if they are the very best possible, but I've tried them and they are truly delicious. 'Narragansett' also makes rich, full-bodied cereal and polenta. But you can't buy the seed from commercial suppliers. If you garden in the Midwest, of course, you don't care, because you can't grow the corn anyway. But given our current system for mass-producing food and seed crops, the fact that you can't grow the corn in the Midwest now means that you can't grow it in southern New England, either — unless you can save your own seed.

Some varieties are not available because they have peculiarities with respect to production of the seed itself. If a watermelon produces few seeds, for example, it will not usually be offered commercially. It's simply too expensive to produce the seed. A home gardener, though, might be happy to save such seed. And a market garden might be able to easily produce the handful of seeds needed for a single field's planting.

Being dependent upon seed companies for your seed means being dependent upon random fads in foods as well as other people's choices and preferences. Saving your own seed means independence. It lets you make your own choices and have your own preferences.

When you save your own seed, the seed is always "available." It is common these days for all the seed of even very popular varieties to be produced by just a single grower. If that grower experiences a crop failure, the seed isn't available anywhere.

Sometimes, even if the seed is "available," you can't necessarily find it. There can be a poor correlation between variety names and the material you actually receive. Seed companies often change lines or suppliers, so that what they are selling one year and the next may be different strains, even though they are called the same thing.

I once grew a squash that the packet identified as 'Red Kuri'. It was a scarlet teardrop shape, delicious and uniquely flavored. The next time I ordered the seed the squash was orange-and-green speckled and inferior in flavor. Somewhere, somehow, the variety had been crossed up. And many companies were selling the crossed-up seed.

I ordered a packet of seed from each of a half-dozen companies, growing a few plants from a new company each year. Some companies were selling an inferior-flavored orange thing as 'Red Kuri'. I have yet to find that wonderful scarlet variety again. Maybe all the material I've tried in recent years is really something else, not 'Red Kuri' at all. Maybe the variety I liked so much in the first place was something else. All I can say at this point is . . . I wish I had saved the seed.

I like to produce my own seed even of varieties that are readily available commercially. My own seed is usually bigger, fatter, and more vigorous. I can plant it earlier than commercial seed. I also have much more of it, so I don't have to skimp. I can sow generously and then thin, instead of sowing thinly, then having gaps that have to be replanted later and less optimally.

And with my own seed, the price is right.

When you save seed, you become a plant breeder. You are choosing which germplasm to perpetuate. This means that you are both deliberately as well as automatically selecting for characteristics that are important to you, for plants that are fine-tuned to your needs and growing conditions and region. After you have saved seed of a variety for a few years, you have your own line of the variety that is slightly different from anyone else's, and it is usually better adapted to your needs.

Knowing how to save your own seed also means that you can take advantage of genetic accidents, ideas, and dreams. Last year, for example, I noticed one squash plant in perhaps a hundred that was resistant to powdery mildew. I saved the seed from it. Perhaps I can use it to develop new powdery-mildew-resistant varieties. Powdery mildew after the first fall rains is what ends the squash grow-ing season in my region. Resistant varieties could be very useful. Many new varieties got their start when some gardener or farmer simply noticed something that was different and special — and saved the seeds.

We gardeners and farmers care about our direct relationship with soil, plants, and food. To grow plants from seed bought from others is one level of relationship. To grow plants from our own seed, to save seeds from our own plants, goes to a deeper level. It is fulfillment and continuity — plants and people maintaining each other, nurturing each other, evolving together. It completes the circle.

Saving Seed from Hybrids

Hybrids don't breed true to type from seed. Some hybrids are even sterile, though most will produce seed. This seed can be used to derive a pure-breeding variety by the methods described in chapters 9 and 10. Such a variety derived from a hybrid is a new variety and should be given a new name. It is not the same as the hybrid from which it was derived. In other words, you can save seed from hybrids as the first step in creating a new open-pollinated variety, but you cannot reproduce a hybrid by saving its seed.

This section on seed-saving practice, then, refers to *pure-breeding,* not hybrid, varieties.

Seed-Saving Overview

Saving seed is easy. Plants want to make seed. They cooperate fully. To save seed, all you have to do is let the plants produce seed, then grab it quick before the birds or squirrels or bugs, and before it gets rained on and molds or sprouts in the pod.

Saving seed of pure varieties is another thing entirely. Plants don't care at all about pure varieties. The outbreeders would all rather cross with that strange inedible ornamental variety down the street in the yard of your neighbor. Even the inbreeders outcross far more often than they are "supposed to," especially under organic growing conditions.

To save seed of pure varieties, we need to know something about the outcrossing tendencies of the crop so that we can isolate it sufficiently from other varieties or wild plants of the species that it could cross with.

Finally, every variety contains genetic variability. Some of this is desirable and even essential to the vigor and adaptability of the variety. Some of it, though, is undesirable. So, we need to grow an appropriate number of plants in order to maintain the amount of genetic variability that we want. At the same time, we must select and rogue to eliminate the genes associated with specific kinds of variability that we don't want.

Given the genetic heterogeneity in most varieties and the greater vigor of the more wild-type forms, the natural tendency of most varieties is to deteriorate quickly to something that is far less useful to its human associates. To maintain a variety we must actively breed in order to counter this tendency.

There is actually no such thing as "saving" a pure variety. There is only further breeding, either deliberate or accidental. We either select in order to hold the variety in its current form and to eliminate undesirable types, or we select in order to change the variety in some preferred direction. Both processes involve exactly the same principles.

Roles and Purposes

"What's my role with respect to this variety?" That's the first thing I ask myself about every seed-saving project. Am I the sole savior or creator of the variety, the one person without whom it would be lost forever? Or is my line better than everyone else's, and especially worthy of preserving and distributing?

Am I planning on building up the precious stock, then giving or selling it to seed companies or others? Will I be distributing it through the Seed Savers Exchange? Will many or even all future plantings of this variety all over the country be descendants of these seeds I hold in my hands today? If so, I will want to be pretty careful and rigorous. I will use serious numbers of plants, and serious isolation distances.

Often, however, I'm saving seed just for myself, and I know others have the variety as well. In that case, I can be quite casual about most nearly everything. Numbers of plants? I grow what I need for the table, and use special tricks (see Chapter 19) to deal with maintaining heterogeneity.

Isolation? It's often minimal. I usually plant so as to be able to recognize hybrids, which is much easier than avoiding them (see Chapter 18). If I can recognize hybrids I can eliminate them or not as I choose in future generations. Who knows? The hybrid might be more interesting than the original material. And if the seed is just for my own use, what's an outcross or two among friends?

Growing Seed

I T'S THE MIDDLE of winter. I am sitting indoors in my big easy chair while it rains outside. I'm looking at seeds. I always visually examine all the seed I'm planning to plant each year, and winter is the best time for it. I spread the corn out in a tray in my lap. I remove any weed seeds and cull out any seeds that are off-type — or make plans to plant them separately.

Yipe! What's that white flour corn kernel doing here in this lot of flint corn? Obviously, that's what the seed company ran through the seed cleaning equipment before the flint. Get that out of there! The 'Magenta' corn is supposed to be magenta. Out with any occasional white or black kernels!

Then there's the 25-pound bag of 'Scarlet Runner' beans I bought just to eat. The seed is smaller than I expected, and I suspect they are a bush type. I'll plant some. I think I would like to cross a pure black seed into a runner bush. And it would be nice if I could get the lavender color separate. Different pure colors would probably have different flavors. Let's look at all the beans, pull out some to plant, see if there is anything unusual.

So I am sorting through the 25 pounds of beans, one tray at a time. Never can tell. Maybe . . . ha! A black! I sort a bit more. More blacks. Some different shapes. A nearly solid *brown*. I never even dared hope for that. I thought it should be possible theoretically, but never saw one, not in the runner bean species. What a find! Pure brown line, here we come!

I finish the sorting with 24¾ pounds of beans left to eat and a half pint of unusual types. I'll plant some of the ordinary beans, of course. But it is the "off" types that are the most exciting. To my planting plans I add the "Mutant Ninja Runner Bean Project."

Preparation and Planning

The most important part of seed saving happens before you turn over the first square foot of soil or plant the first seed. It's planning. Saving seed should be part of every garden's yearly harvest. But it *does* require extra planning. In this chapter I give general rules and suggestions about growing seed, a checklist to review as you plan your plantings. Certain topics — isolation distances, numbers of plants to grow, and selection — are only mentioned here, and are covered in separate chapters.

If you are new to seed saving, the planning may seem complicated at first. It introduces new sets of issues. With a bit of experience and familiarity, however, you will take seed saving into account automatically, just as our ancestors did. You'll begin to plan for seed saving just "naturally," and the planning is part of the fun.

I do most of my garden and field planning during the winter and early spring. I move all the seed I'm going to plant to a single big box, with the varieties of each species stacked together. At a glance I can tell how many isolations I need, and can figure out how to arrange them within the context of everything else I'm going to plant. I can also tell at a glance whether my plans and dreams fit the amount of land and time I have. (They never do initially; I always have to remove some stuff from the box.)

Here are some general rules and considerations:

Visually inspect your planting stock. If the corn seed is supposed to be sweet, cull out that field-type kernel! If it is supposed to be white, get rid of that yellow.

Even tiny seed is worth examining carefully. You can detect and eliminate contaminating weed seeds.

Don't, however, eliminate variability in seed type arbitrarily. There is no reason the mustard seed needs to be all one color.

If you're going to go to the trouble to save seed of something, it's very nice to know what it is. Record what you are going to plant — *exactly* what you are going to plant, including the exact source of the seed. If I bought the lot of seed from Territorial in 1997, for example, "T97" will follow the variety name in my notes as well as on the field marker. "98;T97" would represent seed from my own growout in 1998 of seed I obtained originally from Territorial in 1997.

Remember that many very different lines are being sold under the same variety name, and that seed companies may change suppliers from one year to the next. You haven't identified anything very well with just a variety name. You have to write down the source and the year.

Plan, organize, and label things in the field so you can tell which variety is which. Give plants enough space so that you don't confuse different varieties. The plants need the space anyway. Crowded plants may yield decent vegetables, but they don't produce very good seed.

It's better to alternate varieties that are distinguishable. Separate the two green-podded varieties of pole beans with a purple- or yellow-podded one, for example. Or alternate species and accomplish both identification and isolation simultaneously. Alternate sections of common beans *(Phaseolus vulgaris)* with runner beans *(Phaseolus coccineus)*, for example.

If you're going to do hand-pollinations with squash, thin to one plant per hill, and plant hills far enough apart so that you can easily trace a vine to its source.

Rotate seed crops. This is useful for all crops as a means of minimizing pests and disease, but it is critical for seed crops that are capable of volunteering. If you plant squash and volunteers come up all over the patch, you'll be hard pressed to know who anybody is in July (before fruits develop), when you need to hand-pollinate flowers for breeding or seed saving. Plant the squash where you did not have any squash last year.

Gophers, moles, and cats often rearrange plantings gratuitously. All the more reason to try to choose varieties for each row and for adjacent rows that are distinguishable. I still have squash seed from 1998 with labels like "98 big green mystery 6–2." Gophers were especially active in the squash patch that year, and they rearranged several sections of seeds before they germinated.

Then there was the year I entertained myself and the local squirrels by letting them have all the unneeded squash seed as we ate the squashes during the winter. Between the squash and the oilseed sunflower seeds I put out for birds, those squirrels had a wonderful winter. I enjoyed watching them stuff their little cheeks. It was amazing how much squash and sunflower seed they ran off with.

That spring, when I went down to the garden to plant some ornamental sunflowers I was planning to continue breeding, I noticed that there were already things that looked quite a lot like sunflowers coming up in the area I wanted to plant. In fact — they were sunflowers. Oilseed sunflowers.

That year, oilseed sunflowers and cull-squash-seed squash came up all over my garden and compost heap as well as those of my next-door neighbor. I had forgotten something important. Humans are not the only creatures that plant things. The squirrels planted all the excess seeds I gave them. They planted them in all the soft, easy places to dig — gardens and compost piles.

All varieties outcross to some extent. The so-called outbreeders simply do more of it than the so-called inbreeders. Furthermore, there is much more outcrossing if you garden or farm using organic methods. Living land is full of pollinators of many species.

Every variety will need isolation from other varieties as well as from wild plants of the same species if you want to save seed and keep the variety pure. How much isolation you need depends upon the specific variety, the situation, and your goals and purposes with respect to the seed. Isolation is covered in Chapter 18. Organizing your plantings so as to provide appropriate isolation is a major part of the planning.

How many plants do you need to save seed from? This issue is covered in Chapter 19.

In some cases you can eat and save seed

from the same patch. In other cases you have to choose one or the other. You can't both eat a green bean and save the seed, for example. But you can eat and save seed from the same winter squash.

You'll often need to use one part of the patch for seed saving, or mark the plants you want to save.

Are the needs of the seed crop going to be different from those of the crop if you were just growing it as a vegetable? If plants of the seed crop must be allowed to dry out, you may want to put the planting in a separate bed or at one end of your garden. That way you can stop watering the seed crop when the plants begin to dry without interfering with the watering of the rest of your garden.

Seed crops are also good candidates for growing using drip irrigation methods.

Do you want adequate seed or the very best possible seed? In many cases, adequate seed is all you really need. If you want to save the best possible seed, you'll probably have to give the plants extra space, extra nutrients, and optimal watering.

Almost all seed crops need to stay in the ground somewhat beyond eating stage in order to produce seed. They often need more space too, because they are going to be much bigger than regular vegetable plants before they're done making seed. In many other cases, however, you grow the plants the same way you would for just a vegetable crop, and simply use certain plants or a certain section for seed saving.

Some plants, the biennials, produce seed only in their second year. So, if you are growing most types of onions, carrots, or cabbage, for example, you will probably need to lift and store plants during the winter, then replant them in spring. See Suzanne Ashworth's book *Seed to Seed* for more information about growing various biennials for seed.

Some crops are very easy to save seed from. Self-pollinating annuals such as beans, peas, and tomatoes can be handled with very little more work than it takes to grow the vegetable crop.

Unfortunately, you probably can't save seed of everything you might like to grow. You can grow some things as vegetables that you can't do as a seed crop. For example, any crop that is late-season for your area is unlikely to produce a seed crop reliably, since the seed crop requires extra growing time. Even if you can produce seed some years, the seed is likely to be inferior to what you can buy commercially.

You can seed-save early and midseason sweet corns easily, for example. The latest variety you can produce as a sweet corn crop will need to be maintained by others who have a longer or warmer growing season.

Planting

The following are a few simple rules related to planting.

1. Inspect all seed before you plant it. Eliminate weed seeds. Remove off-types and either eliminate them or plant them separately.

Remove any seed that has a bumpy skin or atypical blotches. Such seed is usually from diseased plants.

Smaller or lighter seed is often from immature plants or pods, and doesn't germinate as well as typical seed.

2. Never plant all the seed at once or in the same year if it is material you can't afford to lose and can't replace.

3. Don't experiment on seed you can't replace. When the seed is rare or unique, it is tempting to give it extra care and treat it specially. Resist the urge, and use your normal methods. Try out any new methods on seed you don't care about. Don't presoak seed for the first time with a rare lot of seed, for example.

4. Plant plenty of seed if you have it available, then thin to the desired number of plants. That way you will be selecting for material that germinates vigorously under your conditions. If you plant minimal amounts of seed and keep every plant, you won't be doing much selecting.

Tending Plants

1. Watch plants for disease. Some diseases are seedborne. Don't save seed from diseased plants or plants that are stunted, or that produced mostly sterile pods, or that failed to thrive. If anything looks diseased, and I don't recognize the disease, I either get a professional diagnosis fast, or I eliminate the plants fast, to be on the safe side.

Universities with agricultural colleges usually have faculty or staff who can help you identify an unfamiliar disease. In many cases they can do it from just a good description. If they can't help, they usually can tell you whom to take or send specimens to.

2. Practice selection and roguing, as described in Chapter 20.

3. If establishing isolation depends upon detasseling (corn) or removing flower scapes on some plants, remember to do that during the season.

4. Mark plants or sections of plants that are reserved for seed saving. Mark them really obviously and prominently. Bright orange surveyor's tape is good — I use big swaths of it. Tell everyone involved in harvesting what the markers mean. After all, it's hard to save seed from a fruit someone has already eaten accidentally.

Harvesting seed is covered in Chapter 21.

Isolation

CONSIDER Buffalo Bird Woman, a Hidatsa Indian woman, working in her garden in North Dakota in the mid-1800s. She doesn't know anything about the role of pollen in reproduction of plants. She has never heard of isolation distances. Yet she saves all her own seed. She even raises extra seed corn to trade to others. A string of her seed-grade ears of corn is so valuable, it can be traded for a tanned buffalo skin. (See *Buffalo Bird Woman's Garden: Agriculture of the Hidatsa Indians,* by Gilbert L. Wilson; Minnesota Historical Society Press, 1987.)

Yet, today, in order to save seed, we need to talk extensively about isolation distances. How did Buffalo Bird Woman get away with not needing to know about all this?

Imagine a pioneer family with their fields and gardens. They save seed from their own corn, soup beans, pole beans, and other crops. They don't know any more about plant genetics than the Indians. Yet, they, too, are "getting away with it." How is this possible?

Traditional Seed Saving and Isolation Distances

I believe the explanation lies in a number of factors, a combination of social and cultural habits, as well as a few useful myths:

1. Distinct varieties were planted separately. Different corn varieties were each planted in plots of their own, never mixed together.

2. Sizes of the plantings were large enough so as to provide automatic isolation.

3. Plantings were often separated from those of neighbors by pastures, woods, or

other kinds of plantings.

4. Often, each family grew just one variety of a species. Grandma had "a favorite bean" that had been passed down in her family, not several favorite beans.

5. Where the family grew two or more varieties of a species, the different varieties often had different growing niches, maturity times, or purposes that dictated that they would be grown in different areas.

6. If a family grew two or more varieties of a species, the plants and the seeds looked distinctly different. (A family would not grow two varieties of white beans that looked the same, for example.)

7. The gardener or farmer was thoroughly familiar with her varieties, so that off-type plants or seeds were easily recognized.

8. Gardeners and farmers had gardening and farming parents, and very traditional patterns of planting that had been passed down through the generations. Seed savers had seed-saving parents, and inherited the patterns as well as the seeds.

9. Where scientific knowledge did not suffice, useful myths filled in the gaps.

10. Maintaining pure varieties was sometimes not important.

11. Even when "pure varieties" were important, the concept of "pure variety" was much looser, making people more tolerant of less-than-perfect seed saving.

Buffalo Bird Woman planted five different corn varieties: hard white (flint), hard yellow (flint), soft white (flour), soft yellow (flour), and sweet. She planted each in a separate field. The corn varieties had maturity times ranging over the entire season, with the soft white being earliest. Differing maturity times helped provide isolation.

Buffalo Bird Woman knew that "corn travels." That is, she knew that growing different types in fields adjacent to each other didn't work, that off-color and off-type kernels would be found in the ears. She had a cultural myth that the kernels of one type could "travel" and get into ears of another type if the fields were too close together. The myth of "traveling" helped her keep varieties pure by encouraging appropriate isolation.

Sometimes, Buffalo Bird Woman tells us, women with adjacent fields made arrangements so that the adjacent corns would be the same variety. They also tried to separate fields of different types. They learned from experience, for example, that corn traveled more over "soft ground." In other words, the corn was better isolated by untilled ground, which probably contained trees and brush that broke up wind patterns.

Finally, Buffalo Bird Woman saved only the best, biggest, most true-to-type ears for seed. In this way, she would recognize and eliminate many of the crosses that did occur. She and her neighbors grew only five varieties of corn, each distinctly different in appearance, and each thoroughly familiar. Recognizing crosses and eliminating them was relatively easy. It didn't matter very much how many crosses occurred if she could eliminate them from her seed crop each season.

Buffalo Bird Woman also had five kinds of beans, which appear to have been bush drying beans. Each had a different color. Her pattern for planting beans and corn put each in hills. She alternated rows of corn hills with rows of bean hills. Nine rows of corn with alternating hills of beans made a block that was separated from the next block with a row of hills of squash.

Buffalo Bird Woman kept beans of one type together in planting, threshing, and storing. It's a good guess that different types were planted in different blocks. These cultural and agricultural traditions gave pretty good isolation distances. If your mother was a successful seed saver and you learn and always follow her traditional planting orders and patterns, you don't need "book learning" about isolation. Your cultural pattern provides it.

Buffalo Bird Woman's squash were yellow or green or mottled, and of various shapes. She seems to have practiced no special methods to isolate different varieties, and apparently didn't have distinct varieties. Rather, she had a very diverse population of squash (*Cucurbita maxima*), and used all the shapes and colors as summer squash, either cooked immediately or dried in slices for winter use.

Now, imagine a pioneer family in a small settlement. Pioneer Man has two varieties of corn. One is early season and yellow. It will make a crop no matter what. The other is a full-season white that provides the best possible yield in an average or good year. In a bad year, though, it doesn't make a crop at all.

The maturity times are so different that the corns are largely isolated automatically. The early-season corn is yellow. The biggest, plumpest ears are the first ears on each plant, and they all form before the later, white corn is flowering. Just saving the biggest, prettiest ears for seed is enough to guarantee purity of the yellow corn.

The late tassels on the yellow corn can contribute pollen to the first ears on the late white corn. However, when this happens, pale yellow kernels appear on the white ears. Merely choosing fully white, true-to-type ears of white corn for seed is enough to eliminate nearly all crosses with the yellow.

Furthermore, the yellow corn is a small plant, and the white is big. The hybrid plants are intermediate, and are regularly eliminated.

A small creek with marshy land and alders separates Pioneer Man's farm from one neighbor. A woodlot separates the corn from that of the other closest neighbor, who is growing the same two varieties anyway. These two varieties are fine-tuned for excellence in this particular region. Everyone grows them.

Pioneer Woman has her favorite variety of pole bean for green beans. She grows just the one variety. She has a bush bean for shelly beans. It's planted fairly far from the pole bean, because it's short. The pole beans are the tallest thing in the garden, and are the last row.

There are bush drying beans, but they are a field crop. They're a long way from the pole beans.

Pioneer Woman has both hot and sweet peppers in her garden. Her mother taught her that if you grow hot and sweet peppers side by side, the sweet peppers will be hot. So Pioneer Woman grows hot and sweet peppers in opposite corners of the garden.

In fact, hot peppers cannot influence the flavor of sweet peppers. But if you grow the two types adjacent to each other, and save the seed, you'll have crosses. The crossed, hybrid seed, when grown up the next generation, will generally produce hot peppers.

The idea that hot peppers can influence the flavor of adjacent sweet peppers is a myth, but it's a useful myth. If you believe this myth, you isolate your peppers. You don't need to isolate them just to grow good sweet peppers, but you do need to if you are saving seed from sweet peppers.

The larger your plantings and the fewer the varieties you grow of any one species, the easier the seed saving. Seed saving is easiest for farmers and people with *large* gardens. With large plantings, you just plant different varieties in separate blocks and save seed from the middle. That alone gives you pretty good isolation for seed for home or farm use. Seed saving is also easiest if your plantings are far from those of neighbors.

All the added detail about isolation for seed savers of today is necessary because most of us these days don't have plantings that are very large, let alone far from those of our neighbors. Nor do very many of us have seed-saving parents whose traditions we could learn directly. Further, many of us want to grow lots of varieties of each species, not just one. And some of us are growing seed for distribution or sale, not just for home or homestead use.

We need to know more about isolation distances because we need to know what we can get away with. And we need to learn various tricks so that we can get away with as much as possible.

Isolation Distances for Organic Farmers and Gardeners

In order to maintain pure varieties, we have to isolate them from other varieties of the same species. If they are outbreeders, we have to isolate them. If they are inbreeders, we still have to isolate them, especially if we have healthy organic land that is alive with myriad species of pollinators. If the crop is "highly self-pollinating," it just means that the percentage of hybrids will be lower, not that there won't be any.

When I wrote the first edition of this book, I accepted conventional wisdom about isolating inbreeding crop plants. I said: "With the plants that are very highly inbreeding, only a little isolation is required — often just enough to prevent physical mixing of the plant material." This is probably valid for the corporate and university fields where data on isolation distances are most often determined experimentally. However, I have come to the conclusion that it is not true at all for those of us who grow garden or farm crops organically.

In retrospect, I should have become suspicious just by the "official" listings of pollinators. If you look in Table I, you'll see that for most of the insect-pollinated crops the listed pollinator is usually honeybees, or bumblebees, or both. Go to any healthy organic garden or farm, though, and look for the pollinators there. I'm talking about a long-time organic garden, one that includes or is flanked by permanent unmowed and unsprayed areas where complete insect life cycles take place.

If it's a farm-scale operation, there may be a little creek with wild banks. There are natural, unmowed corners and strips along fence lines. The less-fertile, rougher ground has been left as a woodlot, or is being allowed to regenerate whatever is natural for the area. A huge blackberry thicket, more effective than any fence, separates the property from the next-door neighbor. It's a haven for wildlife of all kinds, including insects.

When you examine plantings flowering in such healthy circumstances, you find that they usually have a dozen or more different pollinators working them. There are honeybees and bumblebees, of course, but also many types of small solitary bees, and dozens of different species of flies. I tried once to count the different species pollinating a few square feet of flowering mustard. I tallied at least a dozen flies and half a dozen bees of various kinds within just a few minutes of watching.

I didn't realize the implications of all this healthy-land "pollinator pressure," however, until I crossed up eight varieties of soup beans and thirty lines of chickpeas.

Beans are often considered one of the more highly inbreeding species. Backyard seed savers frequently give them no isolation at all, or only a few feet. There is controversy in the professional literature as to how inclined beans are to outcross. Most researchers claim they are highly inbreeding, and that little isolation is needed. A small minority report frequent outcrossing. I knew that such things have a lot to do with the climate and the insects of a region, so I took the precaution of consulting Jim Baggett.

Jim Baggett had been the vegetable breeder at Oregon State University for decades, and beans were one of his major crops. OSU is right here in Corvallis, in the Willamette Valley of western Oregon, just a few miles from my home and various farms where I raise crops. So I figured that whatever isolation distance Jim used for his beans would be good enough for me.

Jim told me that beans are really very highly self-pollinating here. He had no trouble with cross-contamination of adjacent plantings. In fact, he said, the rates of outcrossing are so low that you are more likely to get off-types from new mutations than from an outcross to an adjacent planting. All you need for isolation is enough distance to allow you to keep from physically mixing the seed when you harvest.

So I planted my soup bean varieties using three feet of isolation in between different varieties. I planted 'Jacob's Cattle', 'Hutterite', 'Gaucho', 'Pinto', 'Black Coco', and others. In between the varieties of common beans I planted beans of other species such as soybeans or teparies so that the "three feet of isolation" didn't actually cost me any space. I wanted to see which varieties would do well in my region, and to get a taste of each. I would continue to grow the ones I liked best.

All went well that first year. When I harvested the beans they looked exactly as I had expected; that is, each batch was uniform and typical for the variety. That year, I got a seed increase on the better producers, but not enough to eat any batches of beans without wiping out my planting stock.

Some of the beans were so late in maturing that they proved marginal for me. Others didn't yield well enough to bother with. Several thrived and yielded quite well though. 'Black Coco' was especially impressive. It's a large, black, round bean that's very early and very high-yielding, and it has upright, vigorous, purplish bushes. 'Gaucho', a small, deep golden bean with a long shape, also did very well.

The following year, I planted longer rows of the half-dozen varieties that had looked most promising the first year. This time I harvested enough beans to eat a good batch or two of everything and have plenty of each pure variety to continue on with.

But . . . *pure variety?* What were these long beans doing in among the round 'Black Coco' seeds? And why were some of the 'Gaucho' beans speckled gold instead of solid gold?

My beans weren't pure varieties anymore. The 'Black Coco' was all black, but not all the beans were round. Some were big and long-shaped, like a black 'Jacob's Cattle'. And some were shaped quite a bit like a black 'Gaucho'. And there were intermediate forms as well. The 'Gaucho' beans were mostly the gold color they were supposed to be. But there were a few that were the size and shape of pintos, and they were speckled like pintos. They looked just like pintos, except they were gold pintos.

And so it went. Every batch of seed was crossed up. In every batch of beans was stark, unambiguous evidence of crosses — multiple crosses in fact. Every lot of beans had crossed with several of the other lots. It was fascinating to just look at the beans and figure out who had been up to what with whom. It was not, however, much of a success as far as maintaining pure varieties.

The first year that I had grown beans, everything had crossed with pretty much everything else in the whole patch. There was no sign of it on the beans I harvested the first year, however. This result is typical. The shape of a bean is determined mostly by the shape of the pod, which in turn is determined by the genes of the mother plant, not those of the seed. Some of the outer layers of the seed coat are also determined by the mother plant. The result is that you usually can't tell a cross has happened by looking at the beans that result from the cross. Instead you find it out one year later when you plant those beans and see what kind of plants and what kind of seeds they produce.

If Jim Baggett, Plant Breeder Extraordinaire, doesn't need to isolate his bean varieties, I wondered, how come mine were all crossing? Jim has created new varieties, increased the seed up to a hundred pounds or more, released it to seed companies, and so on. There is no possibility of his being wrong about *his* need to isolate bean varieties. Why did my need for isolation seem to be so different?

I had a similar experience with chickpeas (garbanzo beans). I was unable to find any hard data on isolation distances. Rich Hannan, at that time curator for the USDA chickpea collection, told me that chickpeas are one of the most highly inbreeding of all species, that they needed only about three feet of isolation — just enough to prevent physical mixture.

I planted about thirty different accessions from the USDA collections. I gave them five feet of isolation between varieties. As usual, I didn't "waste" the isolation space. I planted heirloom soup peas in between the sections of chickpeas.

Most of the chickpeas didn't perform well. This is not surprising, since chickpeas are not normally grown in western Oregon. Even where they are grown (in the more arid climate of eastern Washington, for example) the seed is treated with fungicides. I was not treating anything. I wanted beans that could be grown organically.

About six accessions did quite nicely. They yielded reasonable quantities of seed that, as expected, looked just like the seed I had planted. One of the most productive varieties was, delightfully enough, the black-and-brown-mottled popbean I talked about in Chapter 3.

The next year I planted the half-dozen varieties that had done best the year before. This time, however, I noticed something funny about the plants. They didn't seem to be very uniform.

Pure varieties of chickpeas have either green plants and produce white or light seeds, or they have pinkish or purplish plants and make darker-colored seeds (black, brown, tan, or green). The first year, each variety had been uniform for seed type and plant color. The plants were either all green or they were all pink/purple. But this year the green-plant varieties weren't all green. About 5 to 10 percent of the plants were pinkish/purplish. And blocks of pink/purple plants didn't look very uniform either. Some plants were much lighter pink than others. Uh oh.

Several uh ohs, in fact. Uh ohs in every single one of my varieties. Each variety had a number of crosses. I harvested plants with atypical color or leaf size and threshed them separately, and they showed segregation for seed types and colors. Off-type seed also showed up in the main threshing. That is, the plants I could recognize as crosses were, indeed, crosses. But I couldn't recognize all the crosses. Back to the drawing board. I sent off to the USDA for replacement samples and started over.

In conversations with both Jim Baggett and Rich Hannan, I've found that their university fields have very little pollinator activity. Commercial agribusiness-style tilling and spraying is the norm. The fields are, relative to organic plantings, pretty sterile.

My observations about outcrossing of the inbreeders under organic conditions may also explain some or even most of the contradictory observations in the professional literature about pollinators, outcrossing frequencies, and necessary isolation distances.

I suspect that those growers with healthier, more species-rich, vital lands — those who are closest to growing organically, in other words — are the ones who are noticing more pollinators and hence more outcrossing, and who require a greater isolation distance between varieties than that which is usually specified.

In fact, even those farmers who use sprays and till fencepost to fencepost may have an organic-growing situation with respect to pollinators if they have nonagricultural land such as woods or marshes or permanently untilled ground nearby.

Isolation Basics

In order to maintain pure varieties, we need to prevent their becoming contaminated by outcrossing with alien varieties. There are two methods involved: isolation and roguing.

We prevent or minimize accidental crosses by isolating the variety from foreign pollen. That is, we grow the variety in such a way that it is not exposed to foreign pollen, or the exposure is limited.

We rogue for a variety of reasons, but one of the main purposes is to eliminate hybrids created by accidental crosses in the prior generation. In other words, we rogue partly because we usually do not have perfect isolation. Whenever possible, we rogue before the plants begin to flower, so that the rogue contributes not at all to the next generation, either as female or male parent.

Maintaining pure varieties usually requires both isolation and roguing in every generation.

Accidental crosses between two varieties often produce very vigorous hybrids. This means that if we cull out all the slowest-growing material in a planting of a variety, we may be eliminating all the pure material and keeping all the hybrids. To maintain pure varieties that are true to type, we have to eliminate the rogues and keep the plants that are true to type. To do this, we need to know what "true to type" is for the variety. This is easy if we are familiar with the variety, and much harder if we aren't.

When I am growing a variety for the first time, I usually plant just a small amount and let it grow without roguing the first year, and I don't save seed. I've found that it usually takes me one season of growing a variety before I know what "true to type" is. I need to see whether the biggest corn plants produced ears that are true to type, for example. Or whether the vining squash plants or the bush type ones yield the good tasting fruit. If I try to save seed the first year, I am too likely to rogue the wrong plants.

Becoming familiar with the variety before trying to save its seed helps in other ways, too. If I'm not saving seed, I don't need to isolate at all. I can put a known variety I need to save seed from in one end of the garden or field, for example, and the dozen others of the species that I am "trying out" or just growing for the table at the other end.

Furthermore, more than once I've gone to heroic measures to isolate something that was already thoroughly crossed up before I got it. If the material is already a mess, and I still want it, I have to "clean it up." In that case, the first couple of years I grow the material, I don't need to isolate it much. The genetic mishmash that's already in the variety is larger than the additional mess introduced by a few more gratuitous crosses. Only later, when the variety is purer, does serious isolation become more important.

If I am growing seed just for my own use, I often use roguing instead of isolation to maintain pure varieties. I'll arrange adjacent varieties so that I can recognize the F_1 hybrid plants.

When I save seed, I always write on the label card specific information about which other varieties of the species I grew that year and how large and how close the plantings were. That way, when I plant such seed, I know which hybrids to be watching for, and I can make a good guess about how many to expect.

If I'm growing seed for distribution or ultimate distribution to others, then I practice "serious isolation."

Isolation Overview

Isolation involves protecting our pure variety from foreign pollen of three kinds.

First, there can be wild plants and/or weeds that are close relatives of our crop and able to cross with it. In some cases they are the same species. In a few cases they are technically a different species but are in the same genus, and can still cross enough to create problems.

For example, the common weed known as Queen Anne's lace and the cultivated carrot are both members of the species *Daucus carota*. If there is Queen Anne's lace flowering in or near the patch with our carrots, we'll get crosses. To save carrot seed, we need either to eliminate the Queen Anne's lace or to make sure none is flowering at the same time as our carrots.

Wild mustards often cross with various domestic mustards or other brassicas, even when they are not the same species. In my garden, all the brassicas seem to be able to cross with each other. The hybrids, however, are often almost completely (though not totally) sterile.

Second, we need to protect our variety from pollen of other varieties of the same species that we might be growing.

Finally, in many cases, we also need to protect our variety from wild or cultivated material being grown by neighbors and others in the area.

There are three basic ways to establish isolation distances. We can look them up. (See Table I and/or Suzanne Ashworth's book *Seed to Seed*.) We can establish them ourselves. (See Chapter 9, page 120.) Or we can guess.

I use all three approaches. I usually start by looking them up. If no information is available, I usually guess based upon flower anatomy and pollinators. Even when there is information, we must still do some checking, watching, and guessing, because our conditions and pollinators aren't identical to those of others.

Differences in regions, growing conditions, and pollinators are part of the reason why different sources sometimes give different information, and why I give multiple numbers and sources in Table I where they are available. If you grow using organic methods, the larger outcrossing frequencies and isolation distances are more likely to be the relevant ones.

I always watch an unfamiliar plant as it flowers and note the flower anatomy and the pollen vectors. Flower anatomy often gives us clues as to whether the plant is primarily an inbreeder or primarily an outbreeder. If the pollen is loose and blows around in the wind, then wind is at least one of the vectors, and some isolation is undoubtedly required. The plant may be primarily inbreeding or primarily outbreeding depending on whether pollen from the same flower or from other flowers usually reaches the stigma, and whether there is an incompatibility system.

If something unfamiliar is being pollinated by bees, we can make a good guess about isolation distance by considering what is necessary for other bee-pollinated species. For the bee-pollinated outbreeders, though, we'll often find that the distances are in the range of $\frac{1}{4}$ to 1 full mile. There are ways of isolating plants other than by distance, however.

Factors that Affect the Need for Isolation

1. For most practical purposes, plants that are generally inbreeding need less isolation than those that are deliberate outbreeders. The primary inbreeders include tomatoes, common beans, and peas, for example, but not runner beans or fava beans, which out-cross regularly instead of only occasionally. The term "inbreeding" is, at best, relative, however.

2. With insect-pollinated crops, what's the pollinator pressure? Under organic growing conditions, even the "inbreeders" may cross and may need more isolation than is neces-sary just to prevent physical mixture of the seed.

However, many people grow some of the more inbreeding crop varieties side by side for years without experiencing enough cross-ing to be noticeable.

3. Do you need absolute purity? Can you tolerate no outcrosses whatsoever? If so, you need very much greater isolation than if you can tolerate, say, a fraction of a percent of outcrosses.

Most seed obtained from seed companies has a fraction of a percent of contamination in it already. If you rogue out four or five out-crosses per thousand seeds that were in the material when you got it and introduce one or two new ones, that may be good enough.

4. The size of the plantings matters. A single 50-foot row of each of two varieties of beans separated by 50 feet in a garden, with, say, a corn patch in between, will be more than well enough isolated for most purposes. But if you have twenty 50-foot rows of each

variety, with the blocks still separated by 50 feet, you'll get many more crosses. If your bean rows are 50 feet away from a 200-acre commercial planting, you may get so many crosses that trying to save pure bean varieties in that garden becomes impractical.

5. Arrangement matters too. Bees tend to fly up and down rows, for example. But they often skip sections of rows. They are much more likely to skip 20 or 50 feet down the same row, however, than to fly over 50 feet of corn patch to a different row.

6. If you have two different varieties sepa-rated by 50 feet, it matters a whole lot what the 50 feet is. Visual and wind barriers are better than bare ground or low plantings. Flowering plantings that insects can "clean themselves up on" may be better than non-flowering plantings.

7. How much space do you have? For most of us, isolation is a matter of compro-mises between what might be optimal and what is actually possible.

Isolation Distances, Absolute and Practical

The isolation distances recommended by Suzanne Ashworth in her book *Seed to Seed* take all the available professional literature into account and give numbers that are strin-gent enough to establish *absolute* isolation under virtually all conditions. These distances are quite large. Most of the distances for the legumes, for example, are in the range of ¼ to 1 full mile. For chickpeas, for example, Ashworth notes that some researchers say that crossing is common. She recommends a distance of ½ mile.

Where absolute purity of a variety is required and no crosses at all can be tolerated, I believe isolation distances for the so-called inbreeders need to be about the same as those for the deliberate outbreeders. In most cases, distances ranging from about ¼ to 1 full mile.

Most of us these days don't have the ability to isolate most plantings by ¼ to 1 full mile. Fortunately, however, only rarely do we need to. In very few cases are we working with absolutely pure varieties.

Most home gardeners will find that they can raise different varieties of most insect-pollinated crops and keep them acceptably pure with isolation distances of about 20 to 100 feet between separate rows or blocks of different varieties.

The 20- to 100-foot minimum works best when the plantings are not too large, and when the grower combines the basic isolation distance with other tricks to maximize isolation effectiveness.

Alan Kapuler, for example, gardens and grows seed crops here in western Oregon on 2 acres of organic garden that is alive with pollinators. Yet he can grow dozens of different varieties of common beans and runner beans at once and save seed, and maintain and do seed increases on pure varieties.

He does this by alternating rows of trellises of runner beans and common beans. Usually each trellis is separated from the next by a small block of corn as well. This gives good enough isolation to save and increase seed for sale, even though there may be only about 40 feet between one variety of beans and the next variety of the same species. Alan gets occasional crosses, but they are rare.

Generally, the seed he sells is much "cleaner" than what he starts with, and it is clean compared to other seed that is available commercially.

Growing alternate rows of different species of beans on trellises provides both a visual barrier and material for the bees to "clean themselves up on" between different varieties.

Unless you have absolutely pure varieties and need absolute isolation, I suggest the following Basic Rule for Everyday Seed Saving: Isolate things as well as you can with the space you have. If you don't have much space, use any of the various tricks and methods given later in this chapter and make the most effective use of the space you have.

Even if your isolations are less than optimal, with a little luck, most of the time you'll get away with it. And, of course, sometimes you won't. Sometimes you'll get crosses. I crossed up eight varieties of beans and thirty varieties of chickpeas, however, and the world did not end.

In some cases, absolute purity is desired or required, that is, no outcrossing at all can be tolerated. Here's an example:

My black-and-brown-mottled popbean, a chickpea, outcrosses. The plants are of a size, color, and form that represents primarily dominant characteristics, so crosses are usually not recognizable.

The characteristics I'm most interested in, flavor and popping, seem to be destroyed totally by outcrossing and are not easily recovered by backcrossing. That is, an outcross destroys the variety, but invisibly. Further, no signs of outcrossing show up in most cases until a full generation after the hybrid has contributed pollen enough to contaminate everything.

In addition, one off-type in a thousand could break one tooth per off-type. At a rate of one off-type per thousand, almost everyone who ate a few pounds of popped bean per year would break at least one tooth. With ordinary beans, the off-type, if eaten, has trivial consequences. At worst, the bean cooks, looks, or tastes different from the rest. An off-type popbean, however — one that didn't pop and wasn't soft when you were expecting it to be — can do you serious personal injury.

Finally, I am breeding material that I believe (on my more optimistic days) to have the potential to be the start of a whole new snack food industry. If so, thousands of pounds of chickpeas might someday be growing from those I'm breeding and maintaining right now.

I isolate this chickpea from small garden plantings by at least ¼ mile. I simply don't grow it within miles of any large commercial planting. Given the situation, in this material, no crosses at all can be tolerated.

Isolation Tricks and Methods

1. Where large isolation distances are required, we can grow just one variety to seed in any given year. Note that in many cases we can grow numerous varieties. We just can't have more than one flowering at the same time. (Consider what else is growing in the region or area, though. The air in summer is full of corn pollen, for example, whether we plant any or not.)

2. We can isolate in time instead of with distance. We can grow an early, a midseason, and a late variety, for example, if the flowering periods don't overlap.

3. We can hand-pollinate. Squash and other cucurbits, for example, are easy to hand-pollinate for small-scale seed saving.

4. Visual barriers help minimize outcrossing performed by insects. A 20-foot block of corn in between two different bush bean varieties is *much* more effective in isolating them than 20 feet of open space or space with other low plantings.

If I have two kales that flower at the same time, and I want to save seed of both, I'll often plant them on opposite sides of the house. Just watch the pollinators and see what they do. In one 4x8-foot patch, individual bees will work the kale until it is "used up" for the moment (until more flowers open and anthers dehisce). Then they go to the sage flowers within the same patch, not to the kale in a patch out of sight on the other side of the house.

5. Barriers such as tall corn or hedges or trees also break up wind patterns. Wind carries both windborne pollen as well as pollinating insects.

6. We can grow as many varieties as we want, but remove flower scapes or flowers from all but one for the relevant period. With brassicas, for example, I frequently cut the flower stalks off the varieties I don't want seed or contaminating pollen from. But I keep the plants so that I can continue eating from them. What matters isn't how many varieties I have in the same small space, but how many varieties I have *flowering simultaneously* in that space.

7. We can cage the plants or plantings so

as to eliminate or restrict pollinators. Reemay (or some similar floating row cover material) can be used to exclude many insects. Screens or mesh are sometimes used. With outbreeders, though, we may have to introduce clean insects into the cage to get pollination.

8. We can bag individual plants or inflorescences and either hand-pollinate or introduce appropriate insects.

9. We can eat the edges of the patch and harvest just the center. Rich Hannan tells me, for example, that 250 yards of isolation is needed for fava beans under his conditions but only 50 yards if it is planted with material the bees can clean up on. In other words, he could plant many fava varieties in adjacent 150-yard plots if he harvested just the central 50 yards of each block for seed.

10. For seed intended for personal use, it is often easier to depend upon being able to recognize and eliminate hybrids than to prevent them through heroic isolation measures. I often isolate brassica varieties by just 20 feet or so, if I know I can recognize the hybrid. The official isolation distances are usually in the range of ¼ to 1 mile. I always get a few hybrids under such conditions.

Pure pink/purple plants crossed with pure green ones give pinkish plants that are obviously different from the green parent (but which may or may not be distinguishable from the pink/purple parent). Varieties that have leaves with deeply indented margins, when crossed with those having smooth-margined leaves, usually yield hybrids with leaves of intermediate type. Tall varieties crossed with short varieties give hybrids that can be intermediate or tall depending upon the specifics. So we can detect the cross in

seed of the short variety, but not necessarily in seed of the tall one.

With corn you can just detassel hybrids to eliminate their pollen; then you can keep the plants and eat the ears. I usually eliminate flower scapes instead of whole plants so that I can eat the plants instead of wasting them.

11. Instead of excluding foreign pollen, we can just swamp it out. There is always corn pollen in the air during spring and summer, for example. But we may be able to get away with maintaining a variety by open pollination if we have a substantial patch or field of a single variety and harvest for seed only from the middle. There is foreign pollen in the air in the middle of the patch, but it is outnumbered by thousands or millions of times by pollen from the plants in the immediate vicinity.

12. We don't need to save seed of everything every year. It's easier to save enough seed for several years. That way we need fewer good isolations in any given year.

13. We don't have to plant everything on our own land. Friends and neighbors will often be willing to let you grow a variety in their gardens when you need the isolation — especially if you're in the habit of giving them seed.

14. On a somewhat larger scale, we can lease land elsewhere to grow some of our seed crops, or contract with a grower to grow certain seed crops for us to our specifications.

Official numbers with respect to isolation distances are largely derived from open fields. I believe that the distances needed, even for crops such as corn, are not nearly as large when there are lots of houses, trees, bushes, fences, and hedges to break up wind patterns.

My yard and the surrounding area have lots of trees, hedges, and flowering plants of all kinds that provide ample food for pollinators. There is enough for bees to do so that they don't very often go farther than between adjacent yards on one trip. Small flies are easily blown great distances by wind, but the wind patterns are broken up by houses, trees, and hedges.

I routinely examine the yards of my neighbors within two houses of my own in all directions to see what they are growing that might contaminate seed crops of mine. With houses, trees, bushes, and complicated wind patterns, it usually doesn't seem to matter what more distant neighbors are doing with respect to insect-pollinated crops. Wind-pollinated crops are another matter, though. (And corn is an extreme case.)

In short, look up isolation distances to give yourself a basic idea. If you see isolation distances that are larger than the land you have available, though, don't get discouraged. Look at your specific situation, notice your wind patterns, watch your pollinators — and get creative.

NINETEEN

How Many Plants?

HOW MANY plants of a variety do we need to save seed from in any given generation in order to maintain the variety properly?

There are times when one plant is not only adequate, it's the exactly optimal number.

There are other times when saving seed from just one plant will destroy the variety in a single generation.

In many cases one plant is adequate for some purposes, but not others, or adequate, but not really optimal.

There are four reasons why we usually save seed from more than just one plant in any given growout of a variety.

First, we don't want our entire year's seed ruined by an accidental cross or an undesirable new mutation. If we save seed from just one plant, and it is involved in such an accident, the entire year's growout might be affected.

Second, we usually want to maintain some or nearly all of the genetic heterogeneity that might exist in the variety. The term "genetic heterogeneity" refers to the fact that the individual plants in the variety have different repertoires of genes in them. In order to preserve the entire range of genes present in the variety, we save seeds from an adequate number of individual plants each generation.

Third, some plants have incompatibility systems. If you actually grow only one plant, it may be sterile or nearly sterile. If it produces seed, the seed is likely to be of poor quality or to give rise to inferior, wimpy seedlings. If you grow only two or three plants, they may happen to have the same incompatibility genotype, giving the same results as if you were growing just one plant.

Fourth, many outbreeding plants are subject to inbreeding depression. The greater the extent to which a species tends to suffer from inbreeding depression, the more important it is to maintain as much genetic heterogeneity as possible in the variety — and therefore the more important it is to save seed from many plants instead of just a few.

So the answer to the "How many plants?" question depends upon whether the crop is a basic inbreeder or a basic outbreeder, whether it has a self-incompatibility system, and the extent to which it is subject to inbreeding depression. You can find this information about the crop you're interested in by looking it up in Table I and Appendix A. (Table I is organized by scientific names; if you don't know the scientific name, look up the common name in the index.)

(Note: In Chapter 9 I discuss the basic genetic nature of inbreeders and outbreeders and give the reasons behind the pragmatic suggestions featured in this chapter.)

Inbreeders and Numbers

Inbreeders give us the most options. Many a bean and pea and tomato variety has undoubtedly been saved through just a single seed, or just a few. The inbreeders don't mind inbreeding (even though they may outbreed to various extents too). They don't suffer from inbreeding depression.

Most seed savers try to save seed from at least twenty good plants per generation for most inbreeding crop varieties under most circumstances. "Twenty good plants" means that you grow out more than twenty so that you have twenty left after selection and roguing. Twenty plants per generation is enough to preserve most of the genetic heterogeneity in most varieties. Forty plants per generation will allow you to maintain nearly all the heterogeneity in most inbreeding varieties.

In a few cases, where the plant material is very heterogeneous, larger numbers might be warranted. To maintain a very heterogeneous landrace of an inbreeder, for example, a USDA curator might choose to use a minimum of one hundred plants per generation.

With inbreeders, however, maintaining genetic heterogeneity is optional. The individual plants are vigorous whether we maintain the genetic heterogeneity in the overall variety or not. The main reason we want to maintain the genetic heterogeneity is that it is the basis for future selection, adaptation, and evolution of varieties.

Most seed savers save seed from at least twenty good plants per generation for crops like beans or peas; the plants are small, and we're growing lots more than that anyway. However, many seed savers save seed from just a single tomato plant, for example, or just a few.

When we save seed from just a single tomato plant, what we are preserving is actually a single subline of the variety, not the whole variety (see Chapter 9). That subline is often all we need, though. And when many other people are also saving seed of the same variety, many sublines are being preserved overall. Anyone who needs more genetic heterogeneity in the material can obtain the variety from several different sources and simply combine the lines.

Sometimes saving seed from just one plant is optimal. Whenever we notice a superior individual plant of an inbreeding crop type, we usually save its seed separately and use that seed as the beginning of a new line or new variety. When we do this, we are eliminating most of the genetic heterogeneity of the original line in return for the advantage of the superior characteristic we noticed.

There is another situation in which saving seed from just one or a few individuals of an inbreeding crop is best: when the material has been crossed up. Then, we can "clean up" the material by saving seed of several individuals separately so we can identify one plant that is still pure. We use the seeds of that plant to "restart" the line.

Outbreeders and Numbers

Corn, kale, and sunflowers are examples of outbreeding crops. Kale and sunflowers have incompatibility mechanisms to prevent self-pollination. Corn is capable of self-pollinating, but it is very subject to inbreeding depression.

With most outbreeders, maintaining the genetic heterogeneity of a variety is not optional. When a variety has lost too much genetic heterogeneity, the individual plants are wimpy. That is, they germinate poorly, grow poorly, yield poorly, and die (or yield nothing at all) at the slightest excuse. Even if they yield something, it is too small and/or too inferior to be worth the trouble. The loss of vigor caused by loss of heterogeneity in a variety is called "inbreeding depression."

For most outbreeders, with just a few exceptions, we never try to save seed from one or just a few plants. Instead, we save seed from at least forty to two hundred plants in each generation, depending upon how subject the crop is to inbreeding depression.

Corn is very subject to inbreeding depression. Most seed savers try to save seed from at least one hundred good plants every generation. Many prefer two hundred. (See Appendix A for seed saving information on corn.)

Most professional germplasm curators would want to save seed from at least two hundred corn plants, and possibly many more — up to one thousand if it is desirable to maintain nearly all the genetic heterogeneity, or if the variety or landrace is very heterogeneous.

For outbreeding crop plants that are not as subject to inbreeding depression as corn, we usually choose a number between forty and one hundred as the minimum.

Squash is an outbreeder that is an exception to the rules. It doesn't have an incompatibility system and isn't subject to inbreeding depression. We treat it as an honorary inbreeder. We're usually very happy to save seed from just twenty squash plants. Often, we save seed from a lot fewer than twenty because the plants are so big, and hand-pollination is involved. (See Appendix A for seed saving information on squash.)

Avoiding Inbreeding Depression

In order to avoid inbreeding depression you need to do two things. First, you have to save seed from adequate numbers of plants in every generation so that genetic heterogeneity in the variety is maintained.

Second, the plants need to be grown so as to facilitate cross-pollination rather than self-pollination.

If I plant sixty kale plants all together in a patch and space them two feet apart in all directions, for example, the bees will frequently move from flowers on one plant to flowers on another. If I space the plants six feet apart, the bees will mostly move from flower to flower on a single plant. If the variety is capable of self-pollination at all, it will be mostly self-pollinated under these conditions and there will be inbreeding depression.

It doesn't suffice to plant sixty plants if you arrange them so that each is basically all by itself with respect to pollination. Remember that even one generation of inbreeding causes a loss of half the heterozygosity in each plant of the next generation (see Chapter 9). Even one generation of forced inbreeding of a plant subject to inbreeding depression can result in a severe loss of overall vigor and yield.

If a variety is sufficiently strongly self-incompatible, and spacing is such that pollinators don't cross-pollinate the plants, the result will be little or no seed.

If you interplant the kale such that they are even farther apart and separated by other flowering plants, nearly all of the bee-trips will be from flower to flower on one plant. Interplanting has its virtues, but consider the cross-pollination aspect.

If you plant all the kale in a single row, plants will usually be pollinated by their immediate neighbors. In small plantings of outbreeders, patches or blocks are preferable to single rows.

If inbreeding depression results from failure to allow or provide for proper cross-pollination, the variety may still have enough heterogeneity in it, but the individual plants don't. They are too highly homozygous and thus not vigorous. You can regenerate the variety completely in a single generation, though, just by planting with proper spacing and in a block. That way, the next generation of seed is mostly from all kinds of crosses instead of mostly from selfings.

If, however, the variety doesn't contain enough heterogeneity (because you saved seed from too few plants, for example), you can't regenerate it just by planting it properly for one generation.

Normally, we do not try to save seed of an outbreeding crop such as corn via seed from just one plant. There is an exception, however.

An F_1 hybrid plant between two fairly different varieties is heterozygous at many loci. It has a tremendous amount of heterogeneity within just the single plant — more than enough for an entire variety. We can save seed from a single plant of such a hybrid and use it to breed a vigorous new variety.

What If I Don't Have Enough Room?

With most outbreeders, it's important to save seed from adequate numbers of plants in order to maintain the vigor of the material. But you don't have to grow them all in one year. You can grow a few each year, for example, and pool the seeds from different years when you draw seed for planting.

In addition, there is a difference between optimal and practical. You may be able to get away with just twenty kale plants per year, for example, instead of the minimum of forty recommended — especially if you grow them in a block instead of a row so as to maximize cross-pollination of the plants in all possible combinations.

With inbreeders, if you don't have enough space for the optimal number — well, life is full of compromises. Just grow as many plants as you have room for (or need for the table). You might not be able to preserve all possible genetic heterogeneity in the material, but you don't need it all to have a good line.

TWENTY

Selection

IN SOME WAYS selection is obvious: we save seed from the best plants. Yet in other ways selection is complex, subtle, and full of surprises. It is amazingly easy to select for exactly the opposite of what we want.

To use selection most effectively, we need to know how to choose as well as how to evaluate selection criteria.

Here are some guidelines.

Selection Basics

1. You must practice selection, culling, or roguing as part of every generation when saving seed of a particular variety. There is really no such thing as variety "maintenance." New mutations and outcrosses always occur. Varieties evolve. You have to practice selec-

tion in order to hold a variety in the form that you consider desirable. Unselected or improperly selected varieties deteriorate rapidly.

2. Varieties are often rogued more than once in the season. That is, you go through and remove inferior or atypical plants. In a patch of brassicas, for example, I might go through once and eliminate any premature bolters. Then I might examine the rest of the plants as they come into flower and eliminate those that are smallest.

3. Roguing of outbreeders should be done before flowering if at all possible, so that their pollen as well as their seed is eliminated from the next generation. This isn't always practical or possible, however. Sometimes we can't identify the rogue until later.

In corn, for example, we may not recognize

a rogue until we see the mature, off-type ears. We can easily eliminate the rogue plant's ears, but by then it's too late to call back the pollen it contributed to all its neighbors. When we select or rogue based upon ears, we are selecting using the mother plants only. We eliminate about half of the undesirable effect of a rogue, but not all of it.

Selection based upon just mother plants is practical and effective, but not as effective as selection based upon both parents.

4. Consider the possible virtues of a new mutation, accidental cross, or rogue. You must eliminate it from your original variety. But that doesn't mean you have to discard it. Perhaps it deserves consideration as the basis of a new variety.

I often put the interesting rogue at one end of the row and save seed from the other end.

5. You can usually arrange to eat rogues and culls instead of wasting them. You can chop the flowering scapes off the inferior brassicas, for example, so you have more time to eat the leaves.

6. Any time you select, rogue, or cull, always look at the whole plant, not just the individual fruits. You must select the best mother and father plants, not the best individual fruits. Being best involves more than producing a single impressive fruit. (See Chapter 6, pages 81–82.)

7. In choosing the best parents, remember edge effects — that is, the fact that plants at the ends of rows or on the edges of plantings have more sun, water, nutrients, and space than the rest. If a centrally located plant looks impressive, that's more meaningful than if an edge plant thrives. Examine the edge plants and figure out how being on the ends plants and figure out how being on the ends or the edges affects things so you can mentally correct for this factor.

Did the superior-looking plant have the same amount of space as the others? Or did it have more than the others because its nearest neighbors died?

8. In choosing the best parents, remember environmental effects. Was that superior-looking plant the only one that wasn't shaded by the house? Or is it the plant that is closest to the fertile spot created by last year's burn pile? Maybe the plant really isn't genetically superior. Maybe it is only lucky.

9. Hybrids caused by an accidental cross may germinate quicker and grow faster than the pure seed. Examine your material for trueness to type before you do any culling. If you presoak seed, don't plant just the fastest-germinating seed. Plant it all.

10. Do your roguing only after plants are big enough so that you can tell what should be rogued.

11. With unfamiliar material, plant sparsely enough so that you don't have to thin until plants are large enough to evaluate. You might thin out all the little, slow-growing pure plants and keep all the accidental-cross hybrids.

12. If you plant under conditions that are atypical and save the seed, you may be selecting for material that is inferior under your normal conditions.

If, for example, I plant 'Oregon Giant' pole bean in late spring or early summer, I would not want to save seed from the material. I would be saving seed from the plants that did best under warm-weather conditions. Such material might do poorly under ordinary conditions. When you select, think about what you're doing. What is the essential purpose of the variety? Why are you growing it?

The "purpose" of 'Oregon Giant' in my garden is to thrive and produce from earlier plantings than anything else, to yield earliest from early plantings, and to grow happily in cold mud — not to produce earliest or best from warm-weather plantings.

13. Select for natural resistance to soil-borne fungal pathogens by planting untreated seed, and plenty of it. (See 14.)

14. Select for a variety's ability to germinate vigorously under your growing conditions by planting much more seed than you need for the number of plants desired. Then thin to the best plants. Selection for many important but invisible characteristics will happen automatically.

15. Select for the ability to grow and thrive under organic growing conditions by growing plants under those conditions and saving seed. You don't have to know what diseases you're selecting for resistance to, or what genes or genetic characteristics matter. Just select as usual based upon the various visible qualities and performance characteristics that matter to you. The additional selection for organic-growing capability will happen automatically.

Selection Complexities, Subtleties, and Surprises

16. If you plant too early, so that the early-germinating material is killed by frost, you may be selecting for slow germination, thicker seed coats, or seed dormancy mechanisms.

17. If you eat the early produce and save seed from later material you may be selecting for lateness.

18. If your variety has an extended harvest because various plants mature at different times, and you want to maintain that characteristic, you must save seed from plants with the entire range of maturities.

It's easy to save seed from just the earliest plants, because you notice them, and the seeds are ready first. If, however, you want to develop or maintain a variety that will have an extended harvest period, you'll probably need to collect seed more than once during the season. If you let the early-maturing seed sit around until the late material is ready, the earlier seed may shatter and get lost, or be rained upon and ruined. This would mean that you would be accidentally selecting for lateness.

19. In selection, there are usually tradeoffs. For example, selection for more fruits per plant usually means that the individual fruits will be smaller. Selection for two pods per node on peas generally gives pods that are about as large as those that are alone at the node. But more than two pods per node is usually associated with smaller pod size.

20. Ears from early varieties of corn are generally smaller than those from main-season varieties. You can select for larger ears on an early variety. But you can't expect ears on an early variety to be as large as those of the best later varieties. Later varieties have longer to develop. Generally, they have bigger plants and more leaf mass to support bigger ears.

21. If you select the biggest ears of corn from your harvest as seed ears, you might be selecting for lateness. (The biggest ears are often produced by bigger, later plants.) Or

you may be selecting for plants that have only one bigger ear instead of two or three. The single bigger ear may be more desirable. Or you may prefer the greater total productivity usually associated with more ears per plant.

22. Selecting for high productivity sometimes results in a lower sugar content. That's because the same energy is going into making sugar for a larger amount of fruit.

23. When you are creating new varieties, you can't select for everything desirable simultaneously. If you try to select on the basis of six different characteristics, for example, you will probably not be able to select very effectively for any of them. Decide what your priorities are. You can select most effectively if you concentrate on just one or two characteristics.

24. On the other hand, when you rogue established varieties, you're usually eliminating just a few plants for any given reason. You can rogue for many reasons simultaneously, as long as you have enough plants left for a good seed crop.

25. Saving seed from a volunteer? Consider *why* it is a volunteer.

For instance, that volunteer fava bean may represent a seed from a plant that shattered its seed earlier or more easily than the rest. Save its seed and you could be selecting for shattering, which makes seed saving much more difficult. Shattering is especially inconvenient in a grain crop.

Or perhaps that volunteer mustard is from a seed you planted last year that didn't germinate. Instead, it sat dormant all season, overwintered, and sprouted the next spring. That seed may represent a plant that has mutated back to wild-type seed dormancy mechanisms. If you save its seed you may be dealing with erratic, unpredictable germination from there on out.

26. Usually, you should not eliminate variability in an arbitrary manner. Often certain characteristics go together in ways that aren't apparent. You might eliminate all the different colors of seed in a variety of mustard, for example, so you have just one color. But maybe resistance to a disease important in your area was linked to one of those genes that you just discarded. Now you can't select effectively for disease resistance.

We need consistency for certain characteristics for the crop to be useful to us. We usually try to eliminate variability for flavor, for example. But we don't need the mustard seed to be all one color. With soup beans, though, color is a component of flavor. So we usually do want a soup bean variety to be one color.

27. Some plants produce more seed than others. If we pool the seed from all the plants we save seed from, the plants that make more seed will contribute more to the next generation. We will be automatically selecting for seed production. This may be good if the crop is a dry bean, for example, where the seed is also the useful edible portion. It isn't necessarily good if the plant is a mustard, where a plant that produces a lot of seed may have been one that produced a small amount of edible leaves.

Sometimes we need to correct for differences in seed production. One method is to save a constant amount of seed from all the plants we've chosen.

Selection for the Purpose of Germplasm Preservation

Sometimes germplasm preservation is our primary goal in saving seed. In that case, we try to keep all the genes and genetic heterogeneity in the material, whether this is useful or optimal for us or not.

When USDA curators maintain varieties, for example, they try to maintain all the heterogeneity. If they were to select for material that yields best at their location, they might be discarding genes important to the potential grower who lives in another area. If they selected a certain color or flavor or plant form, they would be eliminating the genes needed by someone who wants the opposite characteristics.

When the primary goal is germplasm preservation, we try to avoid automatically or deliberating adapting the material to our personal preferences or specific situation. We use our knowledge about selection to *avoid selecting*.

We keep seed from all the plants that are true to type, not just the best. (We still eliminate the rogues that represent outcrosses, however.) We may even keep equal amounts of seed from all the plants so as to avoid accidentally selecting for plant size, vigor, yield, or seed production under our conditions.

Evaluating a Selection Program

It's useful to evaluate your selection program and selection criteria. Sometimes the variability that we see isn't genetic. If the variability for a characteristic isn't genetic, selecting based on that characteristic doesn't accomplish anything. It doesn't change the variety.

For example, my black-and-brown-mottled popbean produces seed that is either black-and-brown-mottled or black. The difference isn't trivial. The black beans have a more intense flavor when popped.

So I hand-sorted the beans, then planted the mottled seeds in one row and the black seeds in another. I found that both rows gave me the same proportion of mottled beans versus black ones in the next generation. This meant that the seed color differences in this variety were not heritable.

Here's another way of stating the same thing: Something other than genes is responsible for the variability in seed color in my popbean variety.

Yet another way of saying it: There is no genetic heterogeneity for seed color in this material.

If the beans could talk they might describe the situation this way: "These color differences are *not our fault!*"

All the above versions of the statement have the identical implication: I could select from now till doomsday without having any effect on the seed color of my popbean.

Of course, one then wonders *why* I am getting two colors of beans. Threshing out individual plants provided the answer. It's differences in harvesting that matter. Seeds from pods that dry fully before the plant is cut are always mottled. Seeds from pods that are somewhat green when the plant is harvested are green initially, but they dry out black.

If I harvest a plant only after it is completely dry, all its seeds are mottled. If I cut a plant that has half dry pods and half greenish ones and leave it to dry, when I thresh I get a combination of types.

Knowing that the color difference isn't genetic, I don't waste time, space, or seed by selecting on the basis of seed color. Instead, I focus on other characteristics which are highly influenced by genetic differences in this material, such as plant form, yield, and disease resistance. At the same time, I realize that I can always get more black seeds, if I want them, by modifying my harvesting appropriately.

Many of the differences that we see between various plants of a variety have more to do with the environment than with genes. Some plants landed on better soil, weren't nibbled on as much by insects, and so on. Selection works only on genetic differences, not environmental ones.

There are three simple ways to evaluate a selection criterion. One is the approach I took with chickpea seed color. Plant two different batches of seed that are selected differently. Do they give you the same range of types? If so, the variability is not genetic. It's environmental in some fashion.

A second method involves planting both selected and unselected seed. Instead of discarding the seed from the inferior plants, you save at least some of it. Then you plant some of that seed side by side with the selected seed. Then you compare the two plantings. Is the seed of the rejected plants actually inferior? Or is it yielding plants that are just as good as those from seed of plants you're selecting? If both batches of plants are equivalent, your selection criterion isn't working.

Sometimes the seed of the plants you are rejecting produces the superior batch. If so, your selection is effective, but it is backfiring. (The next job is to figure out why.)

Sometimes the selection is working but has negative side effects you will want to know about and consider. Red color in corn seed, for example, is often associated with less vigorous plants, lower yield, and smaller ears.

The third method involves planting seed from different years side by side. Then you compare. Is your selection working? If so, is it having any unexpected effects as well? The results can prove interesting and instructive.

Does Selection Always Work?

In a word, no.

Sometimes selection is effective and sometimes it isn't. That's why it is often worth our while to evaluate whether any given selection criterion is working in a given situation. Avoiding ineffective selections allows us to focus on effective ones. I can select much more powerfully for disease-resistant popbeans, for example, if I don't discard half the beans every generation over differences in color that are not genetic.

Selection works by genes, not by magic. The effectiveness of selection depends upon underlying genetic heterogeneity for the characteristic in our material.

TWENTY-ONE

Harvesting, Processing, and Storing Seed

MANY WONDERFUL heirloom pole bean varieties have been saved one pod at a time. Whenever the gardener was in her garden, she picked the pods that had started to dry, and tucked them in her pocket. Back in the kitchen, she tossed the pods in a basket with others of their kind to finish drying. During winter, she brings out the baskets full of different types of pods and starts shelling them out by hand in the evening, while talking and relaxing with the family.

The kids, even the littlest ones, are allowed to help. And they want to. They feel honored. They consider it a privilege. "Look at this one!" they exclaim over each unusual type, every new discovery.

I think that our fascination with seeds is hardwired into our genes as a species. Watch young children opening bean pods and finding the seeds. You can see that they know they are doing something special. You can see that they sense the magic.

Harvesting

When should you harvest plants for seed? I let the plants proceed with their natural process of making and dispersing seed for as long as possible. I intervene only at the last possible minute, so that the seed disperses into my hands instead of all over the ground, or into the bellies of self-appointed sharers.

My simple all-inclusive rule for fruits is to allow them to become as mature as possible without actually rotting, falling off, or being eaten by bugs or slugs, or stolen by birds or squirrels.

Adequate seed of some crops can be obtained from fruit that is at eating stage, but the best-quality, largest, most vital seed comes from fruits that were harvested way past prime eating stage. You can save adequate seed from eating-stage tomatoes, for example. But when the tomato has softened way past edibility the seed is bigger and fatter and will germinate more enthusiastically, giving rise to more vigorous seedlings.

You can save good melon seed from the fruits as you eat them. But the best seed comes from fruits that you have allowed to mature beyond eating stage in the field. I tie a large chunk of bright surveyor's tape around the stems of fruits I've chosen for seed so that I don't harvest them for the table by accident.

For plants that produce dry seed, such as beans or mustard or most flowers, I prefer to let the plants or stalks mature and dry out completely in the field. I harvest earlier only if I must. Generally, this is because of wet weather or the presence of competitors — birds or other critters that could beat me to the seed.

Some seed crops are harvested before the plant is completely dry because the seed shatters, that is, pods release the seed to fall out on the ground. With buckwheat, for example, the earliest, biggest seed starts to drop long before all the seed is mature. Getting a good yield of a crop that has shattering seed requires guessing at the best harvesting time. The goal is to recover most of the early, big seed and as much additional seed as possible. Usually, the best harvesting time is just a bit after the earliest seed begins to shatter.

For small amounts of seed for my own use, my basic seed harvesting equipment includes plastic freezer containers, paper bags, index cards, and a pen to write with. Whenever I harvest for the table, I take baskets for the fresh vegetables and the extra containers for seed.

In my main garden I water regularly, and water can ruin seeds once they have started to dry. So I snip flower heads and clip pods here and there as they start to dry, and put them in containers. I record what each seed-head or pod is on an index card and toss it in the container with the seeds. When I get back to the house, I tuck the open bags and containers away on a shelf to let the seeds finish drying.

I often write quite a lot on the index card that accompanies the seed. Seed type, harvest date, whether anything else of the same species was growing in the garden, and, if so, what it was and how close — all this goes on the card, as well as anything unusual that prompted me to save seed from this particular plant. When I clean the seed, it goes into a new container, and the card goes along. With the card, I don't have to recopy all that information each time I change containers or clean seed.

For growing larger amounts of seed, the basic harvesting equipment includes more paper bags, some 5-gallon buckets, and tarps. You will often plant an entire row or section of a row or bed of the crop specifically for seed production. And you should try to arrange harvesting so as to optimize the quality of seed and minimize the labor.

As an example, I grow a lot of brassicas such as kale and mustard. I plant them in beds away from the vegetables that need water in the summer. I don't water them at all. In spring they are growing actively, and we have plenty of rain here in Oregon to cheer them along. In May, when the plants start to want to dry down, the rains stop. Then I need to water the rest of the garden, but the brassicas I allow to dry out completely.

Oregon, where we have little or no summer rain, is ideal for such seed crops. Rains (or overhead watering) damage seed quality once the seed begins to dry. So if you have summer rains or your plants are exposed to overhead watering, you will usually let the plants dry out only partially. Then you'll cut the plants and stack them loosely somewhere under cover to let them finish drying.

When I grow mustard in fields away from home, for example, I leave it until the plants are completely dry. Then I clip the stalks near their bases and toss them on a handy tarp. I transport the plants back home as well as store them on the tarps. I set the tarps in the shade somewhere, under cover if rain threatens, and fluff the plants up a bit so that air continues to circulate through them. That's a good description of how it works when everything for this seed crop is optimal, including the climatic pattern and the weather.

I used to handle mustard in my home garden exactly the same as that in the field. In my home garden these days, though, I can't leave the seed mustard so long anymore, because of LBBs (my acronym for Little Brown Birds). As soon as the mustard starts drying, hordes of LBBs descend upon it and have a wonderful time stripping the seeds out of the pods.

When the flock descends, each drying mustard bush quivers all over, every branch alive with dozens of tiny twitchings and rustlings. It would be enchanting, except for the fact that those birds are capable of eating every single seed of the crop in just a few days, leaving the bushes with nothing but stripped, empty pods. On a plant that one day earlier had thousands of pods, each containing several seeds, you can search for minutes without finding a single good seed.

The first season I got LBBs, I lost most of the mustard seed from my earliest-maturing lot. When a later-maturing batch started to dry, however, I was ready. I watched the drying plants carefully for the first signs of bird damage. As soon as I saw those first stripped branches, I harvested the plants and stacked them loosely on a tarp on my back deck to finish drying. The result was a back deck and tarp full of bird droppings, and still no seeds.

I had overlooked the fact that birds fly. The LBBs flew to the garden, found their breakfast missing, and required all of about one nanosecond to spy it up there on the deck. So up they came. Right in front of my very eyes, just five feet away on the other side of the glass sliding door, they ate all my seeds.

These days, I pile my homegrown mustard plants on a tarp in the driveway on the other side of the house. The LBBs haven't seemed to notice it there. At least not yet.

Other seed-saving friends of mine in the area, when told about the birds, said, "Aha! So you have those too, huh?" Apparently, you can get away with major seed-saving activity in one location without being molested for a couple of years, but a couple of years only. Soon the LBBs discover what you're up to.

I have plans for the future that involve whole fields of mustard. I hope to avoid the LBBs by the ordinary measures of crop rotation. Crop rotation is usually practiced in order to promote soil fertility and minimize plant diseases. However, it's also very helpful in avoiding buildups of pests, or in minimizing the extent to which existing levels of pests focus their attentions on your crop and develop inconvenient habits.

Bush bean plants and pods are usually allowed to dry as much as possible before harvest. If you have summer rain or overhead watering, you may need to harvest partially dry plants and allow them to finish drying under cover. A general rule is to let half the pods dry, then clip and harvest the plant.

If you don't have to worry about water, you can leave the plants until they dry fully in the field. In that case, you still need to harvest them before the pods begin to shatter and dump the seed on the ground.

Pole beans are sometimes grown on the ground for a seed crop. (This is how it is done commercially.) But usually a home gardener will be growing the pole beans on a trellis or on corn plants in the main garden and eating most of the beans. Seed saving consists of just choosing one section for seed production, and picking the pods as they dry. Once a plant matures and begins to dry off the first few pods, it doesn't produce many more green beans. So it's best to use one section for seed saving and keep the main planting well picked so that the plants continue to produce properly for the table. (The same applies to peas.)

When you harvest dry plants such as mustard or beans, you want as little shattering to occur as possible. When you thresh plants, however, you want as much shattering as possible; that is, you want the plants to release nearly all their seed readily. I often harvest dry plants in the morning, when the air and plants are moist from dew. At this time, there is little loss of seed from shattering. I usually thresh in the afternoon or early evening, when the plants are as dry as can be, and release their seed most readily.

Even a fully dry pile of plants is easiest to thresh in the afternoon of a sunny day. The plants rehydrate just slightly each night and morning and dry out again each afternoon. If you thresh in the morning, or on a damp day, you not only get a significantly smaller portion of the seed released, but you have to work much harder for what you do get.

Many people harvest dry plants such as brassicas or beans by pulling up whole plants and tossing them on tarps. I never do it that way. Instead, I always clip the plants and leave the roots and dirt in the field, where they belong. I find the root balls contribute rocks and clumps of dirt that are much harder to separate and clean out of the seed than is the dry plant debris.

Equipment useful in harvesting includes gloves to protect your hands from sharp, dry plants, containers or 5-gallon buckets for fruits, and tarps of various sizes for holding and carrying dry plants. The 5-gallon plastic buckets are a standard in the food industry. You can get used ones for free from restaurants. My local food co-op sells the used buckets for $1 (cleaned).

The tarps can be bought for prices ranging from $5 to about $20, depending upon size. You'll want several of various sizes for both harvesting and transporting plants as well as threshing seed. I have handy little 4x6-foot tarps, as well as sizes ranging up to 15x20 feet or so, which hold about as much as I usually want to carry.

There are such things as one-row combines and small-scale threshing machines. One of these years, I keep telling myself, I'm going to need to get a small used stationary thresher. Even the smallest versions of such equipment, however, are likely to cost more than you'll want to spend unless you are selling seed — and fairly large amounts of seed.

Generally, people who are growing up to about one hundred pounds of brassicas or beans, ten pounds of tomato seed, and a few hundred pounds of dry corn, for example, are harvesting it by hand using the simple tools I have mentioned: gloves, 5-gallon buckets, and tarps.

There are specialized tricks for harvesting certain crops. For example, I harvest seed from bunching onions (*Allium fistulosum*) by clipping the flower stalks as they dry and turning them upside down inside open paper bags. When the seed heads are completely dry, most of the seed will fall out spontaneously or with just the gentlest shaking.

The plants don't dry all their scapes simultaneously. If you wait until the last scapes dry before clipping them, most of them will have already dropped most of their seed. To prevent this, you need to go through the patch every few days and pluck the dry scapes.

Furthermore, if you break up the seedheads by rubbing or stomping on them as we do with many dry-seed crops, the debris from the dried flowers is of such a size that it is almost impossible to separate from the seed.

I give some of the tricks associated with specific vegetables in Appendix A. *Seed to Seed*, by Suzanne Ashworth, treats seed saving species by species, and is an excellent source of such information. In addition, whatever the crop, with experience you will develop your own additional repertoire of techniques, tricks, and methods.

Threshing and Cleaning

One fall evening I went over to the Kapuler family home for dinner. There was a big pile of dry bean plants on a tarp in the middle of the living room floor. Alan and Linda Kapuler, their kids, and the invited guests all shelled beans onto the tarp as we visited. The task is simple. You pick up a dry plant or chunk of plant, shell out the pods, and toss the empty plant (empty pods all still attached) onto a separate pile. It's very companionable.

By dinner time, there was a pile of plant debris and a tarp full of dry beans that were clean except for a few stray dry leaves.

Once upon a time, people always worked with their hands as they talked. They repaired tools, mended, spun, knitted. And I doubt not that they hand-threshed the smaller lots of seeds. It's a healthy pattern. I often listen to classical music when I hand-thresh seeds. I don't have a TV, but if I did, you can bet I would be shelling out beans while I watched it.

There are two basic kinds of seed-processing situations. "Dry processing" refers to seeds associated with dry pods and other dry plant debris. Beans, peas, and mustard are examples. The goal is to separate the seed from all the rest of the dry plant material. In most cases this is necessary because the junk takes too much room to store, and it is full of bugs or eggs of bugs that would eat most of the seed during the winter.

"Wet processing" refers to seeds that are associated with fruits. The seed has to be separated from the fruit, washed, and dried. Often, a fermentation process is involved. Tomatoes, squash, and cucumbers are examples of crops whose seeds require wet processing. To wash seed you simply put the seed in some water and rinse it. To fermentation-process seed you put the seed in some water and let it sit around for a while. *Then* you wash it.

When you wash the seed, you remove various germination inhibitors that prevent the seed from germinating inside the wet fruit. So, after the washing, it's important to dry the seed quickly to keep it from sprouting. Those are the basics.

I don't always clean small lots of tiny dry seeds if they are just for my own personal use. It depends upon how much space the uncleaned seed takes and how hard it is to remove the debris.

For example, I don't clean shungiku (ed-ible chrysanthemum) seed. I don't produce much of it, and it is just for my own use. The seeds are tiny and of such a shape that they don't separate from the debris very easily. So I don't bother to clean it.

I free the seed from the dry flower heads by rubbing them between gloved hands. This breaks up the flower heads and produces a seed/debris mix that is alive with tiny insects of many sorts. These insects would be happy to eat all the seed during the winter. So I put the seed/debris mix in a quart jar and put it in the freezer. This kills the insects and insect eggs. I broadcast the whole thing when I plant. It isn't elegant, but it saves time and it works.

Dry Processing

There are two stages in dry processing. First we use mechanical force of some type to release the seed from the pods or plant material. Then we clean so as to separate the good seed from the bits of leaf, pod, stem, chaff, and other junk.

If you have just a few pods or seedheads, you can rub them between gloved hands and separate the seeds. Or you can put them in a tray and run a rolling pin over them. Just try a few methods until you find one that works. With beans or peas, for a few ounces of seed you can simply open the dry pods by hand over a big bowl and strip out the seeds.

A few dry bean plants can be put inside a pillowcase and rubbed around or stomped on a little to release the seeds. This method is especially useful for threshing individual plants.

Supposedly, you can beat individual plants or masses of dry plants against the inside of a barrel. I've never tried it, though. The last barrel I actually ever saw was in a small country store, and it was full of pickles. I don't think the barrel method would work for most plants — only with plants that release their seed extraordinarily easily. I do have a line of big, 6-foot-high, grain-type fava beans, however, that thresh out so easily that I thresh them simply by hitting them once or twice against something. The "something," however, is (surprise!) a big tarp on the driveway with the edges propped up to make a big tray.

For larger amounts of any dry plant and seed material, tarps are very handy. I pile dry mustard or bean plants on tarps in the field and use the tarps to transport them. The plants then dry in the shade on the tarps until they are completely dry and I am ready to thresh. (Direct sun damages seed.) Threshing consists of turning on some marimba music and dancing on the pile of dry plants until most of the seeds have been released.

Loose seed can be separated from debris by screening with sieves of some type. You can buy sieves of various mesh sizes that are constructed specifically for seed cleaning. (See Appendix E.)

My screens are three pieces of hardware cloth I bought in a hardware store. These come in rolls 3 feet wide. You can have the store chop off whatever length you want. I have pieces of each of the three smallest of the mesh dimensions, ⅛ inch, ¼ inch, and ½ inch.

My ⅛-inch and ¼-inch screens measure 2x3 feet and are unmounted. My ½-inch screen is 3x6 feet and is mounted on two bamboo poles wired to the two long edges. With this set of screens, I can separate just about anything. Sometimes I use the mesh to remove all the large stems and plant debris and let the seeds and small stuff through. Other times I use mesh of a size that retains the seed and lets all the finer debris through.

Winnowing involves separating seeds from debris by using wind or air. A traditional Native American method for winnowing is to toss the seeds in a basket in the wind. You are supposed to toss the seeds and small debris skillfully into the air and catch the seeds gracefully, repeating the process until the chaff has all blown away.

I don't know anyone who can actually do this. It has certainly never worked for me. Perhaps our winds here in western Oregon are just too variable. We seem to have either no wind, or variable winds of constantly changing direction. It's hard enough just to avoid losing the seed. Getting good separation of chaff and seeds just doesn't happen. In addition, the wind always changes direction and blows the chaff right back into my face.

I find it more useful to use a window fan during a period of minimal wind. I don't usually try to pour seed from one container into another in front of the fan, though, as is often suggested. I lose too much seed, or retain too much debris, or both. Instead, I prefer to pour my seed in front of the fan onto (of course!) a big tarp.

When I clean mustard seed, for example, I start by hauling the tarp with the dry plants to the concrete driveway, which I've swept to remove any rocks and debris. Then I dance and stomp and shuffle all over the plants on the tarp. The pods mostly separate from the plant as well as splitting open and releasing the seed.

The next step is to remove the stems and large plant debris. I put down another tarp and prop up the lower edge and corners with three big rocks. (My driveway slopes, and round things such as mustard seeds roll.) For this I use my ½-inch hardware cloth, the big piece that is mounted on bamboo poles.

I put this hardware-cloth frame down on the empty tarp and prop up one end with a concrete brick. Then I ball up the tarp with the seeds and debris and dump the mess onto the frame. All the big stems and plant fragments stay on the hardware cloth. All the seed, chaff, and empty pieces of seedpod pass through the screen. I discard the big stuff, bundle up the tarp with the seeds and debris, and dump the seed/debris mix into a 5-gallon bucket.

Next comes the winnowing. I set a window fan on a box above one end of a big tarp (about 10x15 feet, for example) and attach it to a long extension cord. The bottom of the fan is about 1½ inches above the tarp. Much higher and too many seeds bounce off the tarp; much lower and you don't achieve as good a separation. I put rocks in a few places under the edges of the tarp to turn it, temporarily, into a big tray. Then I turn on the fan so that it is blowing lengthwise down the tarp.

Now, using a smaller container, I scoop up seeds and debris, a quart or so at a time, and pour them in a steady stream back and forth above the fan. The wind from the fan hits the debris and pushes it downwind above the tarp. The dust and lightest debris blows away entirely. Somewhat heavier debris goes all the way down to the end of the tarp. Still heavier debris settles nearer the fan. Heavy scraps of stem settle nearest the fan. There will usually be a swath about 2 or 3 feet wide about halfway down the tarp that contains nearly all the best seed, and this seed is nearly completely clean.

This method gives you separation for seed quality as well. Light, small, inferior seed travels farther than the best seed.

After the winnowing, I bundle the tarp so as to dump the debris at the two ends onto handy little 4x6-foot tarps. Then I bundle the big tarp again so that an edge near the good seed is over the edge of a big stainless steel bowl. Into the bowl goes the clean seed.

From the stainless steel bowl, the seed goes into a quart jar and directly into the freezer. That's because there are always lots of insects on the debris, and the seed carries insect eggs. I always freeze-cycle all my brassica seed as the last step of threshing and cleaning.

I often use a couple of large stainless steel bowls for lots of seed that are too small to require the fan and tarp treatment. I put the seeds and fine debris in one of the bowls and shake it a bit. The chaff and junk will migrate to certain points, leaving much cleaner seed everywhere else. I gather the areas of clean seed and move them to the second bowl. I discard the regions containing seed with lots of debris. (Generally, I have lots more seed than I need, but not necessarily lots more time.) After a few passes, most of the seed is (mostly) clean.

One of my favorite tools for cleaning bean and chickpea seed is a large cardboard box-top of the sort that tables and large appliances come in. Mine are about 4x8 feet, with edges about 4 inches high. In other words, they are big cardboard trays.

I put a cardboard tray on the ground, lift one end to tilt it a little, and dump the seed/debris mix into the high end. The seed bounces over the debris and rolls down to the other end. Some seeds roll better than others, depending upon shape. But debris and chaff rolls little if at all, and I can tilt the box to various degrees and jiggle it as well. The result is almost perfectly clean seed after one or two passes.

When I clean large amounts of chickpeas, for example, I first thresh them out by dancing on the plants on a tarp. Then I bundle up the edges of the tarp to form a funnel-shaped bag and shake it a bit. The seed and pod pieces and smaller twigs settle to the bottom of the bag. The big plant stems and pieces remain on top. I gently set the tarp down and open it up, and remove and discard that surface layer of large junk. Then I pour the mix of seeds, pod pieces, and smaller debris into a 5-gallon bucket.

Next comes the cardboard tray treatment. I sit down crosslegged with one end of the cardboard tray resting on the pavement and the other in my hand. I use a quart-sized container to dip up portions of the seed/debris mix with the other hand. One hand dips and delivers a stream of plant material to the upper end of the tray. The other hand controls the tilt of the tray and jiggles it appropriately.

Seed that is roundish and big, like most beans and peas, will bounce right over the accumulating debris and roll right down the cardboard tray and accumulate, as almost totally clean seed, at the bottom end. By dumping the seeds from greater or lesser heights above the tray, I control how hard they bounce. By the angle of tilt and the jiggling, I control how much momentum the seeds have as they continue to roll. Once you get the hang of it, it's an efficient and satisfying method.

Every once in a while I stop to remove the good seed from the bottom end of the tray, and dump out the debris accumulating at the top end.

I also use my big cardboard trays to thresh out individual chickpea plants, something I do a lot of because of my chickpea breeding work. The individual dry plants start off in separate paper bags. That's how I harvest individual plants in the field, collecting the most productive or interesting-looking plants. In each bag is an index card giving the plant's identity and information about why I decided to collect it separately. I leave the paper bags open so that the plants can continue drying. The paper bag retains any pods that might fall off the plant during transportation or storage.

Later, when I get around to it, I thresh out the individual plants. I put a whole dry plant into one end of my big cardboard tray on the driveway, and stomp and dance on the plant to thresh out the seeds. I remove the three or four biggest plant stem chunks by hand and toss them onto a nearby 4x6-foot tarp.

Then I raise the end of the tray with the debris and seeds a few inches and jiggle it a bit, so the seeds roll away from the debris to the other end of the tray. I lift the clean seed out of the far corner of the tray with the index card label, then put both seed and label in an open plastic freezer container to finish drying and/or for later examination (counting, recording seed characteristics, weighing the yield, etc.).

Corners and crevices of cardboard boxes can retain seeds from one batch and dump them into the next. I tape inside flaps and corners of cardboard boxes so that they don't retain seed easily.

Tarps, too, can contribute to cross-contamination between various batches of seeds, especially with little seeds such as brassicas, which can stick to minor flaws in the tarps. I use new or newer tarps for seed work, and patch rough places with tape. I also take precautions to preserve tarps. I sweep the concrete free of rocks and sticks before I put a tarp with seeds down and dance on them. Small rocks underneath the tarp can cause tears or blemishes when danced upon. Dragging tarps over rough ground causes even worse tears and holes. I use older tarps for such purposes.

To avoid cross-contamination between batches of seeds, I shake tarps and boxes thoroughly after I use them. Then I inspect them thoroughly before I use them on the next batch of seeds.

Clothes can also be a source of cross-contamination. Beans can fall into pockets with one batch of seed and fall out when you bend over the next. Rolled up shirt sleeves or cuffs are almost sure to transfer seeds from one batch to the next. Even the seams of long-sleeved shirts can trap brassica seeds. I wear short-sleeved T-shirts with no pockets when I am cleaning seed. And I shake and brush myself off thoroughly between batches of different seeds along with all the cleaning equipment.

For specific information on harvesting, threshing, and cleaning corn, see Appendix A.

Wet Processing

Seeds of many fruits and berries can be cleaned just by washing them in water. Other fruits, such as tomatoes and cucumbers, have seeds that are coated with gelatinous material. The gel makes cleaning and handling the seeds difficult or impossible, and it usually also contains germination inhibitors. So we need to add a fermentation step to the process, to digest the gel and release the seed.

After fermenting or washing, the seeds are wet and their germination inhibitors have been removed. Therefore, to prevent germination or any loss of viability the seeds must be dried promptly. Little seeds can be dried quickly in sieves or on screens, given a wind or a fan to keep air moving over them. Larger wet seeds, such as fermentation-processed squash, require auxiliary drying of some sort. I use a food dehydrator set on 95°F for up to 8 hours. (See the next section.)

Tomato seed, for example, is surrounded by a gelatinous layer that contains germination inhibitors. In processing seed, this layer is either removed by an acid treatment or digested by fermentation. For most gardeners, the fermentation method is easiest, and it also eliminates many seedborne diseases.

Saving tomato seed is especially satisfying because tomato seed is so expensive. In addition, it's very easy to produce much better quality tomato seed than anything you can buy commercially.

I process tomato seed in two different ways, depending upon the amount of fruit and seed involved. I often save seed from just a single fruit or a few fruits from one plant. With just a few fruits, I squeeze the juice with the seeds out of the tomatoes and ferment just that. That way I don't have the problem of separating the seeds from the pulp after the fermentation.

I start by allowing the fruits to ripen way past the edible stage — on the vine if possible, indoors otherwise. To process, I start by cutting each fruit in half across the pattern of internal seed cavities. I use my thumbs to dig and squeeze the juice containing the seeds out into a bowl. I put this fluid in a clear plastic cup of suitable size.

I like clear cups so I can see the fermentation happening, that is, see the bubbles forming and rising. I use very tiny cups if I have just a small amount of juice, because the surface portion will be discarded. I use containers that are small enough so that the tomato juice is at least an inch deep.

I put the seed-cavity juice with seeds in the cup, put an index-card label under the cup, and leave the cup and its contents alone for one to five days. I don't add water; that would just slow things down. I don't stir. How long the fermentation takes depends largely on temperature. Usually, a mat of mold will form at the surface of the juice. At some time after that, the seeds will be "done."

After the fermentation has been proceeding for a day or two, as indicated by the bubbles, or after the surface mat of mold forms, I test the seeds, and I test them every day until they are done. I dip gently into the container,

remove some fluid with seeds, and put it into a small strainer. I hold the strainer under the faucet to wash the seeds. Then I feel the seeds. If the gel around the seeds is all gone, they're done, and I'm ready to end the fermentation. If the test seeds still have gel on them, I wait a day and test the batch again.

If you ferment the seeds too long, viability of the seed and vigor of the resulting seedlings is harmed. Grey or brown instead of bright cream-colored seed indicates overly long fermentation. Most instructions for saving tomato seed give an arbitrary time for the fermentation. I think it is very worthwhile, however, to test the fermentation and choose the right endpoint instead of just guessing.

When the fermentation step is over, I remove the mat of mold at the top of the tomato juice and discard it. Then I pour the rest of the tomato goop into a strainer. I wash the seeds by holding the strainer under the faucet.

I dry the seeds in the same strainer. I use the flow of water from the faucet to spread the seeds evenly out in a layer on the strainer. I then set the strainer somewhere that has good air circulation, such as in front of a fan.

After only a few hours, the seed will be "semidry." Then I "rustle up" the seeds in the strainer with my fingers, spreading them around and abrading them a little against the strainer. The object is to separate the seeds from each other so that they don't dry together in clumps. (I've found this "rustling up" useful in the drying of all wet-processed seeds.) Then I spread the seed out in the strainer and forget about them for a while. (Several days or more.) By then the seeds are dry enough to store.

With small batches of seeds from a few tomatoes, I do the processing indoors. If there's more than a few ounces of tomato glop fermenting, I do the entire process outdoors. Even a few ounces of fermenting tomato glop smells pretty powerful.

For a big batch of seed, I start by harvesting the whole, overripe tomatoes into 5-gallon buckets. Then I stomp them. I have a heavy-duty plastic garden tub that's big enough to hold a person and a few gallons of tomatoes. I put tomatoes in the tub in a layer up to about 6 inches deep, put on rubber boots, and hop in. I stomp the tomatoes only well enough to break each one into a few big chunks. That's good enough to release most of the seeds. It's tempting to stomp the tomatoes much more. It obviously releases a larger proportion of the total seed and it's also really quite a lot of fun. In fact, it's easy to get carried away. If you don't exercise control, in just a few minutes you'll have stomped the tomatoes to smithereens.

Unfortunately, it is very hard to separate tomato seeds from tomato-fruit smithereens. But it's very easy to separate the little seeds from big intact tomato chunks. So restraint in stompery is warranted.

After the stomping I transfer the tomato material to 5-gallon buckets for fermentation. I usually add a little water in order to facilitate the release of the seeds from the fruits and to make it easy to stir the mess. (Not a lot of water; perhaps about 50 percent as much as the volume of tomato material.) I stir the fermenting material every day, and test some of the seed each day until the seed is "done," as described previously.

Next comes screening. I usually add much more water to the tomato glop after fermentation and right before screening in order to help free the seeds from the tomato chunks and pulp. I use my three hardware-cloth screens to separate the pulp from the seeds. I put the biggest-mesh screen over an empty 5-gallon bucket and pour the tomato glop through into the next bucket. Then I grab another empty bucket, and the next-smaller-mesh screen. And so on.

When I'm done, I have one or more 5-gallon buckets full of tomatoish-colored water with tomato seed at the bottom. I pour off the excess water. Then I pour the final amount of water with the seed in it out onto a propped-up old screen door. I use a stream of water from the hose to finish rinsing the seed and to spread it around evenly on the screen. Then I set the screen with the seeds in front of a fan for drying.

I remember to "rustle up" the seeds after they are dry enough (but not too dry) so that, after the final drying, I end up with nearly all individual seeds instead of clumps. That's all there is to it.

One cautionary note: It's very easy to transfer tomato seeds from one batch to another. They tend to stick to your hands and to the buckets and screens. Wash your hands and rinse strainers or hose down buckets and screens between batches of different seeds.

You can process a pound of tomato seed at a time pretty easily using methods involving nothing more than a few 5-gallon buckets, a good solid stomping tub, a fan, hardware-cloth screens, and a good chunk of window screen. Actual seed yields per amount of fruit vary wildly depending upon variety, but an average is about one pound of seed for every 100 pounds of fruit.

I describe wash-processing and fermentation-processing of squash seeds in Appendix A. *Seed to Seed* gives detailed information on processing methods for just about every conceivable vegetable seed.

Drying Seed

Seed dried at room temperature usually has a moisture content of 10–20 percent. That is not dry enough to store in air-tight containers or to freeze. Such seed should be stored in paper envelopes or bags or other containers that allow some air exchange. Seed stored in paper is subject to attack from insects or pests, however. In addition, many kinds of seed have insects or insect eggs in them as they come from the field. Such seed will be destroyed inevitably if we don't take measures to kill the insects and insect eggs. Freezing is a good preventative measure.

Storing seed in glass jars or, for bigger amounts, in 5-gallon buckets is an effective way of preventing infestations of both insects and rodents.

Only very dry seed can be stored in plastic bags, jars, or other airtight containers. And only very dry seed can be frozen.

Most little seeds like mustard and lettuce can be dried well enough without artificial drying. Only if you live in a desert area, though, are you likely to be able to dry bigger seed such as corn and beans to the "very dry" point without artificial methods.

The way you tell whether a batch of seed is "very dry" — that is, dry enough to store

in airtight containers or to freeze — is by testing it.

Test small seeds or thin ones such as squash by bending them. If they are "very dry" they will snap instead of just bending. With bigger seeds such as corn, beans, or peas, do the "hammer test." Take a few seeds outside and put them down on a piece of brick or concrete. Hit each seed with a hammer. If the seed is "very dry," it will shatter like glass when hit. If seeds are inadequately dry, they will smash or mush instead of shattering.

"Very dry" corn or bean seed has a moisture content of about 6 to 8 percent. This level of moisture is optimal for storage. Too much moisture and the seed remains physiologically active and continues to respire; longevity is affected. Too little moisture and viability as well as longevity will be affected.

To dry seed to the "very dry" stage for freezing or storing in airtight containers, you can use either silica gel (see Appendix E for sources) or a food dehydrator that has a thermostat. You can't use a home oven to dry seed. Its thermostat controls don't go low enough. How fast seeds dry over silica gel depends on how much seed you have, how wet it is, and how much gel you're using. Use the "very dry" test to tell when the seed is dry enough. If you use equal weights of seed and gel, a few days or less is usually sufficient.

Silica gel is likely to be most useful for small amounts of seed. Bigger batches of seed take lots of silica gel, which you then need to dry in the oven to reuse. I think it's easier to dry the seeds in a dehydrator.

Many people use small round food

dehydrators to dry seeds. You set the thermostat to 95°F and dry the seeds for up to about 8 hours. Eight hours is enough to take big squash seeds that have just been fermentation-processed all the way from fully wet to very dry.

I've used a couple of the round dehydrators for years. They have several disadvantages. The holes in the middle of the circular trays greatly reduce the potential amount of seed each tray can hold, and the trays are small to begin with. There is also only a constant air stream, which is divided among however many trays you use. So if you have eight trays instead of four, the drying will take twice as long. In addition, you need to rotate the trays during the drying, because the bottom trays get more air than the top ones.

Nevertheless, these round dehydrators only cost about $50, and they can do a good job on small lots of seed. If you get one of the round dehydrators, get one with a thermostat that goes down to the 95°F that you'll need for seed drying. In addition, it's important that the design be such that the air flow is divided and moves across the trays and out. Some dehydrators force air through all the trays. The air goes through the first tray and then through the second, etc. With such a design, only the material in the first tray is dried very effectively.

I recently bought an Excalibur dryer. I needed it for drying summer squash. The little round dehydrators just don't do the job when it comes to drying pounds and pounds of fully wet vegetables. A one-pound summer squash fills a little round dryer to capacity, and one one-pound summer squash is nothing.

The Excalibur quickly became my pre-ferred dryer for drying seed as well. The large square trays with no holes in the middle are a lot more useful. The air stream is constant over all the trays instead of variable. And the dryer has both a thermostat and a 26-hour timer. Furthermore, by omitting alternate trays, one can even dry big items such as whole ears of corn.

Protecting Seed from Insects

There are quite a lot of insects whose major purpose in life is to eat all the seed you have saved. All organic pea seed raised here in western Oregon has weevils or weevil eggs in it, for example. Unless you do something to kill the insects, they will develop during storage and destroy nearly every seed. In some areas of the country, weevils are a serious problem in beans. Nearly everyone has problems with corn. Grain moths and many other insects love it.

There are three basic approaches to killing insects in seed. Some people sprinkle diatomaceous earth in with the seeds so as to kill or discourage insects. I've never used this approach. My seed collection overlaps too much with my food supply. I have received batches of seed coated with the gritty white stuff, though, and I can vouch for the fact that it doesn't look or taste very appetizing.

To kill weevils and other insects and insect eggs I put seed in glass canning jars with air-tight lids. Then they go into the freezer for at least several days. (The seed must first be fully dry).

When I need the seed, I remove the jar from the freezer a day ahead of time. It isn't a good idea to open a jar of seed right after it has been removed from the freezer. Moisture will condense all over the seed, and I doubt the sudden shock does the seed any good.

A third method for killing insects in seed uses dry ice. I'm planning to start using this method on the larger amounts of seed such as the 5-gallon bucket loads full that I seem to be accumulating. Many university research departments and medical clinics use dry ice. You may be able to find one that will sell you the modest amounts you need, or they will be able to tell you the name of a source in your locale.

Use at least 3 ounces of dry ice for a 5-gallon bucket of grain. For larger amounts, use at least 1 pound of dry ice per 30 gallons of volume. Crush the dry ice and mix it with grain in the bottom 2 inches of the container. Add the rest of the grain and cover loosely for 8 hours or more to let the dry ice vaporize into CO_2. The CO_2, being heavier than air, displaces the air. After the 8 hours, seal the container. Insects will die from lack of oxygen.

Once you have killed the insects and insect eggs in seed, you need to store it in such a way that it cannot become reinfested. Paper packets and bags don't hold off a serious onslaught of insects for very long. For seed I care about the most, I use glass jars or 5-gallon buckets.

Protecting Seed from Rodents

Sooner or later, if you have lots of seeds lying around in open containers drying, or protected only by paper or plastic, you will attract mice or rats.

A few years ago in my house, a mouse suddenly appeared and started processing my squash seeds into more mice. I preferred the squash seeds.

Being a believer in biological controls, I did what I considered the logical thing. I started transferring all my seeds to jars and other rodent-proof containers. Simultaneously, I got a baby ferret. I figured the ferret could eat the mice.

I wish to report that a ferret in a seed collection is a whole lot worse than mice. Mice only remove a few seeds here and there for personal use. They don't love to play in seed, turn over or open all the containers, scatter all the seed on the floor, and steal all the empty containers to roll around the room as toys. Mice don't get excited by noise, and push all the glass jars of seeds off the shelves just to hear them go "crash."

Also, mice are *small*. They can't drag a 1-pound package of corn seed away and hide it under the furniture for later investigation.

With a ferret around, if I leave an opened box from a seed company alone for a few hours, when I come back, half the packets of seeds are missing. They just vanish, only to be discovered at random later in one of the ferret nests occupying an empty drawer I hadn't looked in for a while — where the packages of seed are lovingly intertwined with all those socks I was also missing.

Getting the ferret did solve the mouse problem, though. By the time I had "ferret-proofed" my seed collection, it was also mouse-proof.

Storing Seed

The basic rule for storing most seed is "cool and dry." Seed stores best in the coolest, driest place in your house. On the other hand, seeds *look* best and provide the most emotional and visual gratification if displayed cheerfully and gloriously on shelves in the living room. Take your pick.

Most books and articles on seed saving provide you a chart showing the supposed longevity of different types of seed. Other than learning that onion-family seed is quite short-lived, however, I haven't found such charts too useful. They are talking about average commercial squash seed, for example. My squash seed is from the biggest, best fruit of each plant, which was cured optimally before cleaning. Commercial seed comes from all the fruit, which aren't cured at all.

In my experience, prime, handcrafted, homegrown seed keeps much longer than standard commercial seed. I expect my well-grown squash or corn or bean seed to last at least five years at room temperature, for example, with only modest losses in germinating ability, and to be usable for even longer.

For longer-term storage, you can freeze seed and leave it in the freezer.

PART III

Developing Crops for a Sustainable Future

TWENTY-TWO

Genetic Engineering and Genetically Modified Foods

THE FLAVRSAVR tomato was the first genetically engineered food to be released and sold to the public. FlavrSavr contains an artificial gene designed to inhibit the function of the gene that makes pectinase. Pectinase is an enzyme involved in tomato softening. The idea is that, if the tomatoes stayed hard, they could be ripened a little longer on the vine before commercial harvest. Most commercial tomatoes are picked hard and ripened artificially. Tomatoes allowed to ripen on the vine taste better, but are normally too soft for commercial picking, shipping, handling, and storage.

For years both scientific and political controversy raged over whether genetically modified food in general — and the FlavrSavr tomato in particular — should even be allowed to be grown in open fields, let alone

sold to anyone for eating. All through the Stürm und Drang, however, my thoughts ran on a more simple-minded level. What I was thinking was: "But this isn't going to work."

A hard tomato isn't going to seem very much like a tomato, even if it tastes like one. Furthermore, control of fruit ripening is a highly integrated, holistic process. I doubted that you could change fruit softening without also having a serious influence on flavor.

"These molecular geneticists don't know about pleiotropy," I muttered to myself. *Pleiotropy* refers to the effects of genes on characteristics other than the "primary" one. I'll discuss the concept more "genetically" later in this chapter. At this point, let me put it in philosophical terms. Pleiotropy is a genetic version of the ancient Taoist understanding that you cannot do just one thing.

"The way the world works is like a bow," the Taoist sage Lao Tzu pointed out more than two thousand years ago. "When you pull the string, the top comes down, the bottom comes up, and all the parts move."

One day about 2,300 years ago, another ancient Chinese Taoist, Chuang Tzu, went walking in the mountains with his friend Hui Tzu, a famous logician. They sat and rested at the edge of a cliff, their feet dangling over, and they talked philosophy. Hui Tzu argued that everything should be defined in terms of its usefulness.

"What about the ground you're sitting on?" Chuang Tzu asked. "You're only using that little piece under your rump, right?"

"That's right," said Hui Tzu.

"But what would happen if I took a shovel and dug away this little piece behind that?" asked Chuang Tzu.

"I'd fall," Hui Tzu admitted sheepishly.

"The useful depends upon the useless," Chuang Tzu said cheerfully. "Isn't it amazing how everything is so connected?"

Molecular geneticists tend to be mechanists. They altered just one gene involved in softening of the tomato, figuring they would get a tomato just like the original except that it wouldn't soften. I didn't think so. I thought that if you change that one gene, you'd probably affect a whole lot of things, but especially flavor.

I Meet the FlavrSavr Tomato

All doubts aside, however, the idea of the FlavrSavr fascinated me. I awaited its release eagerly. A firm tomato wouldn't seem very much like a tomato, even if it tasted the same. But there are plenty of hard vegetables that taste great. And even if the tomato tasted very different, it might still taste good. It might be a whole new vegetable. I wanted to taste-test that tomato! And I was prepared to give the FlavrSavr every possible chance. I wouldn't even require it to conform to any concept of what tomatoes should be. I wouldn't require it to fit into the "tomato niche." I'd simply try it, and if it tasted good, if need be I'd think of new niches.

And of course, I thought, I might be wrong about flavor. Maybe the tomato *would* taste just like an ordinary tomato, only be firm. Who knows? Certainly the molecular geneticists all seemed to think that was what should be expected.

For years, however, the controversy in the media was as close as I got to a FlavrSavr tomato. Finally, in 1995, FlavrSavr was released and arrived under the brand name of MacGREGOR'S® tomatoes at my local Safeway supermarket. I knew when they were coming, and rushed out to the store the very first day they became available.

There they were, at long last, sitting in a huge bin under a huge sign proclaiming them as MacGREGOR'S®. They looked beautiful. Big and deep red and perfect. They were firm, even though fully red. Each tomato was marked with a little label stuck on it proudly identifying it as a MacGREGOR'S® tomato and boldly stating, though in very fine print towards the bottom of the label, "grown from genetically modified seeds." I bought two tomatoes, bagged them carefully, and scampered home.

Finally came the long-awaited taste test. I cut a small slice from one of the tomatoes and put it in my mouth. The taste was so bad I almost spewed it out by reflex. Instead I deliberately spat the offending morsel into the trash, then examined the tomato carefully. Had something really awful been spilled on it? I couldn't see anything wrong with the tomato.

I cut a small chunk from the other tomato and sampled it. I spat that out too.

I told myself that it really wasn't a taste test if I didn't swallow. So I took another very tiny portion. But I just couldn't swallow. The body rebelled. I spat that portion out too, then rinsed my mouth out with water.

It's easy to describe the flavor of those tomatoes. They tasted like gasoline.

The tomato did not merely taste like a bad tomato. It didn't taste like a tomato at all. There was absolutely no component in the flavor that suggested "tomato." Had I done a blind taste test in which I was asked to categorize substances as food or fuel, I would have classed the FlavrSavr as fuel.

In all I had read about whether the FlavrSavr should be released, no one had mentioned that the tomato tasted awful. The tomato was subsequently withdrawn, in order, as it was put, "to develop cultivars of more acceptable flavor." In other words, the tomato was withdrawn because it tasted awful.

FlavrSavr was, for all practical purposes, withdrawn by my Safeway store within three days. The tomatoes were still there in the bin, but the sign identifying them had been removed. Those tomatoes stayed there for weeks. I doubt if anyone ever bought them more than once except strictly by accident. Apparently, the one bin more than satisfied the needs of the community for this tomato for the rest of the decade.

Genetic Engineering and Genetically Modified Food

A little personal preamble to this section: I am a "classical" geneticist turned molecular geneticist turned "traditional" plant breeder. As a young scientist I got my start in *Drosophila* (fruit fly) genetics, working my way through undergraduate school at the University of Florida in the *Drosophila* lab of Henry Wallbrunn (the man to whom this book is dedicated).

While at Florida I migrated into the molecular biology lab of Arthur Koch, then moved on to graduate work in fungal and molecular genetics with John R. Raper at Harvard University. Thereafter followed a stint on the faculty in Genetics and Cell Biology at University of Minnesota.

I started in the field that was almost the definition of classical genetics. My academic work subsequently was in molecular genetics. During the last twenty years, however, my concern has been with agroecological issues, and for the last ten I've been doing standard plant breeding on a variety of crops.

Most "traditional" plant breeders come from backgrounds in agriculture and have little experience in molecular biology. Most molecular geneticists come from biochemistry traditions, and don't know much about classical genetics, about agriculture, or about growing or breeding plants. I feel that my personal participation in all these fields gives me an excellent perspective from which to see genetic engineering in its full scientific, historical, and technological context.

Genetic engineering involves one of two

types of modifications. In one case, the genetic engineers synthesize a gene deliberately designed to perform a certain function in a certain crop. More commonly, the desired gene occurs in some other, unrelated, organism, and they start by extracting it.

In both cases they proceed by redesigning the desired gene and linking it up to various other genes and pieces of DNA. Some are control genes to make the desired gene function in the right tissues in the target-crop plants. Others are DNA snips that are good at jumping from organism to organism and inserting into DNA. Yet others, such as genes for antibiotic resistance, are included to facilitate identification and manipulation of the cells containing the invented-gene construct after it has been inserted into plant cells.

Then the gene engineers insert this DNA construct into cells of the crop growing in artificial media. The antibiotic resistance helps them select for and identify which cells and cell lines have experienced a successful insertion of the DNA construct into the plant cell's chromosomes. Cell lines containing the construct must then be grown into whole plants capable of reproducing with the new DNA as part of their genetic makeup.

The methodology of genetic engineering is basically biochemical. It involves manipulating DNA and cells in the laboratory. The traditional and modern plant breeding methods promulgated in this book are field methods involving whole plants. They do not require laboratories, do not construct artificial genes, and do not transfer any genes from any organism to any unrelated one.

These ancient as well as modern field methods transfer and manipulate genes only in exactly the same way as the plants themselves do, that is through normal reproductive processes. Traditional and modern plant breeding exclusive of genetic engineering is merely a speeding up and guiding of natural reproductive processes. Genetic engineering though, is based upon a dramatically different process.

Genetically modified food (or GM food for short) is food from plants that have been bred using genetic engineering. Such plants contain laboratory-constructed artificial genes or genes from very unrelated species.

Some spokespeople from the GM food industry are fond of claiming that genetic engineering is an old technique that has been going on for thousands of years. This claim strikes me as dishonest, or at the very least deceptive. Plant breeding using the normal reproductive processes of plants has indeed been going on for thousands of years. But genetic engineering, which involves physically transferring DNA from one organism to another in the laboratory, is *not* merely more of the same.

Plant breeding has certain natural limitations not shared by genetic engineering. Speaking as a plant breeder, these limitations can be frustrating. Speaking as a consumer, these limitations are reassuring. You cannot, for example, take a gene from peanuts and put it into corn using standard plant breeding methods. That limitation is a matter of life or death to a person who is allergic to peanuts.

There is tremendous controversy over the ethics, safety, and social and environmental consequences of genetic engineering in general, and genetically modified foods in particular. These issues are outside the scope of this book. For a thorough and readable appraisal and evaluation of all aspects of the subject I recommend a little book by Luke Anderson, *Genetic Engineering, Food, and Our Environment* (Chelsea Green Publishing, 1999).

In the rest of this chapter I'll focus on a practical comparison of the advantages and limitations of standard plant breeding and genetic engineering.

Standard Plant Breeding versus Genetic Engineering

Genetic engineering involves directly manipulating DNA. You can't do it without a specialized background and a laboratory filled with several hundred thousand dollars worth of specialized equipment. Does this mean we gardeners and farmers are out of the game now that genetic engineering has become a reality?

I don't think so. In fact, there is more need for our contributions and expertise than ever before. The fact is, there are many limitations as to what can be accomplished with genetic engineering. Much of what we care about most as gardeners and farmers is not very amenable to the genetic engineering approach. And almost all of what matters in breeding plants for organic or sustainable agriculture is best approached using standard plant breeding methods.

An additional aspect is the matter of the scope of the vision. All our food crops were originally created with standard plant breeding methods. All genetically engineered varieties represent relatively small changes in

thoroughly established varieties, not whole new crops or whole new crop species.

Genetic engineering could not create corn starting from wild teosinte today. Nor could it have created any of the other food crops we have today. Genetic engineering is very powerful at making single changes in existing crop varieties. But someone needs to breed the crop or the basic variety in the first place.

Genetic engineering excels at manipulating single genes. Standard plant breeding methods work with the whole genome at once. The whole genome is the natural level for evolution, including the human-directed evolution that develops entirely new crop species. I believe that most of the big advances during the lives of those of us now living — most development of whole new crops, for example — will continue to come from standard plant breeding.

Finally, the universities and multinational seed companies have blindly jumped upon a single genetic engineering bandwagon. They're all going in one direction. The main driving force behind the choice of research problems seems to be mostly a matter of whether the problem is one that can be approached by genetic engineering — not whether the goal is important or needed.

Most plant breeders have retired or are retiring and are being replaced with genetic engineers. Wisdom would have added genetic engineers to a full complement of field-experienced plant breeders using standard methods. Instead, the universities and big corporations have virtually abandoned standard plant breeding. In doing so, they have all but abandoned many of the most interesting and important problems.

• • •

Genetic engineering excels at physically manipulating single genes. It is at its best in situations where single genes have a known and definitive effect on the characteristic of interest. Only a few agricultural crops are very well known genetically. Even in the genetically best-known crops such as corn, peas, beans, and tomatoes, only a few of the simplest traits are very well understood.

Genetic engineering requires exact information about specific genes and a thorough understanding of the relationship between various traits and the underlying genes and biochemistry. In only a very few cases for a very few types of plants do we yet know those relationships well enough for genetic engineering to be able to produce desirable new varieties.

One of the great virtues of standard plant breeding is that we don't need to know anything about the genes that determine the characteristic we care about, or about what any of those genes do. We can guide a variety towards our goals without being limited by our understanding of the underlying biochemistry.

An even more important limitation of genetic engineering, however, is that most characteristics of agricultural importance are quantitative genetic traits. That is, they are determined by many genes, not just one or a few.

Flavor, yield, size, shape, earliness, lateness, cold hardiness, heat tolerance, drought resistance, and other components of ecological adaptation are all traits that involve many genes. Such traits are not especially amenable to the techniques of genetic engineering.

These traits include nearly every characteristic that is most important to gardeners and small-scale farmers.

The great strength of genetic engineering is its ability to change a single gene in an established variety while leaving the entire rest of the genome intact. This is very difficult to do with standard breeding methods. The closest we can come to transferring single genes into an established variety is via recurrent backcrossing (see pages 138–140). However, there are always some blocks of genes too tightly linked to the desired gene to be separated.

A second major limitation of standard plant breeding methods is that sometimes the gene we need doesn't exist anywhere in the whole plant species. If there is a useful gene in mustard, and we want it in tomatoes, we're out of luck if we are using standard plant breeding methods. Genetic engineers have the option of isolating and modifying the mustard gene and inserting it into the tomato. In some cases they can even synthesize an artificial gene and use it.

However, once inserted into the tomato, the alien or artificial gene may not have the anticipated effect, and, even if it does, there may be many other effects that were unpredicted and are unacceptable to the farmer or consumer. This brings us back to the phenomenon of pleiotropy.

Pleiotropy refers to "secondary" effects of genes. The classic experiment was done by Dobzhansky in the early days of *Drosophila* (fruit fly) genetics. In *Drosophila*, there are hundreds of mutant strains available that differ from the basic wild-type stock by a single mutant gene. This is possible because the mutations arose in that specific wild-type background originally.

Each mutation has a characteristic effect on the phenotype of the flies that carry it — such as altering the eye color to white, or causing stunted wings, or increasing the dark pigment in the body. When you actually work with the mutant fly stocks, though, you notice that each mutation affects other phenotypic traits as well.

The flies have a characteristic pattern of bristles, for example. I remember noticing how it seemed that every mutation, whatever its primary effect, seemed also to alter the bristle patterns. You work with flies under a low-power dissecting scope. If you're observant, you soon start wondering whether every gene affects everything else.

Dobzhansky's experiment asked the question, "Does every gene affect every phenotypic characteristic?" In order to address the issue fairly, he considered many common genetic mutations, and he examined their influence on an unrelated part of the phenotype. He chose to look at the effects of all the mutations on the dimensions of the spermatheca.

The spermatheca is an internal organ in the fruit fly female that allows it to store sperm between matings. The spermatheca is usually invisible, so Dobzhansky could be assured that he was not choosing it because of preconceived ideas.

So Dobzhansky measured the spermatheca in wild-type flies and in ones that carried *w*, for example (the mutation associated with white eyes), and in many other strains that differed from wild by just a single gene. The results were unambiguous.

Each mutation had originally been identified and characterized by some simple major effect on the external appearance of the fruit fly. Every gene, however, also influenced the shape of the spermatheca.

Everything affects everything. You can't change just one thing. Everything is connected.

If you change any gene whatsoever in a tomato plant, this will affect the flavor of the fruit, the yield of the plant, and every other characteristic. The real issue is, "By how much?" and "Does it matter?" If you change a gene known to be central to fruit ripening, of course, this will affect fruit flavor, and probably profoundly.

Pleiotropy means that when a genetic engineer makes a change in a single gene, the overall result is less predictable than I think most genetic engineers these days are supposing. Concomitantly, a large number of genetically engineered plants will probably turn out to be undesirable or unworkable in the last stage of their development.

All this does not mean that genetic engineering is fatally flawed. It only means that the most powerful advantage of molecular genetics — that of being able to make simple single changes in genes — is actually less powerful than one might suppose, because simple changes in genes will not reliably translate into simple changes in phenotypes.

Practically, pleiotropy means that plenty of the ideas and projects that are theoretically possible and that can be produced technically in the lab are likely to fail at the last stage, when it comes to producing desirable foods from plants with appropriate yields.

Flavor depends upon levels of dozens, sometimes hundreds, of components. For that reason alone, I think flavor will always be one of the most difficult subjects for genetic engineering to address. Yield also depends upon hundreds of genes. I believe that many genetic engineering projects will fail because the genetic modifications will have unavoidable effects on yields.

You can do stupid things with standard plant breeding, but you are more likely to do stupid things with genetic engineering. It took genetic engineering to create and release a tomato as bad as the FlavrSavr.

Whenever you make a cross to get segregants better than both parents, for example, you also get segregants that are worse than both parents. Sometimes *much* worse. I have produced plenty of weird or unpalatable flavors on individual plants here and there in my various breeding projects.

However, the plant breeder works with whole plants and their phenotypes at every stage of the project. Anything bad-tasting or unpalatable is eliminated as soon as it appears. Standard-bred varieties are taste-tested as they are developed. By the time a standard-bred variety is released, it is already an old friend to its creator, who has been eating it and feeding it to her friends and family for years.

In contrast, genetic engineers start a project based upon certain assumptions and ideas, then spend almost the entire course of the project just manipulating DNA and plant cells. Only at the very end of the project do they have real plants. Only then do they get to find out whether their assumptions and ideas were correct and discover any unpredicted side effects.

The standard plant breeder is like a person who builds a go-cart and tests it at every significant stage of development — at each stage discarding what doesn't work and refining what does. But the genetic engineer has to build his go-cart by design and theory alone, without benefit of any testing. He drives it for the first time in the race. Only if every assumption was correct and everything works exactly as it is supposed to is the go-cart likely to run well enough to finish the race.

Imagine further that the genetic engineer is allowed to use titanium in his go-cart, but not wood, screws, or nails. However, he has access to jet engine parts as well as go-cart parts. . . .

We can breed food crops that are actually poisonous, not merely bad-tasting, though these categories overlap. We can do this with either standard plant breeding or genetic engineering. With standard plant breeding, there are two contexts in which we are most likely to produce something poisonous.

Context 1: We have started a project by crossing an edible species or variety to an inedible, poisonous, or wild one. In this case, we're usually trying to get some useful characteristic out of the poisonous or wild material back into an edible variety. Given a poisonous parent, poisonous progeny are expected in the project and are selected against routinely.

Context 2: We are selecting for insect or herbivore resistance. Most natural plant species have toxins of various kinds that protect them from being eaten. Often, domestication of a plant species involves selecting for lower levels of these toxins. If we select for insect resistance, we may be selecting for higher levels of the toxins once again. Higher toxin levels may make the plants unpalatable and/or poisonous to humans as well.

One needs to be cautious whenever selecting for insect resistance or toxin levels or making crosses involving high toxin levels. I don't know of any plant breeder who has actually poisoned himself or herself, incidentally. Most plant toxins are bitter or foultasting. You don't get poisoned because you don't eat the plant.

A much-quoted recent plant breeding incident involves the breeding of insect-resistant potatoes. The potatoes had such high levels of solanin in them that they could make people sick. (Solanin is the toxin found in wild potatoes, and in the skin of green potatoes.) These toxic potatoes were developed using standard breeding methods. They were, as expected, never released.

It is obvious that any genetic engineering involving pest resistance or toxin levels would have to be evaluated carefully, just as would any standard plant breeding involving these characteristics.

Genetic engineering is engineering. The genetic engineer has some ideas and creates a design, then attempts to impose it on the plants. Genetic engineering is limited by knowledge and imagination. Genetic engineers can only create what they can imagine.

Plant breeders also often start with ideas and designs and attempt to impose them on

plants, but that stage is quickly transcended. What soon develops is more like a cooperation between plants and person — a two-way partnership — a conversation. Plant breeding is limited only by what it is possible for the plants to do. The plant breeder can and often does breed things that she never could have imagined.

Everything in This Book Is Illegal with Genetically Engineered Varieties

There is no inherent reason why genetically engineered varieties would have to exist in an entirely different legal framework than standard-bred ones, but the fact is, currently, they do. To make use of the genetically engineered varieties requires accepting an unprecedented legal situation. This new legal frame of reference outlaws garden- and farm-based seed saving and plant breeding.

Genetic engineering, as it is currently practiced, produces patented genes. Plant Variety Protection (PVP) does not patent individual genes. It only gives limited protection to a whole variety, that is, to a specific arrangement of genes. I can legally use a PVP variety in my own breeding program to breed something even better. Breeders may use the material without restriction to create the next generation of varieties.

In contrast, the new genetically modified genes themselves are patented. You are legally not allowed to save seed, not even for your home garden. You are legally not allowed to make any crosses with the varieties or do any breeding with them. There can be no farm- and garden-based plant breeding with these varieties, no adapting them to specific regions or growing conditions.

With patented genes, everything I'm encouraging you to do in this book is illegal.

With the invention of "terminator" technology, companies like Monsanto, owner of the patent on it, can envision making it physically impossible for gardeners and farmers to save seed, not merely illegal. Seeds with terminator technology carry inactive genes that kill plants in the embryo stage. The genes are activated when the seeds are treated with a triggering chemical such as tetracycline. The treated seed produces plants whose seed is sterile. This allows a company to sell seeds that are only capable of producing one generation.

Monsanto, in response to considerable bad press and a furious public outcry, has stated that they will not develop this terminator technology for license to others. However, I have seen nothing in their statements that promises that they will not incorporate terminator technology into their own varieties.

I think terminator technology is so attractive to seed companies that unless we specifically outlaw it, we will soon have it out in our fields. We need laws that make the use or possession of terminator technology a criminal (not merely a civil) offense, and one involving serious prison time.

I view terminator technology as the genetic engineering version of poisoning wells. Those who develop terminator technology may have some legitimate purposes they think they are trying to serve. So, undoubtedly, did most of those who poisoned wells.

Genetic Engineering and Sustainable Agriculture

I believe genetic engineering has tremendous potential to provide tools for breeding for sustainable agriculture. However, I am underwhelmed by its current tally of actual accomplishments. I would be impressed by a new crop that grew so well it outgrew all the weeds. I am not thrilled by a new crop that has a minor change that allows you to dump huge amounts of *this* herbicide on it instead of huge amounts of *that* one.

I believe most breeding for sustainability is better and more easily done by standard methods than through genetic engineering. There are some exceptions. One field of genetic engineering that has great potential for sustainable agriculture is the development of new genes for viral resistance in plants. Viral resistance can also be developed with standard methods, of course. But, as the Chinese say, why not walk with both legs?

In addition, there are other things I can also imagine that would be sustainability-enhancing, and that could be accomplished only with genetic engineering.

However, I cannot imagine any circumstance in which I would be willing to grow a variety whose seed I am not allowed to save. Any advantage to sustainable agriculture viewed at the level of the specific crop would, in my eyes, be more than offset by the sustainability-destroying bigger picture.

Analyzing the virtues or liabilities of herbicide resistance and insect resistance is much more problematic than analyzing viral resistance. You can argue that herbicide-resistant plants constitute sustainable agriculture if it means you'll now be dumping huge amounts of Roundup on the plants instead of huge amounts of something even worse. Similarly, herbicides used with no-till farming methods can be considered sustainable if you're cultivating a hillside where the soil washes away when it is tilled.

In my eyes, these arguments are comparable to those of the man who says he's treating his wife compassionately these days, because he has started beating her with a horsewhip, whereas he used to use baseball bats and chains.

If a farmer is abusing his land, his neighbors, and the environment badly enough, and some of the new genes allow him to continue abusing them, but less badly, he may claim this is a move toward sustainability, relatively speaking. He might even be right. But he shouldn't expect the rest of us to cheer.

• • •

Meanwhile, back in the kitchen, I'm not eating genetically modified foods. I am both allergic to and intolerant of wheat gluten. Most prepared foods contain at least small amounts of wheat gluten. Small amounts are usually not listed on labels. This fact has caused me serious and repeated grief. We need laws that require full labeling of every ingredient in food, GM or not. As it is, our labeling laws are inadequate. Unlabeled GM food very much exacerbates an already bad situation.

TWENTY-THREE

Conversations with a Squash

"SANDWICH-SLICE" is a new class of summer squash. It is unusually delicious as a raw squash, and it is delicious at a stage that is much bigger than optimal for flavor for other summer squashes. It is best when 4 to 6 inches long and 3 to 5 inches across. Slices of the squash are the perfect size to fit into hamburger buns or between slices of bread for sandwiches.

The squash has such an unusual firm, crisp texture, you don't even need the bread for a sandwich. This is wonderful for me, because I love sandwiches, but I can't eat wheat. I can put a slice of cheese between two slices of squash. I can even make an all-vegetable sandwich — a thick slice of 'Brandywine' tomato and a thin slice of 'Walla Walla' sweet onion in between two slices of Sandwich-slice. Or I can cut the squash lengthwise and in half, to get a shape that fits perfectly inside a taco shell.

Small chunks of the squash are wonderful with dips or cheese. They're great in "cucumber" salads. I even eat a lot of Sandwich-slice just plain and out of hand.

The first Sandwich-slice squash plant turned up in my squash patch in 1998 and began producing the first Sandwich-slice squashes in late July. This was just a year after I started the breeding project. Or maybe no years at all. You see, I never started a breeding project to create Sandwich-slice.

It never would have occurred to me that a summer squash could taste so good, especially at such a large size. Had anyone suggested I try to develop Sandwich-slice, I would have dismissed the idea at once on the grounds that it was impossible. I didn't even like summer squash. Up until Sandwich-slice I didn't grow them. I certainly didn't like them raw.

In 1998, the Year of the Sandwich-slice, I wasn't trying to breed summer squash. I was trying to breed a bigger and better 'Sugar Loaf' winter squash.

'Sugar Loaf' is a winter squash that was bred by Jim Baggett at Oregon State University. It belongs to the species *Cucurbita pepo,* which also includes most Halloween pumpkins, most summer squash, most acorns, 'Spaghetti', 'Small Sugar Pie', and the delicatas. 'Sugar Loaf' is a delicata type. 'Delicata' and 'Sweet Dumpling' are other members of this flavor class.

'Sugar Loaf' is sweeter and more intensely flavored than other delicatas, and the flavor is wonderful and distinctive. 'Sugar Loaf' is also more fine-grained, and has drier, denser flesh than any other pepo variety I've tried.

Winter squash is a major staple for me. Properly grown winter squash of a gourmet variety such as 'Sugar Loaf' is grand eating. The problem is, 'Sugar Loaf' and the other delicatas are small. The biggest fruits only run about 2 pounds. By the time you clean them there is just a few bites of meat.

Many people who use vegetables in small amounts as tiny side courses like these small fruits. But the small fruits drive me crazy. I need a fruit big enough to provide the main dose of carbohydrates and calories for the meal, not a tiny little side course.

Furthermore, 'Sugar Loaf' is harder to clean than most fruits, and the flesh is only about half an inch thick. I like the flavor of 'Sugar Loaf' as well as that of any squash

whatsoever of any species. But I frequently end up eating just the biggest few and letting the average ones go to waste because they are just such trouble to clean for the amount of meat in them.

Sometimes some of my 'Sugar Loaf' fruits don't even make it out of the field. It's lots harder to harvest, transport, and store little squash than the same number of pounds of bigger squash. At the end of the season, amid all the harvesting pressure and physical fatigue, I tend to leave small squash in the field.

I am often tempted not to grow 'Sugar Loaf' at all, but I really can't bring myself to do that. The flavor is so spectacular. In addition, these squash have a unique niche.

'Sugar Loaf' and the other pepo "winter" squashes are really "fall" squashes. The big winter squash of good eating quality, such as 'Sweet Meat' and all the Hubbards, belong to the species *Cucurbita maxima*. Maximas mature late and require a full month of curing for optimal quality and storage. Pepo varieties usually mature earlier, and are prime within a few days of harvest. I need a good pepo in order to have squash to eat during the fall while my big maximas are finishing off in the field and then curing indoors.

I really love and need those 'Sugar Loaf' fruits. But I need them to be bigger. A 4- to 6-pound version would be ideal for my purposes. That's small enough to fix in the microwave oven as a meal for one or two serious squash-eaters. Even a 3-pound fruit would be workable if I could get thicker flesh at the same time.

There are other reasons to do some breeding

work with 'Sugar Loaf'. It is not especially early. The seed is small, and it doesn't germinate very well from the earlier plantings I prefer. The plant is not vigorous compared to some. The vines grow out to about 3 or 4 feet. Compared to much of what I grow, either bushes or vines, 'Sugar Loaf' is not very impressive.

It would be nice if I could develop a variety with the 'Sugar Loaf' flavor, but with bigger fruits and larger, more vigorous vines. I'd also like it to be able to germinate in cold mud and to grow better in cold weather. I wouldn't mind thicker flesh, either.

So how do I go about getting a bigger and better 'Sugar Loaf'?

Why Not Just Select?

One classic approach to getting a bigger 'Sugar Loaf' would be to just select within the 'Sugar Loaf' variety for bigger fruit size. Simple selection is both the most basic and one of the most powerful of plant breeding methods under many circumstances.

A couple of observations, though, suggested that the "just select" approach was unlikely to be very effective in this particular case.

First, 'Sugar Loaf' plants don't show much variation in fruit size from plant to plant. What size differences there are seem to be obviously related to the variations in soil fertility or spacing from one plant to the next.

Second, I've saved true-self-pollinations from plants here and there over the years. By a "true-self-pollination" I mean that the flower was pollinated by pollen from the exact same plant. When the plant at the end

of the season seemed to be giving bigger fruits than average, I would plant that seed out the next year along with "straight-run" — that is, unselected seed.

I've never seen any obvious difference. In other words, seed from plants with the biggest fruits doesn't give me any bigger fruits than straight-run seed.

These two observations suggested that there is not much genetic heterogeneity for fruit size in the 'Sugar Loaf' variety. Selection works by changing gene frequencies from generation to generation. It isn't effective when the breeding population is genetically uniform for the characteristic of interest.

I've discussed the basic genetic relationship between selection and genetic heterogeneity in Chapter 10 and elaborated further in Chapter 20. This time, let me forego more scientific explanations and use an analogy.

Suppose you have a basket full of balls, some golf balls and some tennis balls. You pour part of the balls into a second basket through a screen that is of variable size mesh. The screen bounces away or retains most of the tennis balls, but only a few golf balls. The second basket will have a higher proportion of golf balls than the basket you started with.

The first basket is the collection of genes in a variety that has genetic heterogeneity for size. The screen is your selection method. The second basket is the collection of genes in the next generation.

Now imagine instead that you start with a basket that contains only tennis balls, and pour them through the screen into another basket. The second basket will contain all tennis balls too. It doesn't matter what kind of screen you use; you still get another basket full of tennis balls. The original basket that has all tennis balls contains no underlying heterogeneity for ball size or type. Selecting based upon that basket just doesn't get you anywhere.

It's likely that there is a little genetic heterogeneity for size in 'Sugar Loaf'. If so, if I worked very hard at "just selecting" for a number of years I could probably increase the average fruit size a little. But I want *much bigger* fruits. I didn't think there was enough genetic heterogeneity in 'Sugar Loaf' to get a very much bigger fruit via simple selection.

Choosing the Right Cross

To get a bigger 'Sugar Loaf', then, the obvious method is to start with a cross. The purpose of the cross is to introduce genetic heterogeneity for fruit size, and to contribute genes that will enhance fruit size as much as possible.

The simplest scheme is to cross 'Sugar Loaf' to something with much bigger fruits — the bigger the better. If there were something much bigger that had good fruit quality, it would be a logical candidate. None of the *C. pepo* varieties with good flesh quality (as winter squash) are very big, though.

The biggest pepo I'm familiar with that has good fruit quality is 'Small Sugar', which is only about 6 pounds. Like 'Sugar Loaf', 'Small Sugar' doesn't have especially vigorous vines, and doesn't germinate or grow well in cool weather.

In choosing the right partner for a cross, everything you know about plant breeding and gardening and growing plants comes into play. All the varieties you have trialed in your life help. They give you ideas, extend your understanding, and become the primary candidates for involvement in your breeding projects.

I decided to cross 'Sugar Loaf' to a big Halloween pumpkin, then recover good fruit quality by recurrent backcrossing to 'Sugar Loaf'. I had a candidate in mind for crossing to 'Sugar Loaf', one that has all kinds of agricultural characteristics that I would love to have in a new variety. It was 'Connecticut Field'.

'Connecticut Field' interested me for myriad reasons beyond just fruit size. The other factors have to do with my vision of a more sustainable agriculture. I want to breed crops to facilitate the change to a more sustainable agriculture. I want to breed varieties that specifically enhance the viability and profitability of the farms that are practicing sustainability.

In addition, I believe sustainable agriculture requires agri*culture*, not agri*business*. I think true agriculture is usually only possible on farms smaller than those that have become associated with the agribusiness pattern. So I consider that anything I can breed that creates special opportunities for small farmers, organic farmers, or family farmers is a useful contribution.

I can best describe some of these ideas by talking about my concept of an agroecologically "ideal squash" and 'Connecticut Field'.

The Agroecological Virtues of a Squash

'Connecticut Field' is a classic orange Halloween pumpkin. At up to about 30 pounds with no special treatment, the variety is one of the biggest-fruited pepo varieties. There are several orange Halloween pumpkins that are listed as producing fruits of a hundred pounds or more, but these are *C. maxima* varieties. I can't cross them with the *C. pepo* variety 'Sugar Loaf'.

None of the big *C. pepo* Halloween pumpkins have good flesh quality. This is no surprise, since they are sold by the pound, and it is cheaper to produce water than dry weight. The most productive ornamental varieties always have watery flesh compared to the much smaller pie pumpkins. 'Connecticut Field' has fairly thick flesh that is watery, coarse, and bland.

'Connecticut Field' is very vigorous under my growing conditions. Even commercial seed usually germinates vigorously from early plantings. The vines are vigorous. The leaves are huge. The seeds are big. The fruits are early for such a big fruit — often a full month earlier than anything of even near-comparable size.

The variety is an old public-domain heirloom. I can make crosses with it without legal problems or repercussions.

The plants seem to grow better in cold weather than most squash or pumpkins. Here in western Oregon, temperatures fall into the forties every night for the first third of the growing season. Days with highs in the sixties are more common than those with highs in the seventies until well into the growing season.

I like to plant my squash in April. I like to have the plants well established before the cucumber beetle population gets too high. In addition, the early planting allows me to ignore irrigation for the first month or more, because the ground is still moist from winter rains.

With later plantings, the small plants are not able to reach down into the moist layer of soil, which recedes day by day. Later-planted squash need more watering and pampering. They often fail completely because of cucumber beetles, flea beetles, or water problems. Later plantings work better with transplantings than with direct seeding.

Here's my idea of the perfect growing methods and the perfect variety: In early spring before any of the insects are out in significant numbers, I plant the seeds of this ideal variety by direct seeding. Then I go away. I come back at the end of the season and harvest the fruit.

The plants would be so deep-rooted that they would be able to tap into the water table and need no irrigation. The vines would be vigorous and spreading, and would occupy the ground they covered so thoroughly that weeds would be shaded out. No weeding would be necessary.

Actually, I would undoubtedly visit the squash often for breeding work and to hand-pollinate for seed saving. But I enjoy visiting and observing and breeding plants lots more than I enjoy hard physical labor.

In reality, I would have to irrigate some, because western Oregon gets no effective rain in the summer after about the beginning of May. The top several feet of soil dry out completely thereafter, with moisture receding progressively day by day. Even if plants can reach deeper water, this may not be adequate for good growth. It is the process of expand-

ing into and growing in fertile soil that provides nutrients to the plant. This fertile soil is
usually located in the upper layers.

I can imagine squash plants so deep-
rooted and aggressive that they need no water
in the first half of the season, though. And
perhaps these squash would need only a
good soaking a few times in the last half of
the season — instead of the twice-weekly
watering usually considered minimal here.

I can also imagine such vigorous, spreading vines that one weeding around the plants
early in the season and one tilling between
the hills would be all that was needed. After
that the plants would take over and shade
out any weeds.

The agroecologically ideal squash variety
would be able to perform magnificently
under organic growing conditions, of course.
And it would have such an extensive root
system and efficiency at utilizing available
nutrients that it would be able to grow optimally on more modest levels of nutrients
than most squash require.

It's possible that the ideal squash would be
better at establishing useful associations with
root mycorrhizae than most squash.

My ideal squash would have big seeds. The
squash varieties that germinate well and grow
well enough to establish plants from my early
plantings *all have quite big seeds.*

There are a number of factors that might
be involved. First, those early-planted squash
often don't get much direct sun. Sometimes it
rains or is cloudy nearly every hour of every
day for weeks. Under these conditions, a big
seed with plenty of stored food may be

needed to give the seedling a good start.

In addition, my squash get only organic
fertilizers, which are not very high in nitrogen. Nitrogen is a particular problem in early
spring soils in western Oregon because,
unlike most soil nutrients, the nitrogen is
part of compounds that are water-soluble.
The nitrogen is leached out of our soils by
our heavy winter rains.

Organically grown crops are dependent
upon new nitrogen newly released from
organic matter in the soil. The release of
nitrogen may be from old or freshly added
organic matter, but it requires microbial
action. In the cold, wet soil of a western
Oregon spring, however, there isn't much
microbial activity. Nitrogen availability for
organic crops is a problem.

Perhaps big seeds allow my squash plants
to get established based upon internal
reserves of nitrogen, instead of needing much
nitrogen from the soil initially.

'Connecticut Field' has huge seeds, and
germinates very vigorously from those early
plantings. 'Sugar Loaf' has little seeds, and
germinates very poorly from early plantings.

(My observation about the importance of
big seeds applies only to my early plantings.
In later plantings, when the weather is warm
and sunny, there is not any very obvious correlation between seed size of a variety and
germinating vigor.)

My ideal variety would have big leaves. It
would grow well in cold, cloudy weather.

Many varieties that do manage to germinate from my early plantings don't thrive
thereafter. They germinate, but the seedlings
quit growing at about the size that represents

their having used up the stored food in the seed. After a while, the root rots or the plant just disappears. If the plant survives, it becomes a stunted plant that makes no fruit.

Plants that establish themselves well and grow vigorously during that early period *all have big leaves*. I don't know of any small-leaved variety that establishes itself well in my early plantings. I am guessing that this is because most days are cloudy, and the light is of the low levels that come through clouds, not the high levels associated with direct sun. With little actual sunshine, perhaps big leaves are needed in order for the plant to be able to do enough photosynthesis to support its growth.

'Connecticut Field' has huge leaves, up to more than a foot across. 'Sugar Loaf' has little leaves about 5 inches across. When an occasional 'Sugar Loaf' seed does manage to germinate from an early planting, the seedling usually dies or develops into a stunted plant. Later plantings of 'Sugar Loaf' germinate and grow well. If they don't get wiped out by cucumber beetles, that is.

I could speculate that squash plants with deep green leaves ought to do better in those cloudy, early-spring days than those with paler green leaves. Deep green leaves presumably have more chlorophyll. It seems reasonable to suppose that they would be able to conduct more photosynthesis in low light levels.

I've frequently had deep green leaf color and lighter color segregating side by side in crosses, however, and the deep-green-leaved plants do not appear to have any advantage.

Often, deep green color can indicate better

nutrient utilization, especially of nitrogen. So, for example, within a breeding project, I note the greenest corn plants, because these are probably the ones that are obtaining higher amounts of nitrogen from the soil. They may be better rooted, or better at absorbing nutrients, or, possibly, better at establishing mycorrhizal associations that help them acquire nutrients. The pattern doesn't seem to work for squash, though.

I've also speculated that mottled leaves should impair the plant relative to solid green ones (with more chlorophyll, presumably). But this doesn't check out in the field either. When I've had mottled leaves and solid ones segregating from crosses, there is no correlation between the mottling and vigor or growth of the plants from any kind of planting.

My ideal squash would have deep, extensive roots. I think root growth patterns are related to and established by some of the same genes as vine growth patterns. That is, I think varieties with big vigorous vines also have big vigorous root systems, and bush varieties have shallower and smaller root systems. If I'm right about this, the convenience of bush squash in some circumstances comes at the expense of their ability to tap into deeper moisture and fertility.

I've done some work with germinating various kinds of squash seeds rolled up in damp paper towels that confirms my ideas about the correlation between shoot and root growth patterns. I orient all the squash seeds in one direction on a paper towel, cover them with another paper towel, then add just enough water to create dampness, not wetness. Then I roll up the paper towels with seeds, and stack the roll on its end in a sealed container. The roots emerge and head downward between the paper-towel layers. I open the container once a day and blow into it to provide air exchange.

Within a few days, there are seedlings with nice little root systems. If a variety is a vining type it makes a taproot that is much longer than the shoot. The taproot has just a few branches, and they are short compared to the main root. That is, the root displays strong apical dominance.

If a variety is a bush type it makes a root about the same size as the shoot. And the root is very bushy. There is little apical dominance. That is, there are many branches, and branches are almost as long as the bigger root they emerge from.

Seedlings from F_2 seed that is segregating for bush character show roots ranging from strong taproots to bush types. In other words, the seedlings show segregation for root type.

Under my growing conditions, vining varieties are much more likely to survive and thrive than bush varieties. I think this is because I don't start watering until summer, and the moisture content in the upper foot of soil is quite variable. The vine seedlings probably reach below this layer in just a few days.

Bushes cannot be distinguished from vines in the field until the plants are more than a foot across. Both vines and bushes start out looking like bushes. After a month or so, the vines start running while the bushes continue to grow as bushes. Apparently the root systems are different right from the beginning, however.

Vine form correlates with good early estab-lishment of seedlings in my early plantings. If I have vines and bushes segregating in an F_2, for example, and I plant heavily and thin to the biggest 10 percent of the plants, I will have discarded virtually all the bushes before I can even identify them based upon the top of the plant. Instead of getting $\frac{3}{4}$ bushes, I might get none, or say, one in ten.

If I want bushes out of the cross, I need to plant the seed sparsely so I can keep every plant until the plants are a foot or two across. At this point, the vines start to run, but the bushes keep acting like bushes. Only then can I identify the bushes and thin so as to retain the best bushes.

If I want vines, I don't have to fret that only $\frac{1}{4}$ of the plants will be vines. I just plant excess seed in each hill, then thin to the biggest plant per hill when the plants are about 4 inches high or so. They'll almost all turn out to be vines.

I wait till the plants are 4 inches high or more to give them a chance to grow based upon their own root systems for a while. Early emergence of seedlings for bushes seems to be just as good as for vines. The seedlings are using stored food, not their root systems. The extra vigor of the vines only shows up after the plants have had to provide their own support for a while.

'Connecticut Field' generally sends out two long, vigorous, rambling vines per plant. If the root conformation of 'Connecticut Field' resembles the vines, these roots should be able to reach far down into the soil for water.

I prefer vigorous *spreading* vines to vigorous rambling vines, though. A spreading vine makes more branches and more fully occupies a given amount of space instead of just sending a couple of big vines shooting through the vicinity. The rambling vines leave plenty of room for weeds to grow around and between the vines. This is inconvenient. The wandering vines prevent the use of a tiller between the rows, but don't shade the ground thoroughly enough to solve the weeding problem.

'Sugar Loaf' only grows out to about 4 feet or so, but it has more branched and spreading vines that more thoroughly occupy the space they take than do the vines of 'Connecticut Field'.

I hoped that from a cross between 'Connecticut Field' and 'Sugar Loaf' I might develop a variety with the branching pattern of 'Sugar Loaf' and the vine and leaf size of 'Connecticut Field'. This would result in a big, vigorous, spreading vine form of growth.

The vine style of growth facilitates interplanting. Bush squash, the beloveds of agribusiness, make a dense shade and are high enough that they just don't interplant well. I grow a lot of Indian-style corns that work well in hills planted four feet apart or so in all directions, with up to four corn plants per hill. Vining squash can make good use of the space between those hills. (For an article on some of my corn work, see the Bibliography.)

Interplanting is easy to do in small operations that are planted by hand, but much harder to do on an agribusiness level. Bush plants are much easier for agribusiness to deal with than rampant vines.

For someone who plants by hand, big plants can be less labor-intensive than small ones, as long as they yield as well per square foot of area. If I can get the same yield from one big spreading vine that takes up 100 square feet as from ten plants that take up 10 square feet each, the big plant makes for a whole lot less work. That's just one plant that has to be planted as well as weeded around early in the season. The rest of the space can be tilled until the vines spread.

Agroecological adaptation has to be defined in terms of both a specific growing region as well as a specific planting and growing style. If I plant later, that changes everything. So do any changes in growing style, such as going from overhead watering to row drip, or from direct seeding to transplanting.

If I transplanted, for example, bush plants might be optimal. A more restricted, denser root system might be easier to confine and transplant without damage than a sparser, more extensive, "wilder" one.

(However, if I were going to start planting by transplanting, I would *test* this idea and see if it is actually true.)

Whether vines with their (presumably) extensive but sparse root systems make better use of soil fertility than bush plants, with their (presumably) shallower but denser root systems is going to depend upon where the fertility and water is. If all the fertility is in the top couple of feet of soil, the bushes might do better if they do a better job of colonizing that soil layer with their roots.

Furthermore, these thoughts about roots are based upon guesses, assumptions, and looking at some seedlings in paper towels. Perhaps the vines thoroughly colonize the upper layers of soil after they finish shooting their big taproots through. It's amazing how little we understand about roots and water and fertility.

The important thing is to think about and understand your soil, climate, weather, plants, and growing style. Then figure out how to build your understanding into your breeding. Let your mind and ideas and speculations flow free. But always test all the speculations with direct experiments and observation in the garden or field.

What about fungicide-treated seed? Well, I eschew it. I want germinating vigor incorporated into the genes of the seed. And I want natural vigor to soilborne fungi that matter in my region. Treated seed makes it impossible to select for or maintain good natural resistance or vigor.

Treated seed is like a crutch. If you use it, you will have to continue using it. You won't learn which varieties are adapted to your region. You won't learn appropriate planting styles. And you won't be doing seed saving or plant breeding.

I use treated seed of a variety only when nothing else is available, and only for the time it takes to save my own untreated seed. If the variety can't grow without seed treatments from home-saved seed it doesn't belong in my field.

Treated commercial squash seed seems to germinate very poorly in my early plantings, incidentally — much worse than ordinary untreated commercial seed, and very much worse than my homegrown seeds. I don't know why. If I had to depend upon commer-cial treated seed, my early-planting style would be completely impractical.

The Grand Plan

My plan was to cross 'Connecticut Field' and 'Sugar Loaf', then do recurrent back-crossing to 'Sugar Loaf' some number of generations. Then I would look for a plant with 'Sugar Loaf' flavor and quality that could be inbred to establish a new variety.

In breeding for a particular flavor, I usually find it most practical to backcross to the variety with the desired flavor for at least one or two generations. To obtain specific flavors, I think you are usually dealing with a dozen or more genes simultaneously. To recover a specific flavor from an F_2 isn't very likely without growing out thousands of plants. (See Chapter 10 for how to evaluate various plant-breeding tactics.) 'Connecticut Field' and 'Sugar Loaf' probably differ by hundreds of genes that have serious effects on fruit flavor and quality.

An F_1 plant of 'Connecticut Field' and 'Sugar Loaf' would get half its genes from each parent. If that F_1 is backcrossed to 'Sugar Loaf', the offspring will be ³/₄ 'Sugar Loaf'. If some of those offspring are back-crossed to 'Sugar Loaf' yet again, the resulting offspring will have ⁷/₈ of their genes identical to 'Sugar Loaf'. (See Chapter 10.)

In a population that is ⁷/₈ 'Sugar Loaf', it should be much easier to find plants with 'Sugar Loaf' flavor. With a little luck and a little gentle selection along the way, the ¹/₈ genome from 'Connecticut Field' would contribute some fruit size and some advantageous agroecological characteristics as well.

This backcrossing scheme, with a minor hiccup at the beginning, is still in progress. The rest of this story focuses on something else — something unexpected that came out of the project.

Choosing the Cytoplasm

My plan was to cross 'Connecticut Field' female flowers with pollen from 'Sugar Loaf'. The reason for doing the cross in the direction described — that is, using 'Connecticut Field' female flowers, not 'Sugar Loaf' ones — has to do with the cytoplasm. We determine which cytoplasm we retain when we choose which plant to use as the female parent.

The genes we usually talk about are nearly all nuclear genes — that is, they are located on the chromosomes in the cell nucleus. But there are also a few genes located in certain organelles in the cytoplasm, and these aren't inherited in a Mendelian fashion at all. They are maternal. They come from the mother only.

In plants as well as animals, some genes are located in the mitochondria, and are involved in energy metabolism. Plants have chloroplasts, as well, which also have some of their own genes. Chloroplasts are the central organelle involved in photosynthesis.

If I do the cross using 'Connecticut Field' female flowers, the F_1 plants will receive all their chloroplast genes, all their mitochondrial genes, and one dose of nuclear (Mendelian-style) genes from 'Connecticut Field'. In addition, they will receive one dose of nuclear genes from 'Sugar Loaf', the pollen-parent.

If I use 'Sugar Loaf' female flowers, I'll get all the mitochondrial and chloroplast genes from 'Sugar Loaf' instead.

Plant breeders tend to ignore the possibility of serious genetic differences in the genes located in the cytoplasm. They usually make crosses in whatever direction is most convenient technically. I think though, that we should be paying more attention to the role of the cytoplasm.

'Connecticut Field' germinates and grows better in cold weather and under low light conditions than most varieties. It is quite possible that those characteristics could involve genes in the cytoplasm. So I wanted the cytoplasm from 'Connecticut Field'.

Disease resistance is sometimes also known to be cytoplasmic. Whenever I'm doing a cross involving disease resistance, I use female flowers from the resistant parent, just in case the resistance is associated with the cytoplasm.

Whenever I cross a rare variety to a common one, other factors being equal, I try to preserve the cytoplasm of the rare variety. I think we need a greater biodiversity of cytoplasms in our gardens and on our farms as well as a greater diversity of nuclear genetic combinations.

The Reality

In 1997, I planted both 'Connecticut Field' and 'Sugar Loaf' so I could do the beginning cross. The 'Connecticut Field' grew happily. The 'Sugar Loaf' didn't come up and had to be replanted, as usual.

The 'Sugar Loaf' flowered too late to make the cross. I pulled flowers off the 'Connecticut Field' to keep it flowering. Finally, late in the season, instead of losing the year and wasting the 'Connecticut Field', I crossed it to something else I had around that was half 'Sugar Loaf'. The half-'Sugar Loaf' was an F_1 between 'Sugar Loaf' and a bush acorn. It was from an entirely different breeding project.

I used the 'Connecticut Field' as the female parent, as desired. So my cross gave me seed that carried 'Connecticut Field' cytoplasm and a nuclear genome that was ½ 'Connecticut Field', ¼ 'Sugar Loaf', and ¼ bush acorn.

In addition, each seed would be expected to be genetically different from every other. Each represents hundreds of different genes segregating from the 'Sugar Loaf'/acorn F_1 then combined with a haploid genetic complement from the big pumpkin. A row of plants grown from such seeds should be as wild and weird as it gets.

In 1998, I swayed back and forth right up until planting time about whether to plant the "weird stuff," or to forget that mess and start over with 'Connecticut Field' and 'Sugar Loaf' and try again for the simpler cross. The simpler cross would require only a few plants in the next generation, because an F_1 is fairly uniform. The cross I had made, however, would be segregating for 'Sugar Loaf' and acorn characters. That meant I would need more plants in order to retain the 'Sugar Loaf' characteristics. And the material was only ¼ 'Sugar Loaf'.

The day that I planted the squash, I was still undecided. I finally went ahead and planted a whole 60-foot row of the "weird" stuff. There were so many interesting characteristics in that material, I couldn't resist planting it just to see what it would do.

I halfway regretted "wasting" the row almost immediately. I only had about ten rows, and squash is one of my main winter carbohydrate and vegetable staples. A 60-foot row is a lot of space to produce plants I'm not expecting to be able to eat. A cross of something with inferior fruit to something with superior ones usually gives you inferior fruit in the first generation. The acorn/'Sugar Loaf' F_1 was a good-quality delicata type, but the 'Connecticut Field' was not. I expected only inedible offspring in the first generation.

This was one of my standard early plantings. Wherever I had enough seed, I planted six seeds per hill, to be thinned to the best plant. I do this for pure varieties as well as my experimental material. If you plant lots of seed and eliminate most plants, keeping just the best, you get strong selection for ability to germinate under your conditions. If you conserve seed and avoid wasting it, you'll be keeping nearly every plant that manages to come up. That gives you little selection for germinating ability or early growth.

Then we had three weeks of winter. It rained. And it rained. We had temperatures down into the forties every night, often not much better than the fifties during the day. There may have been a few minutes of sun on one day, but not much more. I would probably need to replant everything.

By then I had changed my mind about the weird material. With the late planting, I needed more space for food, since I always lose so much from late plantings. So I headed back out to the field at the first break in the weather, figuring I'd get that weird row redone and out of the way. I would overplant with something else, something I could eat this year, not just curiosity and dreams.

So I drove over to the field, draped my big

peasant's hoe over my shoulder, filled my fanny pack with the packets of seeds, and slogged out to the field. If any plants of the "weird stuff" were up, I had decided, I would hoe them under and replant anyway.

The Squash Speak

When I got out to the field, the situation was as bad as I had thought. Row 1, no plants. Row 2, a few straggling plants here and there, only from one of my breeding projects, nothing from standard varieties. Only at best one seedling per hill of the six seeds planted. Most hills with nothing. None of the few existing seedlings looking very happy. Row 3, nil.

Row 4 . . . .

"Hey, hey, hey! Hello!" yelled a vigorous row of crowded little plants.

"Well, hello . . ." I responded. I had never been accosted by a bunch of squash seedlings before, so I was a bit taken aback.

There was the most beautiful row of squash seedlings I had ever seen. There were five or six seedlings in every hill. The single row of a dozen hills had about seventy vigorous plants. The entire rest of the field, of ten rows, only had perhaps a dozen plants total. And these plants already had their first real leaves.

"Whatcha up to, Person? Whatcha doin' with that big hoe?" So asked the nearest seedling, one in a hill about a third of the way down the row.

"Oh, er, this?" I said. "Heh heh. Nothing. Nothing at all. I just like to carry it around." I quickly set the hoe aside.

I did no replanting that day. Those plants had earned their place in the field.

Carol Falls in Love

For a long time I walked up and down that row of little plants, squatting here and there and just gazing, contemplating. Fascination. Awe. Admiration. Commitment. Bonding. It was a bit like participating in a birthing. Something important is happening. Something is coming into being. I don't know what it is, but I know my role. I will nurture, guide, facilitate.

The rains continued, and turned into six extra weeks of winter. There was essentially no sun. Every day was cold and rainy or overcast. It never actually got to below freezing at night, but it never got very far above freezing either.

Finally the sun came out, and I headed over to the field to replant. Nearly all the seedlings that had come up during the initial cold had subsequently died. The entire squash patch was empty except for the weird material and parts of three other rows, all involving other breeding projects of mine.

The weird row was baby plants now. They had managed to grow with virtually no sunshine at all.

"How did you do that?" I asked the squash in the nearest hill. "How did you grow so big with no sun?"

"Ha ha! There was plenty of sun! There were several minutes of sun back there in the second week. Enough for anybody!"

"But what about the cold?" I asked.

"Very invigorating," said the squash.

"Squash aren't supposed to like cold."

"They aren't?"

"No, look," I said, pointing around at the

empty rows. "Everybody else *died*."

"Good thing those wimps are out of the way," said the squash. "We're going to need the space."

I replanted all the empty sections in the other rows. Then I thinned the weird row. It was painful, but it had to be done.

I was expecting segregation for bush versus vine type. With a heterozygous bush/vine crossed to a vine, I expected half heterozygous bushes and half vines. I wanted vines. As mentioned earlier, I can select heavily for vines simply by thinning to the biggest plants when the plants are about 4 inches or larger.

So my first round of selection was simple. I thinned each hill from five or six young plants to the biggest one. This gave me a dozen plants in the row, just one per hill. A month or so later, the result was eleven vines and one bush, which I eliminated.

Disaster and Opportunity

The ancient Chinese symbol for "crisis" is a combination of the symbols for "disaster" and "opportunity."

The 1998 growing season was a double disaster. First, there was the six extra weeks of winter. Then I replanted. Then, just as my replanted squash were emerging and most vulnerable, there was a plague of cucumber beetles. Every seedling had a dozen or more beetles on it. Nearly every one was eaten to the ground within a day or two of emerging. By then, it was too late to replant again.

So there I was with only about a quarter of a squash patch occupied virtually entirely by plants that were experimental material —

plants whose fruits could not be expected to be edible at this stage in the project. I love growing squash, but the basic idea is to get something to eat. I decided it was time to learn more about summer squash. I had read somewhere that most winter squash make good summer squash too. I had not checked this out, because I didn't especially like summer squash. Now, though, I had too few squash plants and probably no good winter squash. I could take this opportunity to taste-test all my winter-squash-breeding plants to see if they were good as summer squash.

I also learned how to dry summer squash and evaluated all the plants as to their flavor as dried squash. The results were interesting, but are a story for another day.

Meanwhile, the squash were growing, growing, growing. The vines were exactly what I wanted, all of them. They were very vigorous, very aggressive, spreading vines that fully occupied the space they took and shaded out the weeds. They all had huge, magnificent leaves. The vines rooted at the nodes wherever they had good contact with the soil.

There was quite a lot of variability from plant to plant in leaf shapes and colors, but not overall plant form. (I had eliminated the bushes.)

When the plants began fruiting, the fruits were of various types. They ranged from flattened disks to spheres to short loaf shapes like 'Sugar Loaf'. ('Connecticut Field' is variable in shape, sometimes a little taller than round and sometimes a little flatter.) Most plants had fruits that were light green striped and spotted with darker green, like immature 'Sugar Loaf'. Some were solid green of various shades. Some turned orange in storage and some didn't.

I picked most of the fruits to keep the plants flowering, because I wanted at least one backcross to 'Sugar Loaf' and one self-pollination on each plant. The plants responded by acting like summer squash — cranking out flowers and summer squash with abandon. By early August I had backcrosses to 'Sugar Loaf' as well as true-self-pollinations on every plant. The backcross to 'Sugar Loaf' represented the original project I was trying to do. The self-pollination represented the fact that I had no idea what these plants were good for yet, but suspected that they were good for something just as they were.

To put it another way, the backcross represented my attempt to move the material closer toward 'Sugar Loaf'. But the self-pollination was the best way to preserve the genetic combinations of these individual plants who already existed, and who had so charmed and impressed me.

Meanwhile, I was tasting all my squash as summer squash. The cooked flavor of the squash from the twelve different plants was edible, and as good, I thought, as most true summer squash — which is to say, nothing special. They all tasted more or less the same. As I've said, I don't like summer squash.

Then I started tasting the squash raw, starting with the last plant in row 4, the weird row. My first reaction was that I had made some mistake, and picked up something other than a squash. The raw squash was truly delicious. It was sweet, quite firm, and very flavorful,

with a flavor unlike anything else at all. The fruit was a loaf shape of about a pound at this young stage, and was light green with darker green stripes.

I sliced the squash into chunks and ate the whole thing with little bits of aged Cheddar cheese. I, who had always sneered at summer squash, who always picked them out of salads at other people's houses, if they were so crude as to include them. I, whose attitude had been that anything that tasted good with some raw squash in it would taste even better without.

I tried squash from two other sibling plants that I brought home in the same lot. They were every bit as uninteresting raw as most summer squash.

I ran back to the squash patch and harvested a 1-pound squash from each of the other plants and tried them. Nothing. None of them tasted very good. It was just the single plant that had fruit that tasted good raw.

That single plant seemed to be producing an entirely new vegetable. A unique flavor. An intense sweetness for an immature squash. Firm flesh in the stage up to about 6 inches long and 4 or more inches wide. Just a perfect size to slice for sandwiches. Sandwich-slice was born.

By fall, the Sandwich-slice foundation plant and her siblings had spread into huge vines 40 feet across, and had entirely overgrown the adjacent rows. ("We told you so," said the squash.)

I had kept the squash picked except for the hand-pollinations. I wanted to see how they would do as summer squash, and you have to keep summer squash picked. The plants kept cranking out squash. By October the squash patch was mostly dead from powdery mildew, which is often what ends my squash-growing season.

But Sandwich-slice and her siblings were still cranking out squash. The old central portions of the squash were heavily mildewed. But powdery mildew mostly affects older growth. These vigorous vines had massive amounts of vigorous new expanding growth and new roots at many nodes. They kept growing and producing as much as ever.

There was an unusually mild, late fall. We finally had some near-freezing weather. Almost all the squash plants that hadn't already died from powdery mildew were killed by the cold. Sandwich-slice and her sibs merely quit flowering temporarily. After it warmed up again, they cheerfully resumed flowering, and began producing fruit faster than ever on the huge plants.

The last week of November, long after fresh local summer squash had vanished from gardens and markets, I took a Sandwich-slice and a chunk of cheese with me to a germplasm meeting. There were seed company people, USDA germplasm people, and lots of organic market gardeners. Most of them tasted a chunk of the just-plain, raw Sandwich-slice. Everyone seemed to like it and to be amazed.

"You'll have to teach people how to eat this," said Rich Hannan, head of the USDA/ARS Western Regional Plant Introduction Station.

"It's very easy," I said. "First, you open your mouth. Then, you put in the squash...."

What Rich meant, though, was that nobody expected a squash to be good enough to eat out of hand.

The squash was apparently every bit as good as I had thought it.

I harvested the mature fruit from Sandwich-slice and her sibs and evaluated it. They ranged in size, counting the biggest fruit of each plant only, from 6 pounds to 14. That was about what I had expected. Mature Sandwich-slice fruits were blocky and loaf-shaped, and weighed up to 12 pounds. They were cream-colored and light-green-striped initially, then turned to tan and green-striped when fully ripe. 'Sugar Loaf' is also tan and green-striped. The mature Sandwich-slice fruit looks like giant 'Sugar Loaf'.

As a mature squash, Sandwich-slice was quite interesting. It had an intense aroma when chopped open — an aroma quite like that of 'Sugar Loaf'. The flesh also was intensely flavored, and of 'Sugar Loaf' type. But it was also watery and coarse. Because of the texture, it was nowhere near being a good winter squash. The sibling plants all produced mature squash that were watery, coarse, *and* with little flavor or aroma.

At this point, my project split into two parts. I would continue to try to create a better 'Sugar Loaf'-type winter squash. For that I would use the Sandwich-slice backcross to 'Sugar Loaf'. (That project is underway, and is another story.)

I would also develop a Sandwich-slice summer squash variety. For the continued development of a 'Sandwich-slice' variety, I would plant out a row of the true-self-pollination of Sandwich-slice.

Sandwich-slice was already a good Sandwich-slice. But I could not expect the plant to be pure-breeding or stable. I would need to breed further, select, and stabilize it.

During the winter, I wondered and worried as to whether Sandwich-slice was a one-shot deal. What fraction of the next generation would be Sandwich-slice type and quality? All? None? One out of a hundred? A thousand? It all depended upon what genes and what genetic configurations were involved in creating that flavor and texture of young fruit. All I could do was hope. There is a big difference between a single good plant and the start of a new variety.

Sandwich-slice

In 1999 I planted my squash as part of a friend's melon operation. This was a new experience for me. Black plastic and row drip irrigation instead of overhead watering. I worried about the Sandwich-slice material. I gave them 20 feet between rows and 10 feet between plants. But I was concerned as to whether their root systems would have full access to the space underground. It all depended upon whether the row drip irrigated the entire field, or just a 3-foot-wide band underneath the plastic. My normal squash planting involves overhead watering, not row drip or black plastic.

I also wondered whether the vines would be able to root at the nodes. Even if the entire field was adequately moist, the top 6 inches or more of soil could be expected to be dry. Could vines root at the nodes under these conditions?

I love big vines, but as I thought about it, the only time I had used row drip irrigation, I had quite poor production on the big vining types of winter squash. Perhaps with a row drip scheme, big vines would be counterproductive. It's quite possible that bushes or semi-bushes would work better in such circumstances. They might, I speculated, have a root system that took better advantage of restricted growing space, if restricted growing space was the row-drip reality.

I planted a short row of the Sandwich-slice breeding project, a short row of the backcross to 'Sugar Loaf', several other breeding projects, and lots of winter squash for eating. In addition, I planted about thirty different kinds of summer squash. If I was going to be breeding summer squash, it might help if I was somewhat more familiar with them, I figured.

I also wondered about how unique the Sandwich-slice flavor was. Is there anything else with such good flavor raw? Is there any summer squash that is delicious raw at such a large size?

We had another very late spring. I had to do a lot of replanting. The Sandwich-slice offspring, however, jumped up out of the ground and grew, vastly outgrowing all standard varieties and all my other breeding project material as well.

By June, when watering began to be necessary, I had seven vigorous plants, all of them with spreading vines that had taken over a circle of about 10 feet, already way beyond the row drip and black plastic.

They kept growing. They made a few fruits. I ate a fruit from the earliest plant. It was delicious. It was another Sandwich-slice. Elation. Joy. Relief.

After the first few fruits, though, subsequent fruits dried up. The plants continued flowering prolifically, but produced nothing. They didn't look very happy either. And they didn't root at the nodes in the dry surface soil. When I walked the rows, the plants grumbled and growled at me. One plant with beautiful cream-colored blocky fruits (when she made any) cursed at me every time I went by.

Meanwhile, the bush summer squash plants of thirty different varieties, finally established in a later planting, had begun producing. Most of them had already outproduced the Sandwich-slice–project plants, in spite of being much smaller plants with much less space between plants and rows.

After a month of trying to do hand-pollinations on Sandwich-slice plants, only to have all the fruits dry up, I got the picture. The overall growing method gave adequate amounts of water to small melon plants and small restrained vines like 'Sugar Loaf', and to small, well-behaved bush summer squash. But it was not a suitable growing method for my huge Sandwich-slice vines. They did not produce. I was fearful that they were in danger of actually dying. (The bush summer squash, however, were doing fine.)

So I put an overhead sprinkler in to succor the Sandwich-slice. They began growing even faster, began setting buckets full of fruit, and started rooting at the nodes.

"This is more like it!" the cream-fruited plant said. Thereafter, she greeted me cheerfully every time I walked the rows.

Of the seven Sandwich-slice plants in that generation, all had the characteristic flavor of the original Sandwich-slice. One, however,

made vines that only ran out to 10 feet, and its fruits were only up to about 4 pounds. It wasn't a good Sandwich-slice type, because the fruits were already past prime at 4 inches across, being quite near full size.

Another plant made fruits with good flavor, but the texture was coarse instead of fine-grained. The coarse grain made for inferior eating.

Yet another plant had fruits with an unpleasant off-flavor on top of the Sandwich-slice flavor.

One plant was so late that it did not even begin flowering before the end of the summer, even though the plant was as big as any. The fruit was excellent quality, though.

Two plants had round fruits. The rest were loaf-shaped. One plant had cream-colored fruits that matured to light-green- and green-striped. The rest had light-green- and green-striped fruits.

Of the seven plants, three gave me fruits that were really delicious, and had the other good characteristics too: reasonable earliness; huge, vigorous vines with big leaves; excellent growth in cold weather; and fruit of appropriate size for my purpose.

Meanwhile, I tasted all the other summer squash I was growing. I discovered a couple that I liked cooked and a couple I liked dried, but none whatsoever that I liked raw. Furthermore, there was nothing even marginally edible raw at the stage when it was 4 inches across. Sandwich-slice really is something special.

The obvious way to proceed with development of Sandwich-slice is by planting out seed from the true-self-pollinations of the best two or three plants. As is typical with such projects, the good plants undoubtedly still include much unwanted variability. I am cheerfully willing to tolerate a great deal of variability, but not for fruit quality or stage of usefulness.

I can get rid of some of the genetic heterogeneity by selecting against undesirable characteristics. When one does this one never totally eliminates undesirable recessives, however. Undesirable recessives affecting fruit flavor and quality aren't acceptable.

A more effective way to eliminate undesired variability is by simply inbreeding the best plant for an additional generation or more. As explained in Chapter 9, each round of inbreeding causes a loss of half the remaining genetic heterogeneity.

My approach will be to inbreed a few lines from the best plants each generation until I have uniformity and stability for fruit flavor, quality, and size. Then I'll mass-select thereafter, to retain as much unrelated heterogeneity as possible.

I think disease resistance and overall plant vigor might be as much related to the general heterozygosity of the plants in many cases as to specific genes. If I'm right about this, it will be best to inbreed Sandwich-slice as little as possible. Yet some inbreeding will be essential to achieve consistency in fruit quality.

To Market, to Market, to Sell a New Squash . . .

It's Spring of 2000. I'll soon be planting out the Sandwich-slice. My guess is that this year most, but not all, of the plants will give me fruit of prime type and quality. I should be able to begin to sell the fruit in local markets this year, while I continue to perfect and stabilize the variety.

All I need to do is plant hills 4 feet apart, then thin to 10 feet after tasting one fruit per hill. I'm guessing I'll get at least 100 pounds of fruit per plant, and maybe 200 or more.

It is likely that it will take at least another two to four years before I have a stable variety. The years won't be wasted, though. I can eat all the Sandwich-slice I want and sell it to local stores, restaurants, or in farmers' markets.

In addition, it will take at least two to four years before I understand the variety well enough to know how to grow it optimally. What kind of yield will it give? Will it yield as well per square foot as ordinary bush summer squash? What is the optimal spacing?

How will it behave in intercroppings? Most squash can't tolerate shade. But Sandwich-slice can grow in early spring with nothing but cloudy weather. Maybe it will be more shade-tolerant than most squash. If so, that would greatly expand its usefulness in intercropping.

What are Sandwich-slice's water needs? Obviously, overhead watering is basic. But will the plants be able to take less watering than bush types in an overhead watering system, as I originally imagined? The huge surface area of leaves might mean the plants will require as much watering as bushes, whether deeper-rooted or not. There is lots of exploring to do. I can be learning how best to grow Sandwich-slice, however, while I finish breeding it.

I have a workable variety already, though — one that represents a new class of squash, a variety good enough to eat and even to sell, right now — a mere two years after beginning.

Breeding and Seed Saving for Eight Common Vegetables — an Illustrated Guide

This section gives detailed information on breeding some of the most popular vegetables — tomato, lettuce, pea, bean, alliums (onions), brassicas, squash and pumpkin, and corn. The breeding methods for these few vegetables cover virtually the full range of methods. By reading through Appendixes A and B and the general material in Part II and using Table I as a reference, you should have a good start toward being able to breed any vegetable, whether you have any specific information about it or not.

Tomato

The tomato, *Lycopersicon esculentum,* is in the family Solanaceae. Other vegetables in the family include *L. pimpinellifolium* (current tomato), *Capsicum* spp. (peppers),

Cyphomandra betacea (tamarillo), *Physalis* spp. (husk tomatoes and relatives), and various species in the genus *Solanum,* which includes the potato, eggplant, naranjilla, and garden huckleberry.

The tomato (*L. esculentum*) and current tomato (*L. pimpinellifolium*) can be crossed easily. There are many wild species of *Lycopersicon* with which the cultivated tomato can be crossed with more difficulty and which have served as sources of dominant genes for disease resistance via recurrent backcrossing. (See chapters by Rick in *Hybridization of Crop Plants* and Tigchelaar in *Breeding Vegetable Crops.*)

Breeding System

Tomatoes are basically inbreeders. In modern cultivars the pistil never emerges outside of the fused cone of the stamens. The anthers are on the innermost surface of the stamen cone and shed pollen inward. Since the flowers are pendant (hang downward), the pollen drops over the stigma, thus causing self-fertilization.

Many heirloom varieties or varieties with *L. pimpinellifolium* or wild species in their ancestry have pistils that extend beyond the stamen cone and are exposed to pollinators. These varieties are much more likely to be cross-pollinated.

Tomato stigmas become receptive about a day before the flowers open. Pollen begins to dehisce somewhat later than the stigma becomes receptive, but still in the bud stage. The stigma remains receptive and pollen continues to shed during the entire time the flower is open, which may be one day to a week, depending on conditions.

Tomatoes grown outdoors shed their pollen when jostled by the wind. Greenhouse tomatoes need to be jostled to release their pollen and set fruit. Shaking the plants regularly will suffice.

Isolation

Ashworth, in *Seed to Seed,* summarizes the situation: "The extent of cross-pollination in tomatoes has been a controversy among seed savers for a long time. Some say that crossing is rampant, while others have never seen crossing after years of growing different varieties next to one another.

The specific repertoire of pollinators present is undoubtedly a major factor.

The other major factor that affects the isolation distance is whether your varieties have retracted or protruding stigmas. If all your varieties have retracted stigmas and you don't see any insects working the flowers, you may be able to use just a few feet of isolation — the distance involved when you alternate rows of tomato varieties with rows of other plants, for example. Varieties with protruding stigmas are more subject to being crossed by insects and will need more isolation.

Inbreeding Depression

The tomato, a natural inbreeder, does not display inbreeding depression.

Hand-Pollination and Crossing

Tomato flowers are emasculated and pollinated in the late bud stage. At this stage the bud has reached full length, the sepals have started to open, and the petals have started to change color from light to bright yellow, but the petals are still tightly shut. Once the petals start to open, it is too late; anthers are already dehiscing. (See illustrations on pages 306–307.)

Once the buds are emasculated, they can be pollinated immediately or in a separate step the same day or later. Sometimes repeat pollinations are used to increase the success rate. George describes emasculating and cross-pollinating at separate times, with timing dependant on the region and the weather: "For example, in Californian field production, flowers are emasculated early in the morning (from about 06:30) on the same day as pollination; whereas in North European greenhouse hybrid seed production, flowers are emasculated up to two days in advance of pollination."

Note that you can *see* when tomato (or other plant) pollen is released by using a hand lens. Use this general information to help you, but the exact timing depends on the variety and the weather, and the final authority is the plants. Ultimately, you establish whether dehiscence has occurred at a certain stage under your conditions not by reading, but by looking at the plants.

Emasculation of tomato buds is performed as follows. Choose buds at the appropriate late bud stage. One or more buds on a cyme may be at the right stage at once. You may have to figure out a right time of day to catch the buds in the right stage. It will usually be sometime in the morning. The highest success is with bigger buds, the first few on a cyme. Fruits and uncrossed flowers are usually removed to eliminate competition with the cross.

At the appropriate late bud stage, the corolla — the ring of petals (yellow) — is folded up around the stamen cone (also yellow), and the two are attached at their bases. Thus there is a two-layered cone encircling the pistil. In emasculation this double cone is lifted off as a unit, leaving a bare pistil with a ring of sepals (green) at its base.

To remove the double cone, insert one tip of a fine forceps between the anther cone and the pistil and the other between the calyx (sepals) and the corolla (petals) and gently pull the cone off, taking care not to injure the pistil. You may need to dissect away one petal and stamen first and/or to remove the remaining stamens afterward.

You can transfer pollen by using pieces of anther cone or by scraping pollen from the cone off anthers with a triangular dissecting needle. Just-opened flowers are the best and easiest source of pollen.

As is true for most Solanaceae family members, tomato pollen is long-lived and stores well. It will remain viable for weeks at room temperature. If desiccated and refrigerated, it will last a number of months.

Emasculated flowers do not need to be covered to prevent contamination, as there is no pollen to entice insects and no bright parts to signal to them. Under hot, dry, or windy conditions, emasculated flowers may need to be covered with glassine bags to prevent their drying out. Outdoor crosses are most successful if done under cool, relatively wind-free conditions.

Hybrid Seed Industry

There is an extensive hybrid seed industry for tomatoes, even though they don't display inbreeding depression and thus there is no special biological advantage to the hybrids. The best open-pollinated varieties of tomatoes are as vigorous as the best hybrid varieties. The major reason for the industry is the financial advantage to seed companies in producing and promoting proprietary varieties that no one else can multiply.

Commercial hybrid seed production depends on hand-emasculating and hand-pollinating the plants. Genetic male sterility has started to be used in making hybrids. This saves the emasculation step, but hand-pollination is still required. Hybrid tomato seed is economical to produce because the plants and produce are high-value ones for which people will tolerate high seed costs, because each successful pollination produces two hundred or more seeds, and because hybrid seed is mostly produced in Taiwan with inexpensive but skilled labor.

Genetics and Breeding

Much of the recent professional work has been aimed at developing machine-harvestable determinate varieties for processing, as well as those with good shipping and holding qualities for the supermarket trade. Another trend is toward developing specialized varieties associated with every use, growing method, and growing region. Such specialized cultivars are replacing general ones commercially.

Tomato varieties for greenhouse producers need to be indeterminate so as to maximize productivity, for example, and need resistance to some pests and diseases that aren't very important outdoors. Field crops need to be resistant to the major diseases prevalent in the area. Dominant genes conferring resistance to many diseases and pests have been discovered in wild varieties and transferred to commercial cultivars by recurrent back-crossing.

The tomato is one of the most popular vegetables for genetic experimentation, and many of the genes of commercial importance have been identified and mapped. (For chromosome maps of a few hundred of the genes, see Tigchelaar.) Some of the genes of greatest agricultural significance are given in the following paragraphs. Symbols that start with small letters are recessive; those that start with large letters are dominant. (Most of the information is from Tigchelaar; the rest is from Mike Courtney's review in the 1989 Summer Edition of the Seed Savers Exchange.)

Genes that affect plant growth habit include *sp, br,* and *d,* which are self-pruning, brachytic, and dwarf, respectively. Self-pruning means determinate. Nearly all determinate varieties carry it. A few extreme bush types, especially those for pot culture, carry brachytic or dwarf genes. 'Redbush' carries *br;* 'Epoch' and 'Tiny Tim' carry *d.*

The gene *c* is associated with the potato-leaf phenotype.

Two different genes, *j-1* and *j-2,* can cause jointless pedicels. In plants with jointed pedicels, the tomato stem breaks in picking so that the tomato has a section of stem on it. When tomatoes with jointless pedicels are picked, the tomato separates from the stem at the fruit.

Dominant genes for disease resistance have been transferred from wild species by recurrent backcrossing. These genes could be transferred into heirloom and home garden favorites by simple recurrent backcrossing.

Four different genes confer resistance to leaf mold — *Cf-1, Cf-2, Cf-3,* and *Cf-4.* Fusarium wilt immunity for race 1 and race 2 is associated with genes *I-1* and *I-2,* respectively. Resistance to verticillium wilt, Septoria, late blight, Alternaria, and Stemphylium is conferred by genes *Ve, Se, Ph-1, Ad,* and *Sm,* respectively. Tobacco mosaic virus resistance can be conferred by *Tm, Tm-2,* or *Tm-2².* Nematode resistance is conferred by *Mi.*

There is genetic variability for insect resistance of various kinds, but most professional breeders and growers depend on insecticides instead. Organic gardeners and farmers could be pursuing the genetic approach. Tigchelaar includes a summary of work that has been done and a list of wild species resistant to various pests from which the needed genes could be obtained. See Rick and Tigchelaar for information about crosses with wild species.

Male sterility is associated with a number of genes. The ones I've seen listed are all recessives.

The *u* locus is associated with the uniformity of ripening. Many older cultivars have green shoulders when the rest of the fruit is ripe. Modern cultivars usually ripen uniformly.

The gene *nor^A* is nonripening and is found in 'Longkeeper'. Other genes that affect ripening include *rin* (ripening inhibitor), *Gr* (green ripe), and *Nr* (never ripe), among others.

The gene *pat-2* causes parthenocarpic fruit — that is, plants carrying it can produce fruit without setting seed. The cultivar 'Severianin', for example, carries this gene. Varieties with parthenocarpic fruit can set fruit without being pollinated, so they can set fruit in weather that is too cold for pollen release or function. Fruit so set is seedless. Other genes also can cause parthenocarpic fruit.

Mike Courtney provides an excellent summary of tomato color genetics in the 1989 Summer Edition of the Seed Savers Exchange, which I've used for this discussion. Tomato color is determined by a combination of the color of the skin and the color of the flesh. Skin color is yellow or colorless, the difference being determined by a single gene: y^+_ is yellow; yy is colorless. Flesh color involves a number of genes but is basically yellow, orange, or red.

The red tomato of commerce has a yellow skin and red flesh. Tomatoes with a colorless skin and red flesh are pink. Yellow flesh color is determined by the *r* locus, with *rr* being yellow and r^+_ being red. Yellow tomatoes have yellow skin and yellow flesh; white,

ivory, and lemon tomatoes have colorless skin and yellow flesh.

The color of the flesh depends on the content of carotenoids. These include various forms of carotene, all of which are orange, and lycopene, which is red. Of these, beta-carotene is of nutritional significance because the body can convert it to vitamin A. Lycopene has no vitamin A value. Various genes control the total amount of carotenoids; others determine which form of carotenoid predominates. Red tomatoes have most of their carotenoid in the form of lycopene. Yellow tomatoes have no lycopene and only trace amounts of beta-carotene. Orange tomatoes have no lycopene and large amounts of carotene.

Crimson tomatoes carry the gene *og^C* and have somewhat more lycopene and somewhat less carotene. Crimson tomatoes have a yellow skin. When crimson flesh is combined with a colorless skin, purple tomatoes result.

Tangerine tomatoes are associated with the gene *t*. They have no lycopene and most of their carotenoids in the form of zeta-carotene. Among these are 'Golden Jubilee', 'Orange Early', 'Orange Tall', 'Golden Delight', and 'Sunray'. Tangerine flesh and colorless skin result in a muted gold tomato. Two examples are 'Glecker's Gold Glow' and 'Golden Glow'.

Apricot-colored fruits are associated with the gene *at*. They have reduced lycopene levels but normal beta-carotene levels. The gene *Del* has increased delta-carotene. This is one cause of orange fruits. The gene *hp* (high pigment) increases the amount of lycopene and carotene, as well as the amount of chlorophyll in the immature fruit. 'Redbush' has this gene. The gene *dg* (dark green) also increases the amount of chlorophyll in the immature fruit and has more lycopene and beta-carotene in the mature fruit.

The gene *gf* (green flesh) is responsible for mature fruit that retains some of its chlorophyll. In r^+ forms the green and red pigments yield red-brown fruit. In *r* forms, which have no red pigment, the mature fruit is green. 'Evergreen' is an example of the latter.

The gene *gs* is associated with green striping. The fruit is either red and gold striped or green and gold striped depending on the configuration at the *r* locus.

Orange fruits are associated with two dominant genes, *B* and *Mo_B*. Plants carrying just *B* (e.g., 'Caro Red') have no lycopene and five times as much beta-carotene. Plants carrying both *B* and *Mo_B* have ten to twelve times as much beta-carotene as normal. 'Caro Rich' carries both *B* and *Mo_B*. *B* is closely linked to *sp* (self-pruning; determinate) on chromosome 6.

The tomato is second only to the carrot as a source of beta-carotene in the American diet. Thus it is particularly unfortunate that red has been seized upon as the color of choice for tomatoes. Orange tomatoes fare well in blind taste tests; but red tomatoes fare better in tests where participants can see the color. Commercial breeders and growers are limited by this popular bias, but home gardeners can breed and grow any color tomato they want.

Given that *B* and *Mo_B* are both dominant, it would be easy to transfer them to other varieties by crossing them with 'Caro Rich', followed by recurrent backcrossing. One-fourth of each backcross generation would be expected to be deep orange and to have high

beta-carotene content. Perhaps amateur plant breeders should spearhead a drive toward breeding more nutritious vegetables and lead the way with orange versions of all their favorite tomato varieties.

Tomato flavor depends on the presence and the amount of dozens of substances, and so it has complex inheritance patterns. Most of the acid and many of the other components of tomato flavor are in the juice, however, so meaty tomatoes are inherently not as full-flavored as juicy tomatoes. Meaty tomatoes are often preferred by those who like mild-flavored tomatoes, though.

Harvesting, Processing, and Storage

Tomato harvesting, processing, and storage is covered in Chapter 21.

References

Charles M. Rick, "Tomato," in *Hybridization of Crop Plants,* pages 669–680; George, *Vegetable Seed Production,* pages 208–223; Edward C. Tigchelaar, "Tomato Breeding," in Bassett, *Breeding Vegetable Crops,* pages 135–171; Ashworth, *Seed to Seed,* pages 151–155; Mike Courtney, "The Genetics of Tomato Fruit Color," *1989 Summer Edition* of the Seed Savers Exchange, pages 33–36.

Lettuce

Lettuce, *Lactuca sativa,* is in the family Asteraceae (Compositae). It is thought to derive from the wild species *L. serriola* (wild lettuce, prickly lettuce) and crosses freely with it, so much so that the two might be regarded as within one species.

Other genera and species in the family

Asteraceae that may be found in Appendix A are *Arctium lappa* (gobo), *Bellis, Calendula, Carthamus, Chichorium endivia* (endive), *C. intybus* (chicory), *Chrysanthemum, Cynara cardunculus* (cardoon), *C. scolymus* (artichoke), *Dahlia, Helianthus annuus* (sunflower), *H. tuberosus* (sunroot, Jerusalem artichoke), *Inula, Matricaria, Petasites, Polymnia, Scolymus, Scorzonera hispanica* (scorzonera, black salsify), *Silybum, Stevia, Tanacetum, Taraxicum officinale* (dandelion), and *Tragopogon porrifolius* (salsify, oyster plant). (I listed these because of all the interesting wide-cross possibilities.)

Stem lettuce, also called asparagus lettuce or celtuce, is a form of lettuce that has been bred for the stems instead of or in addition to the leaves. It is the same species as ordinary lettuce (*L. sativa*) and crosses freely with it. Celtuce is sometimes referred to as *L. sativa* var. asparagina or as being in the Angustan group.

Flowering

Most lettuce varieties are day neutral. Those used for winter production in greenhouses are usually long-day plants.

In crisphead lettuce the seed stalks need assistance to emerge. One option is to slice through the head to expose the core. Some breeders thump the head hard with the palm of their hand to break the leaves off at the base. Alternatively, the leaves may be removed, or the plants may be deheaded so that axillary shoots develop.

Breeding System

Lettuce is a strong inbreeder. The flower is made up of numerous tiny, perfect florets. All the florets open on the same day. The flowers open, pollen is released, the style finishes emerging, and the stigma becomes receptive and is fertilized all within a few hours. The flowers open just once on one day. Each floret produces only one seed.

Within the individual floret the anthers are fused into a tube that encircles the style. The inner surface of this tube sheds pollen onto the style as it emerges. The style accumulates a coating of self-pollen as it emerges and before it is receptive. As soon as it becomes receptive, the self-pollen fertilizes the flower.

Isolation

The literature is somewhat contradictory concerning isolation. According to George, "Although up to 5% cross-pollination has been observed in lettuce in some areas most authorities regard it as a self-pollinating crop and only specify a physical barrier (e.g. adjacent sections of greenhouses) or a minimum of two meters between different cultivars."

Lettuce will cross with and should be isolated from its wild relative, *L. serriola* (prickly lettuce). Celtuce must be isolated from other lettuce varieties.

Inbreeding Depression

Lettuce is a natural inbreeder and does not display inbreeding depression

Hand-Pollination and Crossing

Lettuce is one of the more difficult vegetables to do crosses with. (See illustration on page 306.) Professional breeders depend on washing the pollen off the stigmas at exactly the right time: after it has all been shed but before the stigma becomes receptive and the pollen has germinated. A fine mist of water is directed at the entire flower. The right time for washing is after the style has grown beyond the anther cone but before the stigma lobes have begun to curl outward. "This method produces about 25%–75% crosses if done properly," Ryder says. The chance of success can be increased by using more than one washing. One still must be able to identify hybrids from selfings, however. That is, even these "controlled" lettuce crosses are actually fertile X fertile crosses.

The florets do not develop with complete synchrony. When the first ones are ready to emasculate by washing depends on the variety and the weather. You have to observe the plants and figure out their timing under your conditions. Usually it is in the morning — early morning on warm days, later morning on colder or overcast days.

Individual florets can be emasculated by hand and then pollinated. The anther cone is removed with forceps early on the morning of flowering. This cone is difficult to remove; the flowers are tiny and delicate, the process is time-consuming, and a successful cross produces just one seed. Usually fertile X fertile crosses, with washing as described, are done instead.

Genetics and Breeding

Much of the professional breeding of lettuce has been oriented toward disease resistance and achieving greater uniformity of harvest time. Other important trends involve shaping cultivars for specific regions and growing methods. Recently, for example, growing lettuce in greenhouses during the winter has become an important production method. Varieties for winter greenhouse production need to be able to grow well with less light and heat than outdoor varieties.

In spite of the popularity and importance of lettuce, it is not very well known genetically. Breeding is done largely by using general principles. Somewhat more than fifty genes have been identified; there are no real chromosome maps. Many of the genes involve disease resistance; usually the resistance is only to specific races of specific diseases. For a description of the genes involved and breeding lettuce for disease resistance, see Ryder, *Leafy Salad Vegetables*. A few of the other genes that have been identified are described in the following paragraphs (from Ryder's chapter in Bassett).

The formation of anthocyanin, which produces red or reddish leaves, is under complex control. In general, hybrids between red and green varieties are red or reddish. Two dominant genes, C and G, influence anthocyanin formation. The distribution of the pigment, however, is controlled by another gene, the R locus. R, R^s, R^{bs}, R^t, and r are all alleles at the R locus. R is associated with full red leaves; R^s is spotted; R^{bs} has red-brown spotting; R^t is tinged with red at the leaf margins.

The gene Sc (scallop) describes the leaf margin.

The U locus controls leaf lobing. U is non-lobed, u^o is oak-leaved; u is lobed.

The gene t is associated with slow seed stalk formation, or bolting resistance.

Genotype ww has white seeds; $W\_$ has black seeds. The y locus is associated with yellow seeds. Genotype $yyW\_$ has yellow seeds. Genotype $yyww$ has white seeds.

Given the difficulty with which lettuce

crosses are made, it would be nice to have a number of known marker genes to help identify hybrids. Unfortunately, there aren't many. The color and leaf-shape genes are most effective because they can be identified in seedlings. Seed-color genes also are often used, with identification of the hybrids being delayed until bolting and seed formation.

Harvesting, Processing, and Storage

Lettuce seed ripens irregularly. Ashworth suggests that the way to get the maximum amount of seed is to shake the plants daily over a container or bag. Alternatively, entire plants are harvested when much of the seed is dry, then put upside down in bags. Seed can be freed from the plants by shaking or rubbing them between the hands. Screens of various sizes are used to separate seed from debris. Winnowing usually isn't effective, as the debris blows about the same as the seed.

References

Ryder, *Leafy Salad Vegetables*, pages 13–94; Edward J. Ryder, "Lettuce Breeding," in Bassett, *Breeding Vegetable Crops,* pages 433–473; Ashworth, *Seed to Seed*, pages 92–94; George, *Vegetable Seed Production,* pages 120–132.

Pea

Garden and field peas, *Pisum sativum,* are in the family Fabaceae (Leguminosae). Field peas used to be referred to as a separate species, *P. arvense,* but there is no justification for this, as garden and field peas can be crossed freely.

Breeding System

Peas are strong inbreeders. They normally self-pollinate before the flower opens. The flowers are protogynous; the stigma becomes receptive well before the pollen is shed. This facilitates crossing, allowing the bud to be emasculated and hand-pollinated in one step.

In moderate weather pollen can remain viable for several days. The stigma remains receptive until a day or more after the flower wilts.

Isolation

According to Gritton's chapter in Bassett, crossing of peas in the United States is normally less than 1 percent. The pea is, in most areas, one of the most highly inbreeding crops, and the isolation distances used are often just a few feet — just enough to prevent physical mixture of plants or seeds.

However, this picture is complicated by the fact that in some areas peas can be extensively cross-pollinated. Reported cross-pollination frequency for peas ranges from 0 percent in New York to 60 percent in Peru, where a number of insects cross-pollinate them.

Official isolation distances in Europe are 100 meters. Ashworth (*Seed to Seed*) recommends 50 feet.

I believe most seed savers in temperate regions can use just a few feet of isolation between varieties — such as is represented by separating different pea varieties by a row of something else — as long as you inspect the plantings and plants (buds and flowers) for insects and insect damage that might suggest more care is needed.

Inbreeding Depression

Pea is a natural inbreeder and does not display inbreeding depression.

Hand-Pollination and Crossing

Open pea flowers have three distinctive kinds of petals. (See illustration on page 307.) The outermost, or *standard,* is the one that opens up wide into two lobes when the flower opens. The next inward petal is the *wing,* which surrounds the center of the flower. Interior to the wing is the *keel,* the innermost petal that protects the pistil and stamens. Peas are hand-pollinated in the bud stage, and in buds these three petals are all folded up, so you must penetrate three layers of petals.

The outer two petals, the standard and the wing, are normally opened by hand and left intact so that they can close afterward and protect the fertilized stigma. The innermost petal, the keel, must be slit open or removed. Most people remove it. Pea flowers need not be bagged after hand-pollination.

Hand-pollination is done when the buds are as big as possible but before pollen has been released. In my experience this stage varies widely with both variety and weather. For many varieties under cool (normal early-spring) conditions, the late predehiscent stage is when the petals of the bud have grown out to just barely beyond the sepals. But plants and buds can grow at temperatures that are too cold for pollen dehiscence. Under these conditions buds grow without dehiscence until it warms up. So in cool weather the buds may be quite large and the petals may extend far beyond the sepals when the anthers dehisce. But in warm weather the same variety may dehisce at a much earlier stage, when the buds are much smaller and the petals are still shorter than the sepals. Exact timing also is influenced by variety.

It's much easier to work with the big, late-dehiscing buds associated with cool weather. Don't try to do crosses when the weather is too cool for them to take, however. (You learn to tell this by noticing what kind of weather pods normally set in.)

Buds lower on the plant also are bigger, easier to work with, and more likely to make successful crosses, and they produce more seed. So whenever possible, make crosses with material grown early in the year, and do them early in the flowering season. Midsummer flowers pollinate when the bud is so tiny that I have not been able to do crosses with them at all. I plan to try again, but with growing the plants in the shade. I also may try emasculating them at a very early stage and then pollinating them a couple of days later.

It takes a little practice to manipulate pea flowers and do crosses. Figure on ruining several flowers before you acquire the skill. I usually start by stripping off the sepal that is directly in the way of opening up the standard and wing. Then I open up the standard and wing and hold them open with one hand. With the thumb and forefinger of my other hand, I grasp the keel about halfway down and tear it off so as to bring most of the stamens with it but leave the style untouched and undamaged. When I'm very clever about it, I get all ten stamens. Usually one to three remain, and I remove them with forceps. Alternatively, you can remove the keel in pieces with forceps and then remove the stamens.

I use pollen that has been shed the same day I do the crosses — that is, from buds somewhat bigger than those I'm using as females or from just-opened flowers, depending on the weather. Fresh pea pollen is bright yellow and moist-looking. Old pollen is pale and dry-looking.

Pea pollen can be stored for up to six days after dehiscence. (I think this means in a refrigerator.) Whole buds or flowers that have just dehisced probably could be wrapped to prevent them from drying out and then refrigerated. Pea pollen is said to be viable for about a year if vacuum dried and stored at 25°C. I think this means that you can store it for an extended time by desiccating it and putting it in the freezer, as described in Appendix B. I intend to try it.

I usually transfer pollen by using the whole male flower with the petals stripped off, by using anthers, or by removing the keel of the male flower and using it. The latter has the advantage that it doesn't sacrifice the male flower, just removes the extra pollen. To use the male keel I pull it off as described for female flowers; it constitutes a little bag of pollen into which the female stigma can be dipped.

The stigma is at the very end of the style. That subterminal brush on the style is not the stigma.

After hand-pollinating, I close the standard and wing back around the pollinated pistil. I am told that Baggett's group uses a bit of tape to hold the entire flower closed for additional protection from drying out. They fold a small piece of Scotch tape in half, then open it again and use it to clamp the flower closed. I plan to try this trick next season.

My own experience is that crosses take best in mild weather. About half my crosses take under optimal conditions; much fewer take late in the season or in hot weather. Perhaps the tape trick will improve the odds on crosses done under these conditions.

Pea seed, if mature enough that it has accumulated most of its dry matter, can be germinated without drying down. Just shell out the fresh, mature, but not dry peas and soak them in water for a few days until they germinate. (Change the water once or twice a day to provide plenty of oxygen and discourage bacterial or mold growth.) This trick can cut about a month off the time required to do a generation of pea breeding. It also works with all the other legumes I have tried.

Labeling pea crosses is trickier than with many vegetables because the stem that supports the bud is small and delicate. I use tiny labels used for pricing jewelry, and the smallest labels in the class. These are about ½ inch long and ¼ inch wide and are supported with a slender thread.

It's very difficult to breed and eat from the same pea plants. The cross labels twist around and hide, so it is easy to eat the crossed pods by mistake. The care needed to examine the plants for labels makes harvesting too much work. And maturing pods cut the production of the plant short; peas must be kept picked to stay in production. For these reasons, I usually plant separate patches for breeding and eating.

Genetics and Breeding

There are three basic classes of pea varieties,

depending on their intended purpose: dry (soup) peas, shelling peas for use as green shelled peas, and edible-podded peas. Some peas also are used for green manure or live-stock feed. Home gardeners usually grow shelling and edible-podded types. Recent breeding efforts have concentrated largely on developing bush types with simultaneous pod ripening for picking by machine and on incorporating disease resistance into these varieties. Much additional work has been aimed at obtaining peas of every given matu-rity for every purpose for each of the major pea-growing areas.

In shellers professional breeders aim for small pea size. This is because consumers have learned to associate large peas with past-prime peas. The smaller peas fool them into thinking the peas are prime. That isn't neces-sarily true. I think home gardeners should give some attention to breeding for large peas, as these produce more peas for the same amount of shelling work. The classic variety 'Alderman' (aka 'Tall Telephone') has huge peas that are among the tastiest.

Professional breeders have been concen-trating almost entirely on dwarf varieties. For home gardeners there are many advantages to tall, pole varieties. We home gardeners will probably have to breed them ourselves. In many cases all you need to do is transfer a gene for tall into established dwarf varieties by recurrent backcrossing.

So far professional breeders have released only green-podded shellers or edible-podded varieties. This is boring. We home gardeners should do something about that, too. Many bean varieties have purple or yellow pods. The genes for colored pods also exist in peas,

waiting for us to use them.

In short, in spite of the fact that the pea is one of the most popular vegetables that receives a relatively large amount of breeding attention, there are still plenty of opportuni-ties for amateurs to work with peas to fill the needs and niches that are being neglected.

Genetically speaking, the pea is one of the best studied vegetables. See Gritton's chapter in Bassett for a list of many of the known genes, chromosome maps, and references to the current professional literature. Some of the genes that are most relevant agronomi-cally are described in the following para-graphs (taken from Gritton).

Seed shape is determined by a number of genes. The two most important loci are R and R_b. The recessive allele at either of these loci causes the seed to be wrinkled instead of round. The round seed becomes starchy faster than the wrinkled seed, but it germi-nates better in cold weather. Round types are usually used for early cultivars; otherwise, wrinkled types are used. Wrinkled genotypes are $rrR_b\_$, $R\_r_br_b$, and rrr_br_b. Round seeds are $R\_R_b\_$.

The round versus wrinkled phenotypes are apparent at the seed stage, since they affect the starches in the cotyledons, which are part of the embryo. Segregation in the F_2 seed can be seen as you shell pods on the F_1 plants.

Cotyledon color can be green (ii) or yellow ($I\_$). You may have to pierce the seed coat to see the color of the cotyledons. In addition, color fades with age and is most obvious in fresh seed. Canning cultivars traditionally have a light green color that resists leaching in the canning fluid. Freezer cultivars tradi-tionally have darker green peas. (There are now dual-purpose cultivars, too.) The genes *pa* and *vim* are associated with dark green seed and foliage color.

Plant height is controlled by three genes: *cry, la,* and *le.* Alleles associated with tallness are dominant.

Two independent genes, *p* and *v,* are associated with reducing the fibrous membrane on the inside of the pod. Modern edible-podded varieties have genotype *ppvv.* Shelling peas are generally *PPVV.* Crosses of edible-podded peas and shellers produce F_1 plants that have shelling-type peas. Plants carrying *Dpo* have tough, leathery pods that dehisce readily when mature.

The gene *n* increases the thickness of the pod wall and causes the overall shape of the pod to be round instead of flat in cross section. All snap peas have the genotype *ppvvnn.*

Synthesis of anthocyanin (the purple, red, and blue pigment) requires the presence of the dominant gene *A.* Plants of genotype *aa* have white flowers; green, not purple, pods; and no purple in the leaf axils or seed coats. Plants of genotype *A_* have some purple, but other genes control where it is distributed.

Purple pods require the presence of three dominant genes: *A, Pu,* and *Pur.* The purple ring in the leaf axil requires *A* and the dominant gene *D.* (Plants that are *dd* have green leaf axils even if they can produce purple pigment.) Genes *b* and *ce* (in the presence of *A*) change the shade of color in the flower to pink and rose, respectively.

The official name of the gene associated with yellow pod color is *gp.* Yellow pods are recessive to green.

Purple, brown, or black color or spotting in the seed coat is controlled by a number of genes (see Gritton). All of them require *A* to be expressed. Dozens more have been identified that control the number of basal branches, stem fasciation, leaf morphology, leaflet margin shape, conversion of leaflets to tendrils or vice versa, and so on. Additional genes have been identified that control the presence, absence, or amount of wax on the pods and the rest of the plant. Other genes control the distribution of the wax.

Two genes, *fn* and *fna,* are associated with the number of flowers at a node. This characteristic is very much influenced by the environment.

Some genes that determine the shape of the pod are *lt* (25 percent wider), *Bt* (blunt apex), and *Con* (affects curvature).

Yet more identified genes are associated with dimples on the seeds, texture of the seed surface, seed shape, and whether seeds stick together in the pod. There are also genes that change the pollen color or drastically alter flower anatomy.

The genes *En, Fnw, Fw, sbm,* and *er-1* are associated with resistance to pea enation mosaic virus, fusarium near wilt, fusarium wilt, pea seed-borne mosaic virus, and powdery mildew, respectively.

I have seen no information on the genetics of pea flavor other than the basic sweet wrinkled-seeded versus the round-seeded characteristic. I also have seen no genetic information on size of pods or peas, number of peas in the pod, cold hardiness, heat resistance, or many of the other characteristics that matter most agronomically. Yet these characteristics are clearly inherited. It is instructive that even with peas, where so much is known compared to most vegetables, large parts of every breeding project still must be guided largely by general principles.

Harvesting, Processing, and Storage

Harvesting, processing, drying, and storage of peas is covered in Chapter 21.

References

Earl T. Gritton, "Pea Breeding," in Bassett, *Breeding Vegetable Crops,* pages 283–319; Suzanne Ashworth, *Seed to Seed,* pages 139–140.

Common Bean (*Phaseolus vulgaris*)

Phaseolus vulgaris, the common bean, is the major type of dry bean used in North America and, in its green bean form, is one of the major fresh, canned, and frozen vegetables. Some home gardeners also grow varieties that are shelled and eaten as immature seeds. Two other species within the genus that are common as commercial beans are *P. coccineus* (runner beans) and *P. lunatus* (lima bean).

Common beans can be distinguished from runner beans, which have similar seeds, by the fact that runner bean seedlings have cotyledons that remain in the ground when the seed germinates. In common beans the cotyledons emerge aboveground when the bean germinates. Fava beans, too, have cotyledons that emerge. However, favas have their hilum at the end of the seed, not on the side as common beans do.

To a large extent the varieties grown for use as dry beans, green beans, and shellies are all different varieties, although there is some overlap. Dry beans, optimally, have a fibrous pod and are stringed; these characteristics make them easy to thresh. In addition, the seeds develop and dry rapidly once the pods reach full size.

The ideal shelly bean also has pods that are fibrous and stringed, and therefore easy to open by hand. In addition, the seeds develop rapidly after the pod reaches full size, but they don't dry down quickly. They stay in the fresh green state for a while.

The ideal green bean should have little fiber in the pod, because the pod is the edible crop. Optimally, it is stringless. In addition, its seeds should develop slowly after full pod size is reached so that it stays in the right stage for green-pod use as long as possible. (Once seeds start to develop, the pods invariably toughen and become more fibrous.)

Green beans for canning usually have white seeds so they don't color the canning liquid. Green beans for use as a fresh commercial crop need to have a bit more fiber in the pods than those for canning or freezing so that they can stand up to shipping and handling.

Among home garden varieties, especially of pole beans, there is overlap in the uses for many cultivars. However, any cultivar that is optimal for one of the three main uses is inherently suboptimal for the others.

Breeding System

Beans are basically inbreeders. Like peas they have three kinds of petals: standard, wing, and keel. Also like peas the style and ten stamens are enclosed within the keel, and dehiscence and self-fertilization occur in the late bud stage. In beans, however, the style is twisted into a coil, the stamens are twisted around the style, and the keel is twisted as well. The anthers are actually pressed against the stigmatic surface such that they shed their pollen directly on it. According to Bliss, the stigma is receptive from at least two days before the flower opens to at least one day after.

Isolation

Isolation theory and practice for the common bean is discussed extensively in Chapter 18.

Inbreeding Depression

As natural inbreeders, beans do not suffer from inbreeding depression.

Hand-Pollination and Crossing

Bean crosses are often done in a greenhouse. This isn't essential, but it appears that mild conditions, adequate moisture, and preventing the crossed flower from desiccating are important considerations. Outdoor crosses are usually done in the morning, when the turgor of the plants and flowers is higher.

It isn't possible to provide a good drawing of the anatomy of the bud of bean because of the twisting of the parts. You can learn the basics by reviewing the material on pea.

According to Bliss, buds are at the right stage for emasculation when they are plump, showing color, and due to open the next day. He notes, "If self-pollination has already occurred, the stigma will be swollen, light green rather than white, dried rather than sticky, and with some remnant pollen attached." I have never done any bean crosses, but I suspect that the situation is

similar to that for peas — that dehiscence occurs in the late bud stage, but exactly when depends on variety and weather. You have to watch for the right stage for the variety and check the weather on any given day.

Apparently it is much easier to injure the style or stigma in a bean cross than a pea cross, and it's harder to get into the bud physically. The standard and wings are opened and held back as with the pea. Then most or all of the keel is removed with forceps to expose the style, stigma, and anthers.

Flowers that have just opened are used as the source of pollen, Bliss says. They may be used immediately or refrigerated for use later. There are two common pollination methods. In both, the pollen is transferred from the stigma of the male flower to the stigma of the female. In the rubbing method, the stigma of the male flower is made to extrude from the keel by pressing on the wings; the pollinated stigma of the male is then rubbed against the stigma of the emasculated flower.

In the hooking method, the pollinated male stigma is removed with forceps, rubbed against the female stigma, and hooked through the style so that it remains near the stigma in the female flower. One report gives success rates of 30 to 40 percent for rubbing and 70 to 80 percent for hooking.

The bud is taped shut after pollination to prevent desiccation. A 4-centimeter strip of Scotch tape is formed into a circle with the sticky side inward. It's placed over the standard. It should encircle only the petals, not the sepals. A small piece of wet tissue may be put inside the tape circle to raise the humidity around the bud.

Because of the difficulty of emasculating

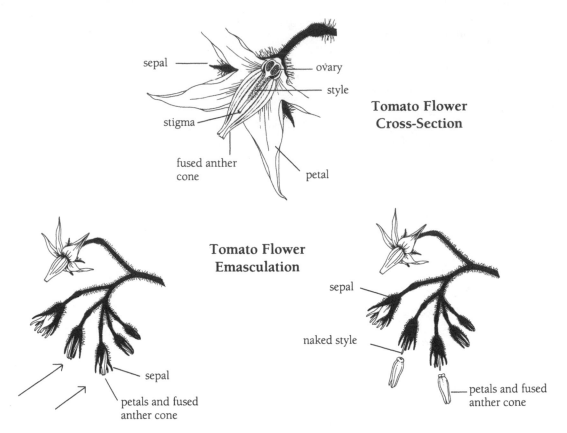

sepal — ovary — style

Tomato Flower Cross-Section

stigma

fused anther cone — petal

Tomato Flower Emasculation

sepal

naked style

petals and fused anther cone

sepal

petals and fused anther cone

A. Unemasculated flowers. The two central buds (arrows) are at the correct stage for emasculation.

B. Emasculated flowers and parts removed.

Lettuce Floret Development

In stage 1, the stigma has not yet emerged from the anther sheath surrounding it. In stage 2 the stigma has started to emerge. It is already covered with pollen, but because it is not receptive, self-pollination has not yet occurred. Stage 3 is the best stage for removing pollen by washing. Stage 4 is too late: the stigma has become receptive, and self-pollination has occurred.

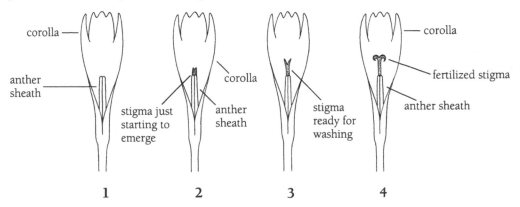

corolla —

anther sheath

stigma just starting to emerge — corolla — anther sheath

stigma ready for washing

— corolla — fertilized stigma — anther sheath

1 2 3 4

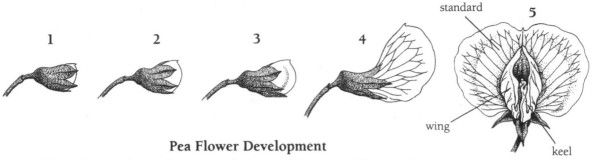

Pea Flower Development

Bud 2 is at the stage that is often optimal for emasculation and pollination. But sometimes bud 1 or bud 3 is the correct stage, depending upon the variety and the weather. Bud 4 is too old; it will have already self-pollinated. It would be a good source of fresh pollen.

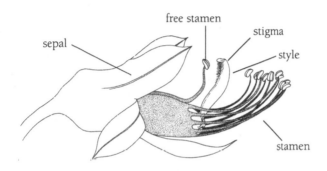

Pea Flower Anatomy

Here is a flower with the petals removed.

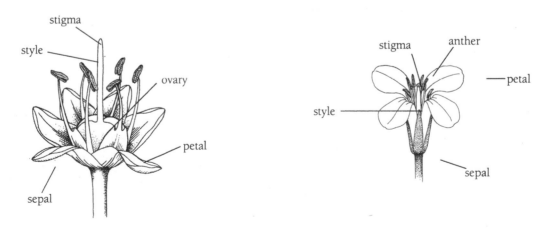

Onion Flower   **Brassica Flower**

beans, fertile X fertile crosses are often done instead. This type of cross is practical whenever the cross will yield recognizable hybrids. It is much easier to do because the bean's stigma can be made to protrude from the keel when the wings of the flower are pressed back. The exposed stigma can be pollinated without permanently opening the flower or removing the keel. The stigma can self-pollinate later, of course, but the applied pollen has a head start. Sometimes the stigma of the male is hooked behind that of the unemasculated female, clamping it in place and preventing the female stigma from returning into the bud, where it can be self-pollinated.

Genetics and Breeding

Extensive breeding work has been done on developing cultivars for dry use and for green-pod use. I know of no work on developing better shelly varieties. It seems arbitrary to me that the major fresh use for peas is as shellies and the use of whole pods is secondary, whereas for beans the reverse is true. I think it's a matter of historical accident. I would like to see home gardeners produce dozens of shelly varieties and turn the shelly bean into as popular a vegetable as the shelling pea.

Most varieties grown primarily for dry use are bush cultivars. Traditionally, green-pod varieties were pole types. The two major green bean classics, 'Blue Lake' and 'Kentucky Wonder', are pole beans. In this century, in the United States, commercial green beans have come to be harvested by machine, and bush green beans predominate. In Japan, where farmland is used very intensively, pole beans are the major commercial green bean type.

Silbernagel's chapter in *Breeding Vegetable Crops* does not give a summary of the genes that have been identified as affecting agriculturally important characteristics, and I know of no such summary anywhere else. Silbernagel does give references to the professional literature that provide gene lists and information about breeding for specific characteristics. You can develop a general feel for the kinds of situations involved by reading the section on the genetics of peas in this appendix.

A number of traits are used as marker traits, however. Anthocyanin (purple or red pigment) inheritance is complex but is often used as a marker to identify hybrids. Hybrids between purple and green varieties are usually purple. That is, purple-podded is dominant to green-podded, purple-flowered to white-flowered, purplish foliage to green foliage, and so on.

Crosses of pole and bush cultivars normally give pole-type F_1s.

Colored seed is dominant to white seed. The seed coat is maternal tissue, however, and as such it expresses the genes of the mother, not those of the seed.

Harvesting, Processing, and Storage

Harvesting, processing, drying, and storage of beans is covered in Chapter 21.

References

M. J. Silbernagel, "Snap Pea Breeding," in Bassett, *Breeding Vegetable Crops,* pages 243–281; George, *Vegetable Seed Production,* pages 192–199; F. A. Bliss, "Common Bean," in Fehr and Hadley, *Hybridization of Crop Plants,* pages 273–284.

Alliums (Onions)

Onions, garlic, chives, leeks, and shallots are all members of the genus *Allium*. For information about which varieties correspond to which species, see Table 1.

Flowering

Many onions are biennials or biennials raised as annuals. The common bulb onion, for example, is an *Allium cepa*. It is ordinarily grown to produce a bulb the first year. For seed production the bulbs are inspected, rogued, and stored for replanting the following year. Many onions need a cold treatment to induce flowering. For most varieties two weeks in the refrigerator will suffice.

Perennial species such as *A. fistulosum* normally don't flower the first year but do flower in subsequent years. Some species and some varieties don't flower at all, or flower only under special conditions. Instead, they make bulbils in the flower heads (top sets) or multiply by division at the base.

Breeding System

Alliums form large clusters (umbels) of tiny flowers. Allium flowers are perfect, but alliums are outbreeders. Bees and flies of various types and other insects are the pollinators. Alliums are self-compatible, but individual flowers usually don't self-pollinate because of protandry. Pollen shedding in an individual allium flower occurs and is often completed before the stigma becomes receptive, thus preventing self-pollination. However, many flowers in a single umbel are opening, pollinating, and becoming receptive at various times, so fertilization of flowers by pollen from other flowers in the same head is possible and common.

According to Pike's chapter in Bassett, pollen begins to be shed shortly after the flower opens, weather permitting, and is shed over a period of three to four days. The stigma becomes receptive when the pistil reaches full size. In some cases this is after pollen release is completed. In other cases the stigma becomes receptive after, but the same day as, the pollen dehisces (Dr. Kalloo's *Vegetable Breeding* vol. I, page 25). The stigma remains receptive for several days.

Cytoplasmic male sterility exists and is the basis for the hybrid onion seed industry (see Pike).

Isolation

For commercial production, isolate alliums by 1,000 meters from all others of the same species. You also may cage plants in small plots, but if so, hand-pollination or introducing insect pollinators is necessary. Fly pupae can be placed in the cages so that they emerge clean and ready to go to work. To collect blowflies, place a piece of meat in an open jar and put it outdoors. Wild flies will lay eggs in it, and the larvae will pupate on the sides of the jar. Caged flies will live longer if water is available.

Most seed savers will want to avoid all these problems by growing just one variety of each allium species. Note that varieties that aren't flowering don't count. You can grow other varieties of *A. cepa* bulbing onions if you, for example, are growing just one variety for seed and the others are making bulbs, not flowering. Usually your neighbors' gardens present little or no problem, because they normally won't have flowering alliums, only bulb-producing first-year material.

Inbreeding Depression

Alliums generally display inbreeding depression, enough so that self-pollinating for even two generations causes problems with vigor.

Hand-Pollination and Crossing

Hand-pollination can be performed by emasculating the flowers and then transferring the pollen several days later. Most of the flowers on the umbel must be sacrificed to prevent contamination, however, and individual flowers produce just a few seeds.

Allium crosses are usually performed as fertile X fertile crosses instead of as hand-pollinations. Commercially, small numbers of plants of the two varieties to be crossed are caged together, and pollinators are introduced. Alternatively, you could transplant two plants to some area away from other varieties of the same species and let the bees do it.

My crosses of alliums have all been wide crosses, and I do them as hand-pollinations. I start by removing all the opened flowers on an umbel as well as buds too small to work with. I open and emasculate the buds that are big enough to work with by removing all their petals, sepals, and stamens. (See illustration on page 307.) I don't try to cover the umbel. I depend on the lack of petals, pollen, and nearby flowers to discourage pollinators. I have not checked this method rigorously, but I've never seen pollinators on any flowers I've prepped this way.

Under my conditions, most species of alliums seem to finish releasing pollen days before the styles finish growing, so none of my prepped flowers are ready to hand-pollinate. I inspect my prepped flowers on subsequent days until the most advanced style reaches full size. Then I hand-pollinate it and all the others on the umbel. (This is probably the right time for the oldest flower and too early for the others.) I repeat the hand-pollinating daily for the next few days. In other words, each flower I prep receives several doses of pollen, some too early, others probably too late, and, hopefully, one absolutely optimally. This takes very little time compared to that required for prepping the flowers in the first place.

Genetics and Breeding

Most of the breeding work has been with bulbing varieties of *A. cepa*. There are three basic kinds of onions raised for bulbs: fresh eating onions, storage onions, and dehydrating onions. In addition, some *A. cepa* varieties are used as scallions. Because onions have a precise photoperiod requirement for bulbing, much effort has been spent on developing all kinds of onions for each specific region.

The breeding work seems to be based largely on general principles. There are very few identified genes in onions. The genotype *vv* has virescent seedlings; *gl gl* has glossy foliage; *pr pr* is resistant to pinkroot.

The inheritance of bulb color in *A. cepa* is controlled by three loci: *I, C,* and *R. I* and *C* control whether there is any pigment. *R* controls whether it is red or yellow. *R_* is red; *rr* is yellow. *I* is a dominant pigment inhibitor. Pigmented plants must be *ii. II* is white, no matter what the other genes are; *Ii* is buff.

The other inhibitor is a recessive. $C\_$ is colored; cc is white. To have colored bulbs, a plant must carry genotype $iiC\_$. If it does, its color is determined by the R locus. $iiC\_R\_$ is red; iiC_rr is yellow. In summary, red varieties are $iiCCRR$, yellow varieties are $iiCCrr$, and whites may be either $II\_\_\_\_$ or $\_\_cc\_\_$.

Harvesting, Processing, and Storage

Allium seed heads begin dropping seed from the earliest flowers as, or very shortly after, the latest flowers dry up. And various seed heads are ready at various times. So for a seed crop of an allium, it is important to watch the plants and harvest the various seed heads as they are ready. I usually harvest at least three times during the season. Once some of the seed on a head has dried up, it should not be exposed to rain or watering. It should be removed and taken indoors to finish drying.

Small amounts of seed can be cleaned by rubbing seed heads between the hands and then winnowing. Once the seed heads have been broken up, though, the debris is hard to separate from the seed. Often sufficient seed can be obtained by shaking whole seed heads without breaking them up; much seed drops free, and the debris isn't generated. When I need a lot of seed, I sometimes break up the seed heads by rubbing but don't bother separating seed from debris. I sow it all.

Allium seed is more easily damaged by mechanical processing that most seed. It must be handled gently. Allium seed also does not store as well as the seed of brassicas, legumes, and many other common vegetables.

References

Leonard M. Pike, "Onion Breeding," in Bassett, *Breeding Vegetable Crops*, pages 357–393; H. A. Jones and L. K. Mann, *Onions and Their Allies*; Dr. Kalloo, *Vegetable Breeding* vol. 1, page 25.

Brassicas (Cabbages and Relatives)

The brassica family, Brassicaceae, includes numerous important vegetables. This section covers the genus *Brassica*. Additional genera and species in the family are covered in Table 1, including *Armoratia rusticana* (horseradish); *Barbarea* spp. (cress); various *Crambe* spp., including *Crambe maritima* (sea kale); *Eruca vesicaria* (rocket); *Hesperis matronalis* (sweet rocket); various *Lepidium* spp., which include maca, garden cress, and Virginia pepperweed; *Raphanus sativus* (radish); *Rorippa* spp. (watercress); and *Wasabia japonica* (Japanese horseradish).

Brassica oleraceae includes cabbage, kale, collards, broccoli, sprouting broccoli, cauliflower, Brussels sprouts, and kohlrabi. *B. campestris* includes turnips, Chinese cabbage, pak choy, Chinese mustard, turnip rape, and a few others. *B. napus* includes rutabaga, swede, rape, and rape-kale. *B. juncea* includes leaf mustards. Condiment mustards and wild mustards account for a few other species. For details see Table 1.

On *Brassica* taxonomy I am following Downey, Klassen, and Stringam, as found in Fehr and Hadley's *Hybridization of Crop Plants*. The abbreviation *gf-* in Table 1 refers to genomic formula. The chromosomes of the various species are microscopically distinguishable, so it is possible to understand the evolution of and relationships between the various species by examining the chromosomes. The genomic formula indicates which genomes are involved.

Example: *A* is the haploid *B. campestris* genome, so Table 1 includes the symbol *gf-AA* for *B. campestris*, indicating it is a normal diploid. *B. nigra* (black mustard) is *gf-BB*. *B. juncea* (leaf or India mustard, of which my much-beloved 'Green Wave' is a representative) is *gf-AABB*. In other words, it is a tetraploid, an amphidiploid, and was derived by a cross of a *campestris* and a *nigra* followed by chromosome doubling.

Flowering

Biennial brassicas must be overwintered to achieve flowering. Whether this can be done outdoors depends on the cold hardiness of the variety and the region. Most brassicas require some period of cold to flower. Brassicas that form tight heads, such as cabbage, may require scoring of the outer leaves to allow the flower stalk to emerge. Cut two slashes in an X into the top of the head.

Breeding System

Brassicas have perfect flowers. Most brassicas are strong outbreeders. Varieties of the species *B. oleraceae*, *B. campestris*, and most others are usually self-incompatible. (An exception is summer cauliflowers, which are usually self-compatible.) The tetraploid species *B. juncea* and *B. napus*, however, are self-compatible inbreeders. The strength of the incompatibility varies from variety to variety and plant to plant, as well as under different conditions. The normal pollen vectors are bees, flies, and insects of various kinds.

In brassica flowers the stigma becomes receptive some days before the bud finishes developing and opens. In some species and varieties the stigma protrudes slightly and is exposed during the last day before the flower opens. The anthers dehisce some hours after the flower opens. When the flowers open and the anthers dehisce depends strongly on the weather.

Vegetative Propagation

Brassicas are easy to propagate vegetatively by shoot or root cuttings. These methods can be used to increase or maintain a specific plant for further evaluation or to overwinter or apply cold treatments to induce flowering.

For shoot cuttings, cut off a vegetative shoot, place it in potting soil, and set it in a moist place to root. Rooting hormone may be used. For root cuttings, use root sections about the thickness and length of a pencil. Make a direct cut across the upper end and a slanting cut across the lower end of each. Place them vertically in potting soil or in the ground.

Isolation

For commercial production brassicas are isolated from all others of the same species by 1,000 to 1,500 meters. Note, for example, that cabbage, kale, collards, broccoli, cauliflower, and Brussels sprouts are all *B. oleraceae* and will cross freely with each other. The other species are equally diverse (see Table 1). Some mustards are *B. juncea*. Other plants we call mustards are actually leaf or stem forms of Chinese cabbage (*B. chinensis*). Which is which is often not obvious.

In my garden, with its myriad of pollinator species, all the brassicas seem to cross to some extent whether they are the same species or not. Hybrids between the species are usually highly, though not totally, sterile. I isolate all brassicas as if they were one big (somewhat confused) species.

There are a number of wild mustards of various species as well, and it usually isn't obvious what species they are. I eliminate any wild mustards growing near a variety I am growing for seed.

Inbreeding Depression

Most brassicas are subject to inbreeding depression. The exceptions are those that substantially inbreed naturally, such as summer cauliflower and the *B. juncea* varieties (see Table 1).

Hand-Pollination and Crossing

Hand-pollination of brassicas is done in the bud stage. If the plant is self-incompatible, emasculation prior to doing crosses isn't required. According to instructions given by Dickson and Wallace for cabbage, buds can be used from one to four days before opening. Self-incompatibility has developed by the day before flowering, however, so if bud pollination is being used for self-pollinating an incompatible plant, earlier buds should be used. (See illustration on p. 307.)

Generally the opened flowers are removed, and the six to eight buds on the inflorescence — representing all those that will open for the next four days — are opened, emasculated if necessary, and hand-pollinated. After hand-pollinating, it is necessary to exclude pollinators. Dickson and Wallace specify a cheesecloth bag around the blossoms.

I used to spend much time making brassica crosses by hand. I never do it that way anymore. I just plant the two different varieties or species side by side. I'll always get a few percent of crosses from the plants at the edges of the blocks where the two varieties are adjacent. A few thousand seeds from these plants can be sown in a few square feet of soil the following year, and the hybrids recognized as seedlings.

Hybrid Seed Industry

Because most brassicas are so subject to inbreeding depression and display such large amounts of hybrid vigor, there is much natural advantage to hybrid brassicas.

My favorite mustard, 'Green Wave', is an open-pollinated type and is more vigorous than any other mustard I've seen, including the hybrids. But 'Green Wave' and other *B. juncea* types are in a very real sense hybrids already, because they are tetraploids containing two entirely different brassica genomes from different species. This apparently gives them the vigor of hybrids, but they still have the genetic stability of any other open-pollinated variety.

The hybrid seed industry for brassicas has been based on the incompatibility system. For information on the incompatibility system and how it's used to produce hybrid seed, see Dickson and Wallace in Bassett. Cytoplasmic sterility also exists and is starting to be used to produce hybrid seed.

Genetics and Breeding

A few of the genes that have been identified in cabbage are given in the paragraphs that follow (from Dickson and Wallace).

T is a dominant gene causing the plant to be tall. Most cabbage cultivars are *tt*; tall cabbage plants tend to fall over. *Ax* is associated with axillary buds.

The *A* locus controls anthocyanin. There are various alleles. A recessive inhibitor, *c*, suppresses anthocyanin formation in all parts of the plant; *cc* has no anthocyanin, whatever the rest of the genotype. *B* is associated with color limited to the midrib. *G* and *H* are intensifiers associated with deep purple color. In plants carrying *M* the color is magenta instead of purple. Other genes also affect color (see Dickson and Wallace).

The gene *En* is associated with entire leaves (renamed from *E*). *W* is associated with wide leaves. The *gl* locus (glossy) has a series of alleles that determine the amount of wax on the stems. With *Hr-1* the plant has hairy first leaves and hairs on the margin, with "sometimes poor expression in heterozygote." Smooth leaves are associated with *sm* along with *wr*. *Pet* controls whether the plant has petioles and leaves instead of sessile leaves. Four genes — *W-1, W-2, W-3,* and *W-4* — are associated with frilling of leaves.

K is a dominant gene associated with heading. Hybrids between heading and non-heading forms are intermediate. Hybrids between varieties with pointed heads and those with round ones have round heads, but the inheritance isn't simple. Crosses between varieties with savoy heads and those with hard heads produce hard-headed offspring. Other such patterns are as follows, with the type the hybrid resembles being listed first: annual habit versus biennial; early maturity versus late; few wrapper leaves versus many; wide core versus narrow.

Core length is associated with two codominant genes with additive effects. Head splitting is determined by three codominant genes that act additively. Crosses of red and green plants are usually pink. Savoy leaf texture involves three or more genes.

For information on genetics of disease resistance, see Dickson and Wallace.

Harvesting, Processing, and Storage

Brassica seed harvesting, processing, and storage is covered in Chapter 21.

References

Michael H. Dickson and D. H. Wallace, "Cabbage Breeding," in Bassett, *Breeding Vegetable Crops*, pages 395–431; R. K. Downey, A. J. Klassen, and G. R. Stringam, "Rapeseed and Mustard," in Fehr and Hadley, *Hybridization of Crop Plants*, pages 495–509.

Squash and Pumpkins

The words *squash, pumpkin,* and *marrow* are applied more or less indiscriminately to fruits of four species: *Cucurbita pepo, C. maxima, C. moschata,* and *C. mixta.* Generally, most of what are referred to in the United States as summer squash are *C. pepo.* Many of the winter squash are *C. maxima.* Some are *C. moschata* or *C. mixta.* The words *pumpkin* and *marrow* tend to be used for some varieties within each of the species but not others.

There are anatomical differences between the species, but most people know which variety is which because the seed pack or catalog gives the species or they look it up. The *C. pepo* species includes the acorns, many of the orange pumpkins, the summer crook-necks, the scallops, the zucchinis, 'Spaghetti', and many others. *C. maxima* includes other pumpkins, the Hubbards, the bananas, 'Arikara', 'Buttercup', 'Golden Delicious', the Kuris, and many others. *C. moschata* includes the butternuts, the cheeses, the melon squashes, and others. *C. mixta* includes 'Tennessee Sweet Potato', most of the cushaws, and others. For extensive lists of which cultivars are in which species, see Ashworth's *Seed to Seed*, *The Garden Seed Inventory*, or a *Winter Yearbook* of the Seed Savers Exchange.

Breeding System

Squash and pumpkins are outbreeders, but they seem to have some genetic characteristics typical of inbreeders. (See the sections on inbreeding depression and on genetics and breeding.) The flowering pattern is mono-ecious; there are separate male and female flowers.

The first few flowers are usually male; then both male and female flowers are produced. Honeybees and bumblebees are the usual pollinators. Squash and pumpkins are all self-compatible.

The female and male flowers are easy to tell apart because the females have a swollen area at the base, which represents the ovary. It looks like a little fruit and will develop into one after fertilization. The female flowers produce nectar only; the male flowers produce both pollen and nectar.

Isolation

The recommended isolation distance for different varieties within a species of cucur-

bits is 1,000 to 1,500 meters. The four different species don't cross with each other spontaneously, however, so you can grow one of each without isolating them from each other. That's only appropriate if you are 1,000 meters or more from any neighbor growing a variety of the same species.

Most seed savers are not 1,000 meters from the nearest squash-planting neighbor and/or they want to grow more than one variety of a species. So they hand-pollinate. The separate male and female flowers, the large flower size, the high success rate, and the fact that you get hundreds of seeds from each hand-pollination make hand-pollination especially easy.

Inbreeding Depression

In spite of their basic outbreeding anatomy, squash and pumpkins display little or no inbreeding depression.

Hand-Pollination and Crossing

The stigmas of squash don't become receptive until the flower opens, at which time pollen also dehisces. Therefore, it is necessary to emasculate at one time and pollinate later.

The normal procedure is to tape male and female blossoms shut with masking tape applied to the end of the flower bud. This should be done in the late afternoon or early evening of the day before the bud is due to open for the first time. These buds are beginning to show color, and the petals may be showing signs of opening at the ends. If the bud is taped earlier in the day, the tape will damage the bud as it grows. It's useful to put a flag or marker of some kind in the ground next to each taped flower; otherwise you might have a hard time finding the flowers again.

Male flowers as well as females must be taped. Otherwise bees can get to and remove the pollen before you do, and the little that is left is contaminated.

Hand-pollinations are done the next morning, as soon as possible after the pollen is released from the anthers. This is usually about 9 A.M. in my garden. Release of pollen happens earlier on warm dry days and later on cool or wet ones. When male flowers are ready to use, the pollen is yellow and fluffy-looking as it clings to the anthers. Pick male flowers to use, as pollen applicators. Untape them by removing the end of the flower with the tape and tear off the petals. It's best to use all the pollen from a male flower for each female flower.

After the male flower is in hand and prepared, untape the female flower by removing the tips of the petals with the tape. Then brush pollen from the male onto the stigma of the female. Retape the end of the female flower closed with a new piece of tape.

Some bees will bite into taped flowers before or after the hand-pollination. Check for holes in the taped flowers or trapped bees.

Cross information is initially recorded on a tag on the stem of the female flower. Once the fruit has developed, make sure to write the information on the fruit before it is harvested. Tags tend to get lost during harvest.

Hybrid Seed Industry

There is enough hybrid vigor in squash and pumpkins so that hybrid varieties can be advantageous. Hybrids may be earlier than parent varieties and may show increased yield.

The hybrid seed industry is based largely on chemical sprays that inhibit the development of male flowers. The variety to be used as the female is planted in rows that are sprayed. For every so many rows of the female variety, there is a row of the desired male variety, which isn't sprayed. A more laborious method is to remove the male flowers from the female variety by hand and provide rows of the male variety as pollinator.

Genetics and Breeding

Although squash and pumpkins and many other cucurbits clearly are basically outbreeders, they seem to have adapted secondarily to being inbreeders.

Whatever the cause, squash and pumpkins have the anatomy of outbreeders but seem to have some of the genetic characteristics of inbreeders. Inbreeding depression is not usually a problem in working with them. Extensive inbreeding is used in developing most squash and pumpkin varieties. After an initial cross, desired offspring are selected and inbred, often for a number of generations. In other words, the breeding approach is very much as it is with peas and other natural inbreeders, except that you do the inbreeding by hand-pollination.

Most breeding of squash and pumpkins is based on the general genetic principles, not on specific information about the genes involved. Not many specific genes have been identified. A few of those that have been follow (see Whitaker and Robinson for more).

Some of the following genes that affect

skin color have been identified in *C. pepo*: *B* (bicolored fruit, deep yellow), *C* (colored fruit, green), *l* (lightens fruit color), *r* (recessive white), *W* (dominant white), *Y* (yellow), and *St* (longitudinal stripes). Two other genes affecting fruit skin color in *C. maxima* are *bl* (blue at maturity) and *Rd* (red at maturity). In *C. pepo, wf* is associated with white flesh.

In *C. pepo, Di, Hr,* and *Wt* are associated with disk-shaped fruit, hard rind, and warty skin, respectively. *Bi* and *cu* affect the amount of cucurbitacin, and thus the bitterness of the fruit.

In both *C. pepo* and *C. maxima* the gene *Bu* is associated with bush growth form.

Harvesting, Processing, and Storage

Squash and pumpkins must be allowed to mature fully if they are to be used for seed. In addition, there is an after-ripening process. Seed quality and viability are improved if the harvested fruit is stored a month or more before the seed is extracted. Most home gardeners remove the seed whenever they eat the squash, as long as it has had at least a month for after-ripening.

Small amounts of seeds can be put in water and worked with the hands to remove flesh and debris. The seeds are then spread in a single layer on a tray and allowed to dry. (They will stick to paper.) It's not always easy to get the seeds clean using this method, however, and it's a lot of work for more than a few seeds.

Alan Kapuler's method for processing cucurbit seeds yields cleaner seeds with higher germination rates and is suitable for small or larger amounts of hand-processed seeds. Alan opens the fruit and puts the seeds and pulp into buckets of water. Both float. He leaves them in the water for one to three days — until the seeds sink. It is important to look at the seeds every day and catch them immediately after they take up enough water to sink. At this point the good seeds with large kernels inside sink, and they are easily separated from both the debris and the seeds with poor germ.

After separating and rinsing the seeds, Alan *immediately* puts the seeds on trays in his food dehydrator and dries them. If you clean seeds using soaking, it is important to dry them promptly and rapidly.

References

Thomas W. Whitaker and R. W. Robinson, "Squash Breeding," in Bassett, *Breeding Vegetable Crops,* pages 209–242; Ashworth, *Seed to Seed,* pages 114–120; Carol Deppe, "Custom-Crafted Vegetables I & II" (Squash seed saving and breeding.) *National Gardening Magazine,* June/July & August, 1998.

Corn (Maize)

Sweet corn and the various field corns — flour, flint, and dent — are all *Zea mays.* So are popcorn and the ornamental corns. The kind and arrangement of starches in the endosperm of the seed determine the basic kind of corn. Flint corn seed has a lot of flinty endosperm and just a little of the floury kind. Flour corn has mostly floury endosperm with just a little flint. Dent corn has a good bit of both types of endosperm. You can tell which class a variety is by cutting open a few of its seeds with wire cutters (longitudinally) and looking. The flinty endosperm looks like flint or glass and is either yellow or white. The floury endosperm is soft, grainy, and white, like chalk.

Sweet corn carries one or more genes that alter starch synthesis, causing the kernels to be sweeter when green but shriveled when dried. The wrinkled seeds have a flinty-appearing endosperm, but its composition is different from that in the flint types.

Breeding System

Corn is an outbreeder. It is monoecious. Male flowers, called tassels, emerge from the top of the plant. Female flowers, called ear shoots initially, emerge from some of the leaf axils on the plant. Once fertilized, they develop into the ears.

Each silk in an ear shoot or ear is actually the very long stigma of an individual flower. The silk is receptive over its entire surface, so if a pollen grain falls anywhere on it, it germinates, grows up the silk, fertilizes the egg, and produces a single kernel. If only a few such fertilizations happen, the corn ear will be mostly blank and have just a few seeds. Corn is self-compatible.

The pollen of most varieties begins to dehisce a few days before any of the ear shoots are ready. The silk is receptive from when it first appears to until it dries up. The pollen is very sensitive to heat and to drying out. It is mostly shed in the morning as soon as it warms up and the humidity drops. Corn pollen is short-lived. It usually lives from minutes to a few hours.

Isolation

Corn pollen is light and is blown long distances by the wind. In areas where corn is common, the air is full of pollen from the fields and gardens in the region. The isolation distance recommended by George is "up to one kilometer." Various factors may reduce the amount of isolation you need.

You may not need as much isolation for a very early variety if it flowers before most of the corn in your area. You may not need as much isolation for a white variety if most of the corn in your area is yellow. Pollen from yellow corn will produce yellowish kernels if it fertilizes white corn, and you can cull such kernels out by hand. You also can reduce the isolation distance if you have a large enough patch so you can save seed from only the plants in the middle. Most home gardeners have too many corn-growing neighbors for any of these methods to be reliable, and they save corn seed by hand-pollinating.

Inbreeding Depression

Corn displays extreme inbreeding depression.

Hand-Pollination and Crossing

Ashworth gives an excellent description of the process in *Seed to Seed,* which also includes photographs of every stage. Basically it involves bagging tassels to collect pollen; covering ears before the silks appear, to prevent contamination; and doing the pollination.

Corn ear shoot bags and corn tassel bags may be purchased from the companies listed in Appendix E. When I first started doing hand-pollinations, I substituted envelopes for ear shoot bags and ordinary grocery bags for tassel bags. But it doesn't usually rain in the summer here. The professional shoot and tassel bags are water resistant.

To bag the ear shoots, you have to learn to recognize them before the silks emerge. Where they develop on the plant (at what node) and what they look like depends on the variety.

To bag an ear shoot, grasp the leaf whose axil it is hiding in and strip it down and off the plant. Then pull the ear shoot bag down over the shoot and firmly into the crevice between the shoot and the stalk. You may need to gently slice the leaf that goes around the shoot and the stalk to be able to pull the bag down far enough to secure it.

You can bag plants over a few days without pollinating them. If you accumulate bagged plants over several days, you will need to cut off the silks occasionally every two or three days so they don't emerge from the bag.

The day before you plan to pollinate, cut off the tip of the ear shoot to expose the silks or a fresh area of silk. The following day, this area will be a brush of dense silks of ideal length for pollinating. (If you bag the ear shoots the day before you plan to use them, you cut off the tips at the same time.)

Obtain pollen by bagging tassels the day before you do your hand-pollinations. Tassel bags are much larger than ear shoot bags. Give the tassel a good shake to get rid of any foreign pollen that may have landed on it. Then position the bag over the tassel and fold the bottom of the bag up so that the bag is tight around the tassel. Staple the bag thus, or attach it with paper clips. In the late morning (or later) of the next day, give the tassel a shake to free as much pollen as possible before removing the bag.

To hand-pollinate, dust pollen onto the brush of silks on the ear shoot. Then use a tassel bag to cover the ear.

Cover the ear loosely enough so that it will be able to grow without being constricted by the bag. This is necessary initially to protect the ear shoot from contamination. Later, what matters is that you've usually written the cross information on that bag, so you don't want it wandering away from the ear. Pull the tassel bag down over the shoot, then fold the lower ends around the stalk and staple or clip them there.

When I'm doing hand-pollinations involving crosses, I plant the corn rows at least 3 feet apart. Otherwise, it's too hard to get into the patch without jostling the plants, causing them to release clouds of pollen, and increasing the probability of contamination. I also walk through the patch with care to avoid bumping plants and making them release pollen.

Corn pollen does not store easily. You can store it for about a week, though, if you collect it soon after it starts to be shed for the day and refrigerate it under conditions of high relative humidity (such as with a damp paper towel in the container with the pollen).

Hybrid Seed Industry

Corn displays extreme inbreeding depression and extreme hybrid vigor. There are many advantages to using hybrid varieties.

There are two main methods for producing hybrid corn commercially: using male-sterile cytoplasm and detasseling. The detasseling method is usable with any variety, not just specialized ones with special cytoplasm. It also can be used on any scale. This is the method I describe here (after the description in George's *Vegetable Seed Production*).

To produce a hybrid, plant alternating rows of the two parents. Usually you can plant two to four rows of the variety you're using as the female for each row of the variety used as the male, depending on how much pollen the male variety produces and how efficiently it achieves pollination under your conditions.

When tassels of the male variety begin to emerge, remove them before they open and shed any pollen. This is called detasseling. You must notice and remove the tassels on tillers as well as those on the main plant. "It may be necessary to de-tassel the crop up to seven times, at two day intervals, to ensure that it is complete," George says. Home gardeners and farmers can easily produce their own hybrid corn using this method.

If you are less than a kilometer from other corn patches and still want to produce your own hybrid corn, plant extra rows of the variety you want as the male parent around the whole patch to saturate the air with the desired pollen and overwhelm any contaminating pollen in the air. Under these conditions you normally hand-pollinate to maintain pure varieties. But you won't be saving seed from the hybrids, so occasional contaminants won't matter.

Genetics and Breeding

Corn genetics has been studied extensively. Coe and Neuffer's 112-page chapter in *Corn and Corn Improvement* describes genes and genetic maps. So much variability is present in corn and so many genes are known to be involved in every trait that I am going to limit my discussion to just a few traits of the seed itself.

The appearance of the kernel depends in part on the chemical composition of the endosperm. The endosperm is triploid tissue. In the fertilization of corn, the pollen tube grows down the silk and contributes two sperm nuclei to the egg. One sperm nucleus unites with the egg to become the diploid zygote, which develops further to become a baby plant inside the seed (the germ). The other sperm nucleus unites with two polar nuclei from the egg to become a triploid. That triploid nucleus divides and gives rise to the endosperm, the storage material that constitutes the rest of the interior of the seed.

The endosperm is normally yellow or white. The outer layer of the endosperm, the aleurone, is often colored. At least fifteen genes are known to affect color in the aleurone alone. The very outer layer of the seed is called the pericarp and is maternal tissue.

The characteristics of the pericarp depend on the genotype of the mother plant, not the seed. Characteristics of the aleurone and endosperm depend on the genes of the seed. But the aleurone and endosperm have two copies of egg genes and one copy of pollen genes instead of one of each.

The *Su* locus is the major gene responsible for all the early varieties of sweet corn. Sweet corns are *susu;* flint and other field corns are *SuSu*. Just that single gene is enough to cause the greater sweetness and wrinkled seed. However, many other genes are involved as well. Most sweet corns have thin, tender pericarps, for example, and other characteristics of texture and flavor.

The other major gene that can cause the sweet phenotype is *sh2* (shrunken). It is a different gene, but the recessive condition also causes sweet flavor and wrinkled kernels. Both the sweetness and the wrinkling are more extreme in *sh2* varieties, enough so that there is more of a problem with seed viability and shattering. The varieties listed as supersweet are usually *sh2* types.

Since both sweet genes are recessive, two varieties that are of different types must be isolated from each other if they are to produce sweet kernels. They differ by two loci, one variety being *susu Sh-2Sh-2* and the other being *SuSu sh-2sh-2*. If the former is pollinated by the latter, for example, the endosperm will be *susuSu Sh-2Sh-2sh-2*. It will have at least one dominant gene at each of the two loci and will be of the field type.

If you have blocks of the two different types of sweet corn near each other, the kernels pollinated by other plants within the variety will be sweet, but those pollinated by the other variety will develop faster and be hard and tough. Likewise, any field corn near a patch of either sweet corn variety will give rise to sweet corn ears that have some hard, nonsweet kernels in them.

Breeders now have other genes and combinations that can be used to generate the sweet phenotype as well. There are also modifiers of the other genes. One especially useful one is *se*, sugary enhancer. This recessive gene, in the presence of *susu*, increases the sugar content and delays the formation of starch. It is especially useful because it gives us supersweet varieties that don't have as many seed viability problems as those generated using the *sh-2* gene.

A number of different genes can cause the floury phenotype. When flint or sweet corn is pollinated by a flour corn, the kernels are usually of the flour type.

Many dozens of genes are involved in the color of the seed. I'll mention only the practical matter that white varieties pollinated by any colored variety produce colored kernels and that black varieties crossed with any other color variety produce black or blackish kernels.

Harvesting, Processing, and Storage

Corn to be saved for seed is allowed to dry on the plants for as long as possible. Ears are then removed and stored in a dry place until there is time to deal with them. Home gardeners usually shell out seed corn by hand.

It's very common to have insects of various kinds in homegrown corn seed, so common that treatment of some sort to kill insects and insect eggs is often essential. I dry my corn seed to "extra-dry" and then freeze it, at least temporarily, as described in Chapter 21.

References

George, *Vegetable Seed Production,* pages 288–293; Ashworth, *Seed to Seed,* pages 187–196; Karl Kaukis and David W. Davis, "Sweet Corn Breeding," in Bassett, *Breeding Vegetable Crops,* pages 75–584; W. A. Russell and A. R. Hallauer, "Corn," in Fehr and Hadley, *Hybridization of Crop Plants,* pages 299–312; E. H. Coe, Jr., and M. G. Neuffer, "The Genetics of Corn," in Sprague, *Corn and Corn Improvement,* pages 111–223, Carol Deppe, "Parching Corn," *National Gardening,* June 1997.

Technical Aspects of Hand-Pollination and Performing Crosses; Overcoming Incompatibility Barriers

This appendix covers the technical aspects of making crosses and hand-pollinations.

Making Crosses

There are two basic aspects to doing a cross. One is physically doing it or arranging for it to happen. The other is recognizing it when it has happened. Usually we need to do both. "Physically doing it" means collecting pollen and pollinating by hand; "arranging for it to happen" means getting the bees (or wind or other pollen vectors) to do the cross for us. The second aspect, "recognizing when a cross has happened" means being able to distinguish the hybrids resulting from the cross from any self-pollinations, or selfings, of the maternal parent.

Sometimes we carefully transfer pollen from one flower to another after emasculating the recipient. (Details on how to collect and transfer pollen and emasculate flowers depend on the species and are covered in Appendix A.) Sometimes, however, the flowers are too small or otherwise too difficult to emasculate, so we do a *fertile* × *fertile* cross, which means that we pollinate the female with pollen of our choice without emasculating it first. This usually means that the resulting seed is a mixture of two kinds — that resulting from fertilization by the pollen that we transferred deliberately and that resulting from fertilization by pollen from the female flower itself. In other words, the seed may represent both the desired cross and undesired selfings. Thus we must depend on our ability to distinguish the hybrids from the selfings.

Even when we do deliberate hand-pollination after emasculation, there are often some contaminating selfings. For that reason crosses are often done in the direction that permits distinguishing the hybrids from the selfings. Here *direction* refers to which variety is used as the female parent. You can cross a short pea variety with a tall one in one of two ways: short female × tall male, or tall female × short male.

In the cross short female × tall male, we emasculate flowers of a short plant and fertilize them with pollen from a tall plant. Since tall is dominant to short, the hybrids — that is, the successful crosses — will be tall. Any accidental selfings will simply be plants of the maternal variety, which will be short. When we grow up the offspring, we can easily distinguish which are hybrids and which are selfings.

The reciprocal cross — the cross done in the opposite direction — is tall female × short male. Again, we expect the hybrids to be tall. And again, we expect accidental selfings to be of the maternal variety. But in this case the maternal variety is tall, so the accidental selfings are tall, and both the successful crosses and the selfings look the same. There is no way to distinguish which is which.

Most geneticists would do a cross of a tall variety and a short variety in the direction that involved using the short variety as the female unless there was some strong reason to do otherwise. That way, the accidental selfings could be distinguished from the crosses and eliminated.

If two varieties look different, the F_1 hybrid can't resemble both of them. If we do the cross in both directions, in at least one direction we should be able to distinguish the hybrid from the maternal parent. Often we can distinguish it from both parents. Anytime we grow up an F_1, we should grow some of each parent variety along with it and compare it to both.

The genes or traits that are most useful for distinguishing hybrids from selfings are called *marker genes, marker traits,* or *markers.* Those that show up in the seed, the seedling, or the young plant are the most useful, because we can discard the mistakes at an early stage. For example, round-podded peas when crossed with wrinkle-podded peas produce a round-podded seed. If the wrinkled variety is used as the maternal parent, the mistakes are obvious. If a grain of pollen from the mother has fertilized one of the eggs that produced a seed, there will be a wrinkled seed in the pod among the round ones.

In most cases tall plants, when crossed with short ones, produce tall or intermediate hybrids. Determinate (bush) plants crossed with indeterminate ones (vine types) usually produce indeterminate hybrids. Plants with purple pigment in the leaves or stems crossed with those without any purple pigment usually produce hybrids that are purple, purplish, or pink. Plants with smooth-edged leaves when crossed with plants with dissected leaves usually produce hybrids with leaves that are intermediate in shape. These characteristics are all distinguishable in young plants or even seedlings.

In some cases the two parents are so similar in appearance that we do not expect to be able to distinguish the hybrid from either parent. In other cases we want to do a cross in a given direction because we want to preserve the cytoplasm of a particular parent. In these cases we emasculate accurately and make exactly the cross we want if at all possible.

Even when a cross is made exactly and properly, the progeny may derive from only the mother. This phenomenon is called *apomixis*.

Apomixis is an aberrant form of meiosis in which seed is produced without any contribution of genetic material from pollen, which means that this genetic material is excluded from the zygote, and hence the next generation. What matters to us is that apomixis interferes with our ability to do crosses, because the plant doesn't use the pollen we give it.

Some species use apomixis as a regular form of reproduction. In many dandelions, for example, the seed is from ordinary-looking flowers but isn't derived via fertilization. In dandelions, though, as in many plants, the extent of apomixis depends on weather conditions, variety, and/or part of the season.

In most plants apomixis represents the rare accident instead of the ordinary form of reproduction. In others it happens fairly commonly, but normal sexual derivation of seed is still the norm. Even where apomixis is rare, we may see it when we do crosses — especially if we have excluded all ordinary pollen and provided only pollen that is somewhat incompatible (such as from a different but related species). Whenever we do a cross, especially a wide cross or a selfing of a plant with an incompatibility system, we need to realize that the rare apomictic events may be more common or easier to achieve than the cross.

We always record crosses to distinguish which parent is maternal and always compare presumptive F_1 progeny with the two parents. When all the progeny of an apparent cross resemble only the maternal variety and show no characteristics of the paternal variety, an accidental selfing or apomixis may be responsible.

When working with outbreeding species, it is often easier to arrange for a cross to happen and then screen the offspring for the hybrids than to hand-pollinate. I often just grow or transplant two plants, one of each variety ("partners," I call them), in a spot where they are mostly by themselves. I plant them very close together — even bend them so that their flowering branches intertwine. That way, as insects pollinate, they will often be going from one plant to the other. They do the cross for me.

Remember that just putting two plants together and letting bees, wind, or whatever do the crossing works only with outbreeding plants. With strong inbreeders the only way to get a cross is to transfer the pollen yourself.

To make a cross between two different varieties, we need to have them both flowering at the same time, or we need to be able to store pollen. If I want to cross an early corn and a late one, for example, I plant the late one as early as possible in the spring and the early one later. I often use the variable spacing I described earlier in Chapter 5. Each variety to be crossed is planted in a row with spacing ranging from 6 inches at one end to more than is necessary at the other. The purpose in this case is to spread out the period during which pollen and plants ready to be pollinated are available. The more crowded the plants are, the later they mature. With a range of spacings I can have plants pollinating and being ready for pollination over a very much longer time period than with any single spacing. This usually gives me appropriate material for crossing, even when the maturity dates of the varieties are significantly different.

You can speed up corn pollen by several days by cutting the plant down and standing it in the shade somewhere, or even by cutting the tassel off. Many kinds of flowers will shed pollen somewhat earlier if you remove them and allow them to start drying out.

Some kinds of pollen can be stored for long periods, and some can't. Where information on storing pollen is available, I include it in Appendix A. In most cases, however, no information is available. When I want to store pollen for just a few days, I usually put freshly shed pollen or even whole flowers that have just shed pollen in a vial or plastic bag to prevent them from drying out. Then I put the container in the refrigerator. Most, though not all, pollen will keep for a few days to a week or two in this fashion.

When I want to store pollen for longer periods, I usually desiccate and freeze it. I put the pollen (or anthers with pollen, but without the filaments) in gelatin capsules (the kind used for drugs and vitamins — many but not all pharmacies sell empty gela-

tin capsules). I put the open gelatin capsules on a layer of silica gel, seal the jar, and leave it at room temperature for a few hours. Then I open the jar, close the gelatin capsules, put them in a vial with more silica gel, cap it, and put it in the freezer.

To do a cross with frozen pollen, I open the vial and remove one capsule. Then I open the gelatin capsule and leave it lying around at room temperature for an hour or two so that the pollen or anther can rehydrate before I use it.

Hand-pollination

For many uncommon vegetables and for wild species, you have to figure out the breeding system and/or how to do crosses yourself, as there isn't any information available. Even with common vegetables, the available information is only a first approximation. Details depend on the exact variety and the weather. You'll have to fine-tune the techniques to your varieties and conditions. So whether you are working with a rare vegetable or a common one, you need to be able to figure out breeding patterns and crossing techniques for yourself.

With peas, for example, the line drawings in Appendix A show that the proper stage for pollination is when the flower-petal part of the bud extends slightly beyond the sepals. This is long before the bud is full size —very long before it opens — and just before the pollen is released. Since the stigma is already receptive then, the flower is opened, emasculated, and pollinated all at once.

In early spring, with large-flowered varieties, I can make pea pollinations very easily, because the buds become quite large before the pollen is shed; the line drawings are more or less accurate. But later in the year (or whenever there is a sudden warm spell) the situation changes. Pollen is shed much earlier, when the bud is much tinier and at a stage when the petals don't yet extend beyond the sepals. The dwarf varieties I've worked with have smaller flowers, and many also shed pollen at an earlier stage.

There are six steps to hand-pollination: (1) obtaining pollen; (2) identifying flowers or buds at the right stage to be pollinated; (3) emasculation where appropriate; (4) transferring pollen; (5) protecting the fertilized female flower (from foreign pollen or drying out) where appropriate by covering or sealing it or using other methods; (6) eliminating competition from other flowers and fruits where appropriate.

First of all, it is important to have the appropriate conditions. There is no point in making hand-pollinations during weather that is too hot or too cold for pollen germination or fruit set, for example. Watch the species and varieties you are interested in and become familiar enough with them that you know when the flowers normally open and when the pollen is normally released (time of day and weather), as well as what kind of weather conditions are required for ordinary pod or fruit set. Among the conditions that may matter are stage of the plant, part of the season, weather during the day of pollination and the next few days, and time of day.

It's usually easiest to get crosses to succeed if they are early in the flowering season but not necessarily the very first flowers. With peas, for example, I wait until the weather warms up enough so that naturally fertilized pods will set. Then I do the crosses. Early flowers (weather permitting) are much more likely to take crosses and will produce more seed than flowers higher on the plant. I don't bother doing any crosses during hot spells; even natural pollinations don't set pods very well then.

With peas I eliminate any other flowers or pods at the node where I am making a cross so as to eliminate competition for nutrients. I usually eliminate noncross pods that are lower than my cross or crosses as well. With some species it's helpful or even essential to remove all fruits that formed before the one to be crossed. In other cases eliminating competition isn't necessary.

Most pollen becomes viable just shortly before the anther dehisces (sheds it). In general, it is normally maximally fertile and viable right after it is shed. Some pollen continues to be viable for several days. Other pollen is dead within a few hours. Whenever possible, I use pollen that has been shed that day. Freshly shed pollen is usually moist-looking and more brightly colored than old pollen. Once you've looked at several flowers of a variety and dissected a few buds at various stages, you'll learn to recognize what fresh pollen looks like.

The information in Appendix A tells you when various flowers dehisce (just after the flower opens, at a particular bud stage, and so on). In reality, though, dehiscence is partly determined by the plant's maturity and partly by the weather. Pollen frequently doesn't shed in cold or wet weather, even if the bud or flower is at the right stage. In such cases the bud or flower usually continues to grow and the pollen sheds at a later stage — usually very promptly within hours of when it warms up or dries out.

For similar reasons most flowers don't shed pollen early in the morning; they shed after it warms up and the fog or dew burns off. As well, most flowers don't shed on rainy days. For some species crosses "take" much better if done at a particular time of day. For most it doesn't matter much as long as the time and weather permit pollen shedding.

You can see when pollen has been shed. Just touch an anther gently and see whether your finger comes away colored. I also pay attention to bees and other pollinators. If they are out working my flowers, it's because pollen or nectar is available.

Bees and other insects remove pollen, so you have to get there first. With brassicas, for example, I often pick the flowers I want as male parents in the morning before the pollen is shed and bring them inside. By a few hours later, the flowers outdoors are releasing pollen, and the bees are out grabbing it anther by anther almost as fast as it is released. The anthers dehisce individually over a period of time, so it's not easy to find pollen once the insects become fully active. In addition, insects, pollinators or not, may have contaminated it.

Books for professional plant breeders usually show a variety of tools for transferring pol-

len. A fine-tipped brush and a jar of 95 percent ethyl alcohol for sterilizing things are common. You can obtain this alcohol in denatured form from a drugstore, but I don't use it. I sometimes use rubbing alcohol (70 percent isopropyl alcohol) to clean my hands and forceps when I have contaminated them with unwanted pollen or am going between varieties. I haven't checked rigorously to see whether the rubbing alcohol kills pollen; I depend as much on using ample paper towels to dry my hands afterward as on the alcohol. Most of the time I don't even use the towels or alcohol. I work with only one variety at a time and go in and wash my hands and forceps in between.

With big flowers I just pick the flower that I want to use as the male, strip away anything in the way, and touch the anthers with loose pollen to the stigma of the female, using the rest of the flower as the handle. Alternatively, I pick individual anthers with a pair of forceps and use them.

My hand-pollinating tools include a magnifying glass, a scalpel with disposable blades (the fine-pointed type), laboratory forceps, flat toothpicks, rubbing alcohol and paper towels (sometimes), a Magic Marker, and little labels.

I use the scalpel and forceps for dissecting tiny flowers when I'm trying to figure out how to work with them. Flat toothpicks are great for transferring pollen when the whole male flower or stamen isn't appropriate. I use a different one for each cross and throw them away. The forceps are my major tool. I use the kind that have fine-pointed tips and that open when you press them and close when you release them, not the reverse. You can purchase tiny labels with strings in an office supply store. For peas, I use very tiny ones, used for pricing jewelry. Forceps, scalpels, and other hand-pollinating supplies are available from Meristem Plants (see Appendix E). Ordinary tweezers and razor blades will do if the material isn't too small.

Identifying female flowers at the right stage to be receptive and then emasculating them can be difficult or easy, depending on the species. Emasculation is not always necessary. The cucurbits have separate male and female flowers; spinach is dioecious — it has male and female flowers on separate plants. Even with perfect flowers, you may not need to emasculate if the plant has a relatively strong incompatibility system or if you're getting in there before self-pollen has shed and you expect to be able to distinguish hybrids from selfings.

When you do need to emasculate, you have to do it before the pollen is released and when the anthers are immature enough that pollen is not released by handling. When possible, try to find a stage at which the stigma is receptive and the anthers aren't, so that you can emasculate and pollinate at the same time. This is easy with species such as peas, where the stigma becomes receptive long before the anthers finish maturing. With species whose male parts mature at the same time as or before their female parts, you need to emasculate at one stage, mark the flower so you can find it again, and then pollinate at a later stage.

Pollen shedding is obvious, so you can easily figure out when to emasculate. Usually you want to emasculate as late as possible so that the bud or flower is as big as possible and easy to handle. It isn't as obvious when the stigma is receptive. With many flowers it seems to become receptive about as the style reaches full size, but it can be earlier or later. If you are working the methods out yourself, start by figuring out the latest stage at which you can emasculate, then try several pollinations of emasculated buds at that stage and see if they work.

After doing a cross, you may need to cover the flower to prevent contamination by foreign pollen. In some cases it is simply to keep the pollinated stigma and style from drying out, because the hand-pollination has destroyed all or part of the flower that usually protects them. Covering the crossed flower isn't always necessary.

With inbreeders like peas, for example, which are not worked by or attractive to insects, there is not much danger of contamination. Covering the crossed flower is not necessary. In other cases, if the petals are removed, there is nowhere for the normal pollinators to land, or they may not notice the "flower" without the petals. So removing the petals may be enough to exclude pollinators. In yet other cases the pollen attracts the pollinators, and if there are no anthers, they won't land on the flower.

Watching the pollinators at work can give you ideas for how to do crosses with a given species, as well as for shortcuts that might be suitable for your conditions and pollinators. Appendix A includes information about covering flowers for specific vegetables. You may be able to get away with less or have to do more depending on your weather and your pollinators.

Overcoming Incompatibility Barriers

Up until now I've been talking about ordinary crosses that take well spontaneously — compatible crosses. But often we want to do crosses that are incompatible. There are two basic kinds of incompatibility. First, there is incompatibility caused by a genetic incompatibility system. This kind is designed to prevent inbreeding. In plants such as many of the brassicas, self-pollination or pollination by plants with the same incompatibility gene(s) or genotype doesn't take.

The second kind of incompatibility is normally associated with preventing crossing between different species. In some cases there are few or no barriers between making crosses between different species in the same genera or even different genera in the same family. The biological barrier between the species in nature may arise from the fact that they flower at different times, have different pollinators, are located in different ecological zones, or are separated geographically, not from physical inability to cross. In many cases, though, crosses between species don't "go" spontaneously, even once a person has done the task of pollinating.

The same kind of tricks are used to get around incompatibility barriers of both types. Usually, when someone gets a rare cross to go, he or she reports the trick used, but in most cases no one knows for sure whether it was the trick that caused the suc-cess. In other words, doing a difficult cross is usually a matter of just trying things.

Crosses will often go in one direction but not the other. When doing wide crosses, it's especially important to try each species as the female and as the male. Even in cases where both crosses can be obtained, the results will be entirely different, since they set a hybrid genome down in the middle of two completely different cytoplasms. Sometimes only one of the combinations is viable. Where both are, there are often differences in appearance, growth rate, vigor, or other properties. And there is no way of predicting ahead of time which combination will be most useful or interesting, even if both are possible. You just have to try them and see.

Different individual varieties and plants may differ in their ability to participate in wide crosses. If I try a cross between a wild perennial mustard plant and 'Green Wave', for example, and it doesn't go, that doesn't necessarily mean I have to start trying the more laborious forcing methods. The cross might go readily between the wild plant and a different variety of mustard, even of the same species.

When trying an incompatible cross, first try to do the cross normally, but with two modifications. First, do the cross as early as possible after the female becomes receptive, to give slow-growing pollen as long as possible to grow before the flower is dropped. Pollinate the flower in the bud stage in cases where the stigma is receptive then.

Second, make sure to exclude all possible sources of pollen other than what you provide. In many cases the incompatible pollen does germinate and start to grow down the style, but its germination and/or growth is much slower than for compatible pollen. Under normal circumstances, where compatible pollen is present, the incompatible pollen is beaten out by the faster-growing competition. If you have excluded all compatible pollen, the cross may go.

"All compatible pollen" includes that of wild relatives, too. When you are doing an ordinary cross, you often don't need to worry about contamination with pollen from the field down the street. But if you are doing a difficult cross in which you would be happy to get just one successful seed from your work, the rare pollinations from far-traveling bees may be much more frequent than the success rate for your cross.

In some cases where slow growth of incompatible pollen occurs, the flower normally dries up and falls off before the pollen grows all the way down the style and achieves fertilization. A number of tricks can help circumvent this problem. The simplest one is to pollinate as early as the bud is receptive. In many plants the stigma is receptive much earlier than the flower opens. By pollinating in the bud stage, slow-growing pollen has much longer to reach the ovaries.

In addition, in some plants, such as the brassicas, the incompatibility mechanism that prevents self-pollination develops late in the maturation of the flower. There is a stage at which the stigma is receptive but not yet capable of recognizing and excluding incompatible pollen.

Many other tricks depend on just giving the alien pollen more time to do its job. Use of hormones that promote fruit set may help, because part of what they do is to slow down flower drop. Some breeders cut off the stigma and upper part of the style of a flower and put the pollen directly on the cut stub. This shortens the distance the pollen tubes have to grow, and in some cases the pollen also may bypass exclusion mechanisms associated with the stigma. Just covering the pollinated flower to keep it moist may allow the flower to stay on long enough for a difficult cross to take.

Many incompatible crosses are much more likely to go at certain periods. Self-incompatible crosses may be rare among fertilizations early in flowering, for example, but common among the latest flowers. Incompatibility of both kinds is sometimes circumvented by weather, either artificial or deliberate. A plant may be willing to accept incompatible crosses late in its flowering period if no other pods or fruits have been allowed to remain on it. Heat or cold shocks can sometimes cause incompatible crosses to take. Self-incompatibility often depends on climate or weather.

In many cases it appears that something in the sticky juices associated with the stigmatic surface may be necessary for germination of the correct pollen or may prevent germination of the wrong stuff. Another idea is that the right pollen might have substances that are necessary for germination. Various tricks are based on these ideas.

Sometimes breeders transfer the stigmatic "goo" from a plant that the pollen would be compatible with to the stigma of the incompatible plant before pollination. I wonder whether it might not be more effective to pollinate a compatible plant, then shortly thereafter transfer the goo with the (hopefully) germinating pollen in it to the incompatible plant. Washing the goo off the incompatible plant also may be necessary.

What about pollinating a compatible plant with the desired pollen and letting it germinate, then cutting the stigma and style off the desired female, slicing off the stigma with the germinating pollen, and putting it directly on the stub? That way the pollen would be stimulated to germinate and start growth by the compatible stigma and not be excluded by any mechanisms, either positive or negative, associated with the stigma of the incompatible plant.

A simpler related method is to mix live pollen from the desired male parent with dead pollen from a compatible plant and use that to hand-pollinate. The idea is that enzymes or proteins on the surface of the compatible pollen may be necessary to set up a reaction in the stigmatic surface or break down the surface. With a huge amount of compatible pollen, the reaction is general enough that foreign pollen is no longer excluded.

You might be able to kill the pollen by heat-treating it, but heat also denatures many proteins. I think the first thing I would try would be to collect the pollen and let it age. Then I'd try some crosses with it to check the killing method.

Henry Wallbrunn, geneticist and orchid breeder, tells me that he often achieves difficult orchid crosses by mixing live pollen. Suppose he wants a cross of a variety-A female and a variety-B male, but that cross is difficult (or up till then impossible) to ob-

tain. He chooses a third variety, variety C, that crosses with variety A readily and whose hybrids with variety A he expects to be able to distinguish from those between A and B. Then he mixes pollen from varieties B and C and uses it to pollinate variety A.

Another possibility that I think might work in at least some cases is to slit open the ovary of the living plant with a razor and push in some pollen. A recent report describes fertilizing excised ovaries with pollen in vitro. Apparently growth of a pollen tube down the style isn't always necessary. My suggestion is a low-tech version.

Sometimes there is incompatibility between the developing embryo and the endosperm of the seed. Incompatible pollen germinates and achieves fertilization and an embryo results, but the endosperm — the seed's food supply — fails to develop or isn't compatible with and can't be used by the embryo. The high-tech solution to this problem is *embryo rescue*. The embryo is dissected out and cultured sterilely in a tissue culture medium that provides nutrients and substitutes for the endosperm. I wonder whether the same thing might be possible in a lower-tech way by transferring the hybrid embryo to a normal seed whose embryo has been removed. I haven't tried it.

In many cases where embryo rescue is required to achieve a cross, the reciprocal cross may be possible without it. The reciprocal cross has the same embryonic genome, but it has a different endosperm.

So far the major "trick" I have used to get wide crosses to go is to look at the weather forecast and do them during a period of ideal weather.

Technical Aspects of Hand-Pollination 331

USDA-ARS Plant Introduction Stations and Germplasm Collections; Using GRIN (Germplasm Resources Information Network)

Collections are listed alphabetically by collection site code followed by site name, address, and phone number. For information about how to use this list and how to approach and obtain material from the USDA system, see Chapter 4.

CLOVER. Clover Collection. USDA-ARS, Department of Agronomy, University of Kentucky, Lexington, KY 40506; (606) 257-5785.

COTTON. Cotton Collection. USDA-ARS-SPA, Route 5, Box 805, College Station, TX 77840; (409) 260-9209.

CR-BRW. National Clonal Germplasm Repository–Brownwood. USDA-ARS, Pecan Research, Route 2, Box 133, Summerville, TX 77879; (915) 646-0593

CR-BYR. National Clonal Germplasm Repository–Byron. USDA-ARS, Southeastern Fruit and Tree Nut Research Laboratory, P.O. Box 87, Byron, GA 31008; (912) 956-5656.

CR-COR. National Clonal Germplasm Repository–Corvallis. USDA-ARS, 33447 Peoria Road, Corvallis, OR 97333; (503) 757-4448.

CR-DAV. National Clonal Germoplasm Repository–Davis. USDA-ARS, Department of Pomology, University of California, Davis, CA 95616; (916) 752-6504

CR-GEN. National Clonal Germplasm Repository–Geneva. USDA-ARS, New York State Agricultural Experiment Station, Geneva, NY 14456-0462; (315) 787-2390.

CR-HIL. National Clonal Germplasm Repository–Hilo. USDA-ARS, c/o Beaumont Agricultural Research Center, 461 W. Lani-kaula Street, Hilo, HI 96720; (808) 959-5833.

CR-MAY. National Clonal Germplasm Repository–Mayaguez. USDA-ARS, Tropical Agricultural Research Station, P.O. Box 70, Mayaguez, PR 00709-0070; (809) 831-3435.

CR-MIA. National Clonal Germplasm Repository–Miami. USDA-ARS, Subtropical Horticultural Research Station, 13601 Old Cutler Road, Miami, FL 33158; (305) 238-9321.

CR-ORL. National Clonal Germplasm Repository–Orlando. USDA-ARS, Route 2, Box 375, Groveland, FL 32736; (904) 787-5078.

CR-RIV. National Clonal Germplasm Repository–Riverside. USDA-ARS, 1060 Pennsylvania Avenue, Riverside, CA 92507; (714) 787-4399.

DBMU. Database Management Unit. USDA-ARS-PSI-GSL-GRIN, Building 001, Room 130, BARC-West, Beltsville, MD 20705; (301) 344-3318.

FLAX. Flax Collection. USDA-ARS, Walster Hall, Room 206A, North Dakota State University, Fargo, ND 58105; (701) 237-8155.

GD. National Plant Germplasm Quarantine Laboratory. USDA-ARS, 11601 Old Pond Drive, Glenn Dale, MD 20769-9157; (301) 344-3003.

IR-1. Inter-Regional Potato Introduction Station. USDA-ARS, Peninsula Experiment Station, Sturgeon Bay, WI 54235; (414) 743-5406.

NA. National Arboretum. USDA-ARS, 3501 New York Avenue NE, Washington, DC 20002; (202) 475-4836.

NC-7. North Central Regional Plant Introduction Station. USDA-ARS, Iowa State University, Ames, IA 50011; (515) 292-6507.

NE-9. Northeastern Regional Plant Introduction Station. USDA-ARS, P.O. Box 462, New York State Agricultural Experiment Station, Geneva, NY 14456-0462; (315) 787-2244.

NPMC. National Plant Materials Center. USDA-SCS, Building 509, BARC-East, Beltsville, MD 20705; (301) 344-2175.

NSGC. National Small Grains Collection. USDA-ARS, Small Grains Germplasm Research Facility, P.O. Box 307, Aberdeen, ID 83210; (208) 397-4162.

NSSL. National Seed Storage Laboratory. USDA-ARS, Colorado State University, Fort Collins, CO 80523; (303) 484-0402.

PEANUT. Wild Peanut Collection. USDA-ARS, Plant Science Research Laboratory, 1301 N. Western, Stillwater, OK 74075; (405) 624-4124.

PIO. Plant Introduction Office. USDA-ARS-PSI-GSL, Building 001, Room 322, BARC-West, Beltsville, MD 20705; (301) 344-3328.

S-9. Southern Regional Plant Introduction Station. USDA-ARS, 1109 Experiment Street, Griffin, GA 30223-1797; (404) 228-7255.

SBMNL. Systematic Botany and Mycology Lab. USDA-ARS, Building 265, BARC-East, Beltsville, MD 20705; (301) 344-2681.

SOY-N. Soybean Collection–North. USDA-ARS, University of Illinois, W-321 Turner Hall, 1102 S. Goodwin Avenue, Urbana, IL 61801; (217) 244-4346.

SOY-S. Soybean Collection–South. USDA-ARS, Delta Branch Experiment Station, P.O. Box 197, Stoneville, MS 38776; (601) 686-9311.

TOBAC. Tobacco Collection. USDA-ARS, Crops Research Laboratory, P.O. Box 1555, Oxford, NC 27565-1555; (919) 693-5151.

W-6. Western Regional Plant Introduction Station. USDA-ARS, Room 59, Johnson Hall, Washington State University, Pullman, WA 99164-6402; (509) 335-1502.

The National Plant Germplasm System holdings and data are now all in a database called GRIN, the Germplasm Resources Information Network. You can access GRIN through the Internet to obtain information or request material. Here are the relevant addresses.

home page
www.ars-grin.gov/npgs/

holdings
www.ars-grin.gov/npgs/holdings.html

holdings by site
www.ars-grin.gov/npgs/stats/site.summary.stats

search
www.ars-grin.gov/npgs/searchgrin.html

accession queries
www.ars-grin.gov/npgs/acc/acc_queries.html

taxonomy index
www.ars-grin.gov/npgs/tax/index.html

taxonomy search
www.ars-grin.gov/npgs/tax/taxgenform.html

germplasm request
www.ars-grin.gov/npgs/orders.html

order form for germplasm
www.ars-grin.gov/npgs/order.html

Plant Variety Protection Office home page
www.ams.usda.gov/science/pvp.htm

PVP search
www.ars-grin.gov/cgi-bin/npgs/html/pvlist.p

Addresses of Seed Savers Exchanges, Seed Companies, and Organizations

The following is *not* a comprehensive list of worthy seed companies or organizations. It is only a list of those referred to in this book. It emphasizes seed companies in the Northwest, and those with special interests similar to mine. For the best available list of seed-saving organizations, seed companies, and other sources throughout the world, see Stephen Facciola's book *Cornucopia II*.

Abundant Life Seed Foundation no longer operates a seed company. The name 'Abundant Life Seeds' was bought by Territorial, but no longer distributes the old AL seed repertoire, which was burned up in a fire.

Adaptive Seeds. The Seeds of the Seed Ambassadors Project. Seed Ambassadors Project, P.O. Box 23, Drain OR 97435.

www.seedambassadors.org. Open-pollinated public-domain seed.

Allan, Ken. 61 S. Bartless Street, Kingston, ON K7K 1X3, Canada. allan@adan. kingston.net. Sweet potatoes, tall peas, organic open-pollinated vegetable seed.

Borman, Dan. See Meristem Plants.

Bountiful Gardens. 18001 Shafer Ranch Road, Willits, CA 95490; (707) 459-6410; Fax (707) 459-6410. Sole U.S. distributor for Chase Seeds of England. Open-pollinated.

Deep Diversity Seed was an incarnation of Peace Seeds in a period in which Alan Kapuler, owner of Peace Seeds, was Research Director for Seeds of Change. Now, as of this 2010 printing, there is no Deep Diversity, and

Alan Kapuler, retired from Seeds of Change, again sells his own breeding work and a collection of germplasm and potentially edible native plants under the name Peace Seeds.

Deppe, Carol. See Fertile Valley Seeds and www.caroldeppe.com for available seeds, news, other books, and further adventures.

Drowns, Glenn. See Sand Hill Preservation Center.

Fedco Seeds. P.O. Box 520, Waterville, ME 04903-0520; (207) 873-7333. (No telephone orders.) Seed for cold climates and short growing seasons.

Fertile Valley Seeds. 7263 NW Valley View Dr., Corvallis, OR 97330. Started 2010 to distribute my own breeding work as well as certain other worthy public-domain organic-adapted varieties. Website catalog only. www.fertilevalleyseeds.com.

Garden City Seeds. P.O. Box 204, Thorp, WA 98946; (509) 964-7016; Fax (800) 968-9210. www.gardencityseeds.com. Northern-acclimated, open-pollinated seeds; perennial vegetables.

High Mowing Organic Seeds. 813 Brook Road, Wolcott, VT 05680, 802-888-1800. www.highmowingseeds.com

J. L. Hudson, Seedsman. Star Route 2 Box 337, LaHonda, CA 94020. Open-pollinated; many rare varieties.

Johnny's Selected Seeds. Foss Hill Road, Albion, Maine 04910-9731; (207) 437-4301; Fax (800) 437-4290. www.johnnyseeds.com. Seeds for northern climates.

Kapuler, Alan. See Peace Seeds, Peace Seedlings, and Seeds of Change.

Meristem Plants and Plant Research Supplies for Cool Climates. Box 254, Deer Harbor, WA 98243. Open-pollinated seeds; perennial vegetables.

Native Seeds/SEARCH. 526 N. 4th Avenue, Tucson, AZ 85705; (520) 622-5561. Fax (520) 622-5591. Preservation of traditional Native American crops and wild relatives of U.S. Southwest and northwestern Mexico. Open-pollinated.

Nichols Garden Nursery: Herbs & Rare Seeds. 1190 North Pacific Highway, Albany, OR 97321-4598; (541) 928-9280; Fax (541) 967-8406. nichols@gardennursery.com. www.gardennursery.com. Seeds and plants. Many rare varieties.

Northern Groves. Four Wands Farm, 23818 Henderson Road, Corvallis, OR 97333; (541) 929-7152. More than 50 hardy bamboos, including a number of edible types.

Oregon Exotics Nursery. 1065 Messinger Road, Grants Pass, OR 97527; (541) 846-7578. Incredible collection of rare, recently collected perennial vegetables, tubers, and trees, including hardy citrus.

Peace Seeds: A Planetary Gene Pool Resource and Service. 2385 SE Thompson Street, Corvallis, OR 97333. www.peace-seeds.com Website catalog only. Alan Kapuler's seed company reflecting his public-domain organic-adapted plant breeding and current interests, which include collecting and domesticating wild food plants of the maritime Northwest. (See also, Peace Seedlings, the next generation.)

Peace Seedlings (Dylana Kapuler and Mario DiBenedetto) 2385 SE Thompson Street, Corvallis, OR 97333. www.peaceseedlings.cn

Peaceful Valley Farm Supply: Organic Growers Source Book and Catalog. P.O. Box 2209, Grass Valley, CA 95945; (530) 272-4769; Fax (530) 272-4794. www.groworganic.com Organic seeds. Organic growers supplies and equipment.

Perennial Vegetables. Seed companies that have especially good collections of perennial vegetables include Deep Diversity Seed, Garden City Seeds, Meristem Plants, Nichols Garden Nursery, Irish Eyes, Peters Seed & Research, Oregon Exotics, Perennial Vegetable Seed Co., Seeds of Change, and Wild Garden Seeds.

Perennial Vegetable Seed Co. P.O. Box 608, Belchertown, MA 01007. Perennial vegetables.

Peters, Tim. Tim Peters was working for Territorial Seed Co. at the time I wrote the first edition of this book. He subsequently out-migrated and started his own seed company, Peters Seed and Research.

Peters Seed and Research. 407 Maranatha Lane, Myrtle Creek, OR 97457; (541) 874-2615. The best collection of perennial grains I've ever seen, as well as many experimental and original varieties, and an excellent collection of Northwest heirlooms. Open-pollinated organic seed.

Pinetree Garden Seeds. Box 300, New Gloucester, ME 04260; (207) 926-3400; Fax (888) 527-3337. www.superseeds.com Focuses on gourmet quality and flavor for the home garden.

Plant breeding for the public domain. As of this 2010 printing, universities and most major seed companies now by and large patent or PVP-protect their varieties. The torch for breeding new public-domain open-access varieties is now being carried by certain freelance plant breeders and small seed companies such as: Adaptive Seeds, Fertile Valley Seeds, Peace Seeds, Peace Seedlings, Peters Seeds and Research, Sand Hill Preservation Center, Seeds of Change, and Wild Garden Seed.

Redwood City Seed Company. P.O. Box 361, Redwood City, CA 94064; (650) 325-7333. www.ecoseeds.com. Open-pollinated. Many rare Native American and other traditional varieties.

Richters. 357 Highway 47, Goodwood, ON L0C 1A0, Canada; (905) 640-6677; Fax (905) 640-6641. catalog@richters.com inquery@ richters.com. Many rare herbs and vegetables.

Ronniger's Potatoes. See Irish Eyes, Inc.

Salt Spring Seeds. P.O. Box 444, Ganges, Salt Spring Island, BC V8K 2W1, Canada; (250) 537-5269. Certified organic, open-pollinated seed. Huge selection of beans and grains.

Sand Hill Preservation Center. Heirloom seeds and poultry. 1878 230th Street, Calamus, IA 52729; (319) 246-2299. (No calls Sunday or Monday.) Open-pollinated vegetable seeds, especially vine crops and tomatoes. Also, many rare and traditional breeds of poultry.

Seeds Blüm. This company, founded by Jan Blüm, was one of the earliest promoters, rediscoverers, and purveyors of heirloom vegetables. Jan Blüm has retired from this phase of the seed business, and sent out her last Seeds Blüm catalog in 1998.

Seeds of Change. P.O. Box 15700, Santa Fe, NM 87506-5700; (888) 762-7333. www.seedsofchange.com. Certified organic open-pollinated vegetable, flower, and herb seed.

Seeds of Diversity Canada. PO Box 36 Station Q, Toronto, ON M4T 2LT, Canada; (905) 623-0353. www.seeds.ca/ This is the major seed savers exchange in Canada. Membership is $20–$35. Includes three publications a year and provides access to more than a thousand varieties.

Seed Savers Exchange. 3076 North Winn Road, Decorah, IA 52101-7776; (319) 382-5990; Fax 319-382-5872. Membership $35–$45. Includes three huge publications per year, the *Winter Yearbook* being the one that lists all the thousands of varieties offered by members.

Seed Savers Exchange in Canada. See Seeds of Diversity Canada.

Southern Exposure Seed Exchange. P.O. Box 460, Mineral, VA 23117; (540) 894-9481; Fax (540) 894-9481. www.southernexposure. com. Huge collection of traditional open-pollinated varieties adapted for the Southeast.

Territorial Seed Company. P.O. Box 157, Cottage Grove, OR 97424-0061; (541) 942-9547; Fax 541-942-9881. www.territorial-seed.com. Specializes in varieties for the Maritime Northwest (west of the Cascades). Distributes one catalog to the Maritime Northwest and another nationally.

Vegetable Garden Research Exchange, The. This publication, run by Ken Allan, provides an outlet for exchanging information among home-gardener researchers and plant explorers and breeders. Receive and exchange information on cultivar trials, culture methods comparisons, plant breeding, and much else. It's scientifically rigorous, and is worth much more than the cost of membership. For information, write Ken Allan, The Vegetable Garden Research Exchange, 61 South Bartlett Street, Kingston, ON K7K 1X3, Canada.

Wild Garden Seed. Shoulder to Shoulder Farm, P.O. Box 1509, Philomath, OR 97370; (541) 929-4068. Unusual salad greens, many farm-bred varieties, and naturalized, native, and semi-domesticated wild edible plants of the Maritime Northwest. Many segregating mixtures from which farm-adapted new varieties can be easily selected. Semi-wholesale. Minimum order $50.

Sources for Seed Saving, Plant Breeding, and Garden Research Supplies

Abundant Life Seed Foundation

Screens of various sizes for cleaning seed, cloth bags for seed, a hand-operated seed thresher, and self-sealing seed envelopes.

Bountiful Gardens

A Seed Saver's Kit, which includes silica gel for drying seed, data forms, and the book *Growing to Seed;* silica gel refills.

Meristem Plants and Plant Research Supplies for Cool Climates

One of the main objectives of this new company is to provide for the needs of amateur plant breeders and researchers. It carries an extensive line of seed saving, plant breeding, and plant research supplies, forceps of various kinds, ear shoot and pollen bags, and dissecting and standard microscopes.

Southern Exposure Seed Exchange

Silica gel, ear shoot bags, drawstring muslin bags (for excluding pollinators while allowing normal flower development), self-seal seed packets, light and heavy Zip-Loc bags, heat-seal pouches and equipment, parafilm, seed saver vials, corn sheller, various kinds of tapes, markers, and labels.

See Appendix D for addresses.

Statistical Predictions
and Actuality

This chart tells you how many plants you need to be 95 percent or 99 percent sure of obtaining at least one of the desired class, for classes of the given probabilities. (Adapted from C. North, *Plant Breeding and Genetics in Horticulture*.)

Probability of the Class	Number of Plants Needed	
	95 Percent	99 Percent
$\frac{1}{2}$	5	7
$\frac{1}{3}$	8	12
$\frac{1}{4}$	11	16
$\frac{1}{8}$	22	35
$\frac{1}{9}$	25	39
$\frac{1}{16}$	46	71
$\frac{1}{27}$	79	122
$\frac{1}{32}$	95	146
$\frac{1}{64}$	191	296

Glossary

accession   The material in a germplasm collection that results from a single collecting event — such as a collector buying beans from one basket in a market. The basket may contain just one variety or a mixture of varieties. A sample of beans from a basket in the next market would get a different accession number, even if they looked identical. They may or may not be the same variety. So an accession may contain more than one variety, and a variety may be represented in a collection by more than one accession number.

alleles   Different variations of one gene at one locus.

allopolyploid   *See* ploidy.

amphidiploid   A polyploid that originates through the combination of two different genomes such that it behaves like a new diploid species.

anther   *See* male parts of the flower.

anthesis   The opening of a flower.

apomixis   The process of producing seed without any union of gametes. Pollen may or may not be necessary to stimulate the process, but it doesn't contribute genetically. The seed produced may or may not be genetically identical to the mother, depending on the kind of apomixis involved.

asexual   Without sex.

autopolyploid   *See* ploidy.

backcross   To cross one of the progeny of a cross back to one of the parents. A backcross is such a cross.

bolt   To produce flower stalks.

bud pollination   A hand-pollination performed when the flower is in the bud stage. Often useful in overcoming incompatibility barriers.

bud sport   A mutant branch or section of a plant that arises when a mutation has occurred in the somatic tissue that gave rise to the bud.

calyx   The ring of sepals on a flower.

chimera   An individual whose cells are of two or more genotypes.

chromosome   The structural units containing the genes.

cleistogamy   The act of releasing pollen and self-fertilizing before the flower opens.

clone   To create genetically identical individuals. Clones are the individuals so created. Vegetative propagation of plants produces clones.

codominant   *See* dominant, codominant, and recessive.

colchicine   A cell-division poison used to double chromosome numbers.

corolla   The ring of petals on a flower.

cotyledons   The first leaves of the embryonic plant within the seed that are used as a food supply for the germinating embryo.

cross   A fertilization involving two different parents, or to make or cause such a fertilization. Opposite of self-pollination.

crossing over   A breaking and reunion that interchanges corresponding parts of homologous chromosomes. It allows genes on the same chromosome to recombine unless they are very close together.

cull   To remove inferior or undesirable individuals. Same as *to rogue* (remove the rogues).

cultivar   A cultivated variety. Professionals use the word *cultivar* rather than *variety*. Common use favors *variety*. In this book I usually use *variety,* but I also use the two words interchangeably.

cytoplasm   The rest of the contents of a cell other than the nucleus.

cytoplasmic inheritance   Inheritance associated with the cytoplasm instead of genes on chromosomes (and in the nucleus).

day-neutral   *See* photoperiod.

dehiscence of anthers   Pollen release.

determinate and indeterminate growth patterns   Indeterminate plants have shoots that remain vegetative at the apex; they produce flowers only subterminally. Determinate plants have shoots that grow vegetatively for a while but then end in flowers. This restricts overall size and often also produces a bushier form. Bush beans and bush tomatoes are determinate. Pole beans and vine tomatoes are indeterminate.

detrimental   Used to describe a gene, allele, or mutation that causes an impairment or disadvantage to the individual that carries it. Often used as a noun. Example: Recessive lethals and detrimentals are common in populations of plants, especially outbreeders.

dioecious   Varieties or species that have male and female flowers on separate plants.

diploid   *See* ploidy.

DNA   Deoxyribonucleic acid, the genetic material.

dominant, codominant, and recessive   If plants of genotype *Aa* look the same as those of genotype *AA*, we say the *A* allele is dominant to the *a* allele or that the *a* allele is recessive to the *A* one. If, however, plants of genotype *Aa* look different from both *AA*'s and *aa*'s, we say that the *A* and *a* alleles are codominant. Strictly speaking, the terms *dominant, codominant,* and *recessive* apply only to relationships between alleles of a gene, not to traits. That is, it's correct to say that *T* is dominant to *t* but not to say that tall is dominant to short. Tall might not be dominant to short in some other cross where different genes are involved. The use of the words to describe traits promotes sloppy thinking but is such a useful shorthand that it's almost impossible to avoid. I usually try to avoid it, but give in occasionally.

edge effects   Plants at the ends of rows or on the edge of a planting are often bigger, bushier, earlier, and/or more productive than the rest because they have more space, get more light, or have less competition for water or nutrients — in other words, because their environment is better. Edge effects often have to be considered when you're trying to evaluate the performance of individual plants or groups of plants.

egg   The female gamete that combines with a nucleus from a pollen grain to make a zygote.

emasculate a flower   To remove the male parts of a flower before pollen is shed.

embryo   The part of the seed that grows into the new plant after the seed germinates.

endosperm   A storage tissue in the seed that is used up by the embryo in the process of germination. *Endosperm characters* are traits associated with the endosperm of seeds.

expressivity   Not all genes or genotypes express themselves consistently. If a single genotype has a variable degree of expression within a given single environment, it's said to have *variable expressivity.*

$F_1, F_2, F_3, etc.$ The progeny of the specified generation after a cross. When two different varieties are crossed, the seed they produce is the F_1 generation. When it is planted, it grows into F_1 plants. A cross of two F_1 plants produces F_2 seed, and so on.

factor   A one-factor cross is a cross in which the two parents differ by one pair of alleles at one locus. (Example: *AA* X *aa.*) The number of factors in a cross is the number of pairs of alternatives. The cross *AABB* X *aabb* is a two-factor cross. The term *factor* dates from before people knew that the factors are really genes. *Two-gene cross* would be ambiguous (Would *AABB* X *aabb* be a two-gene cross or a four-gene one?), so *factor* is still used in this context.

female parts of the flower   The pistil is the whole female unit of the flower. The eggs, ovary, style, and stigma are parts of the pistil. The stigma is at the end of the pistil and is the special surface on which the pollen lands and germinates. The style is the structure that supports the stigma. The pollen tube grows down the style to reach the ovary. The ovary is at the base of the style and contains the eggs.

fertile X *fertile cross*   A cross made without emasculation of the flower, generally because the flowers are of a type that makes emasculation too difficult. In a successful fertile X fertile cross, only some of the offspring are the cross; the rest are the unavoidable self-pollinations.

fix   To fix a gene, allele, or trait in a population or variety is to achieve homozygosity for it in all individuals in the population so that the population is then pure breeding for it.

frankenfood   A derogatory reference to food derived from genetically engineered plants or animals. It is a contraction of *Frankenstein food,* recalling the artificially constructed monster in the novel *Frankenstein* by Mary W. Shelley. Frankenstein killed his creator.

gamete   The haploid sex cells that result from meiotic (sexual) cell division — pollen and eggs.

gene   The basic unit of inheritance. A gene is the section of DNA that codes for one protein or subunit of a protein.

genetic engineering   The creating or altering of organisms or varieties of organisms by direct physical transfer of DNA.

genome   All the genetic material in a cell or an organism. *See also* germplasm.

genomic formula   See ploidy.

genotype and phenotype   The *genotype* is the exact genes in an organism (*TT, Tt,* or *tt,* for example). The *phenotype* is what the organism looks like (tall or short). You identify the genotype by knowing the genetics and parentage involved. You identify the phenotype just by looking at the plant. The phenotype depends on both the genotype and the environment. If I chop the top off a plant of genotype *TT,* its genotype is still *TT,* but its phenotype changes from tall to short.

germplasm   Like *genome,* the term *germplasm* refers to all the genetic material, but the context is usually broader and the emotional value is different. We refer to the genome of individuals or species, but the germplasm of species or genera, or the germplasm in a country or the world. The germplasm of North America would be all the genes in all the species, wild and domestic, that are found or grown in North America. The two terms overlap in formal definition but rarely in practice. *Genome* is a technical, unemotional term. *Germplasm,* with its associations of germ/core/life and plasm/life/blood, has mystical and spiritual overtones. We use the word *germplasm* when we care about it. It is saving germplasm from extinction that biologists and many others are so passionately concerned about.

gibberellin   A plant hormone involved in controlling plant growth, differentiation, flowering, fruit set, and other processes.

GM, GMO, GM foods   Genetically modified. Genetically modified organism. Genetically modified foods. All refer to genetically engineered organisms or products derived from them.

haploid   *See* ploidy.

heirloom variety   A variety that has been around for a while, that has been handed down from gardener to gardener or generation to generation.

hermaphrodite   A flower that has both male and female parts. Same as *perfect flower.*

heterosis   Increased vigor of a hybrid as compared with the two parental varieties. Same as *hybrid vigor.*

heterozygous and homozygous   An individual that has two different alleles at the locus under consideration is *heterozygous* at that locus. (Example: *Aa.*) Such an individual is called a *heterozygote.* An individual with identical alleles at the locus (*AA* or *aa*) is *homozygous* and is a *homozygote.*

hilum   The scar or spot on the seed that marks the area where it was attached to the ovary.

homologous chromosomes   In diploids, the two chromosomes that have the same or similar genes on them and that pair during meiosis.

homozygous   *See* heterozygous and homozygous.

hybrid vigor   Increased vigor of a hybrid as compared with the two parental varieties. Same as *heterosis.*

inbreed   To breed two closely related individuals. In the extreme case, to self-pollinate.

inbreeding depression   A loss of vigor in a variety or population associated with inbreeding. Especially typical of plants that are natural outbreeders. (See Chapter 9.)

incompatibility mechanism   A genetically determined physiological mechanism that prevents or restricts self-fertilization or fertilization by plants or pollen with related incompatibility genotypes.

incompatible   Two plants are incompatible when they are unable to fertilize one another. A plant is self-incompatible when it is unable to fertilize itself.

indeterminate   *See* determinate and indeterminate growth patterns.

inversion   A chromosomal mutation in which one block of genes on the chromosome has been inverted and therefore carries the genes in reverse order.

isolation distance   The distance apart two different varieties of a single species must be to prevent their cross-pollinating (and thus contaminating) each other.

keel   *See* standard, wings, and keel.

lethal   Used to describe a gene, allele, or mutation that kills the individual carrying it in some specific context. Often used as a noun. Example: Recessive lethals are quickly eliminated from populations that are strongly inbreeding.

linkage   If two different genes are on the same chromosome, the genes are said to be *linked.*

locus   A specific position on a chromosome or chromosome pair that is occupied by a specific gene.

long-day plants   *See* photoperiod.

male parts of the flower   The stamen is the basic male part of the flower. Its parts are the anther, stalk/filament, and pollen. The anther is the part at the top that produces the pollen. The stalk or filament is the rest of the stamen, the part that supports the anther.

mass selection   Selection based on working with a population as a whole as opposed to working with various families' or other groupings separately.

meiosis and mitosis   Meiosis is sexual cell division. It results in gametes. It occurs only within the pollen- or egg-producing tissues within flowers. *Mitosis* is the ordinary cell division that occurs whenever a plant grows or reproduces vegetatively. Mitosis results in more ordinary cells.

mitosis   *See* meiosis and mitosis.

modifier   A gene that influences or alters the effect of some other gene with a larger effect.

monoecious   Used to describe varieties or species that have separate male and female flowers on each plant.

mutation and mutant   A *mutation* is an alteration in a gene or chromosome. A *mutant* is the changed thing (DNA, protein, phenotype, gene, chromosome, or individual) that has the mutation in it.

nucleus   The organelle in the cell that contains the chromosomes.

off types   Individuals of a population or variety whose characteristics are atypical or undesirable. Also called *rogues*.

open-pollinated variety   A stable variety, a variety that breeds true from seed. (Normally maintained by allowing natural pollination under field conditions, but hand-pollination is sometimes used to achieve the equivalent.) Not a hybrid.

outbreed   To cross-pollinate instead of to self-pollinate. Alternatively, to cross to a distant relative or an unrelated plant instead of self-pollinating or crossing to a close relative. Opposite of *inbreed*.

outbreeder   A variety or species that naturally cross-pollinates more often than it self-pollinates.

ovary   *See* female parts of the flower.

penetrance   The degree to which a gene or genotype expresses itself. If some individuals of a genotype express a characteristic and others do not, even within the same environment, we say that the gene or genotype has incomplete penetrance.

perfect flower   A flower that has both male and female parts. Same as *hermaphrodite*.

phenotype   *See* genotype and phenotype.

photoperiod   Length of day. A plant that has a photoperiod requirement for flowering is one whose flowering is at least partly under the control of the day length. Long-day plants begin flowering only after the day length exceeds a specific minimum. Short-day plants begin flowering only after the day length falls below a given level. In northern temperate zones, spring-flowering plants are often long-day plants and fall-flowering plants are often short-day plants. Day-neutral plants are plants whose flowering is not dependent on photoperiod.

pistil   *See* female parts of the flower.

pleiotropy   A gene's influence on traits other than the primary one that it is defined in terms of.

ploidy   The number of haploid genomes in a cell, plant, variety, or species. Gametes of most plants are haploid. This means they have one copy of each kind of chromosome. Other than gamete-producing tissue, most plants are diploid — that is, they have two copies of each kind of chromosome. Polyploid plants, varieties, or species have more than two complete haploid sets of chromosomes. They may be triploid, tetraploid, hexaploid, etc.

In autopolyploids the chromosomes are all derived from a single species. In allopolyploids the chromosomes are derived from two or more different species. *Genomic formulas* are used to describe the origin of polyploids. If three diploid species have genomic formulas of *AA, BB,* and *CC,* respectively, then *AABB* would be a tetraploid that arose from a cross of the first two species followed by chromosome doubling. In other words, it would contain all the chromosomes typical of both species (as identified microscopically).

pollen   The male gametes of plants.

polyploid   *See* ploidy.

popbean   Any legume variety whose seed both pops and becomes fully cooked and edible during the quick exposure to heat during popping. The popbeans discussed in this book are all chickpeas, *Cicer arietinum*.

protandry   A situation in which pollen is released before the stigma matures and becomes receptive.

protogyny   A situation in which the stigma matures and becomes receptive before the pollen is released.

pure breeding variety   A variety that is homozygous for everything relevant. This means that all members of the population or variety have the same genotype and give rise to offspring with the same genotype. Same as *true breeding*.

recessive   *See* dominant, codominant, and recessive.

reciprocal crosses   Reciprocal crosses involve the same parents, but with different sexes. If we cross a female from a tall variety with a male from a short variety, the reciprocal is a cross of a male from the tall variety with a female from the short variety.

recombinant and recombination   In a cross of two parent varieties, some classes of F_2 offspring resemble the parents and some do not. Those that resemble the parents are called *parental* classes. Those that do not are called *recombinant* classes, and the individuals in them are *recombinants*. They have different combinations of the characteristics of the parents. In the cross tall green X short yellow, F_2 progeny that are tall green or short yellow are parental types. Those that are tall yellow or short green are recombinants. The process of producing these new types is called *recombination*.

recurrent backcrossing   A breeding approach in

which progeny of a cross are backcrossed to one of the parent varieties, then some of the resulting progeny are backcrossed again to that same parent variety, and so forth. The parent variety that is crossed to the offspring in each generation is called the *recurrent parent/variety*. Recurrent backcrossing is used to transfer one or a few genes to a different variety.

scape   A flower stalk.

selection   The process of choosing the appropriate individuals to use in creating or maintaining a variety. To select is to choose which germplasm to perpetuate or continue with.

self   A cross involving only one parent; a self-pollination. To self is to self-pollinate. Opposite of *cross*.

short-day plants   *See* photoperiod.

somatic   Having to do with the body. Opposite of *sexual*. Somatic cells include all the cells in a plant except those involved in producing gametes.

species   A population of organisms capable of interbreeding in nature. (*Interbreeding* means producing the normal number of fully fertile offspring.)

sport   Old-fashioned word meaning *mutant*. A mutant is a sport with a modern education.

stable variety   A variety that reliably produces similar plants from seed to seed and year to year; a pure breeding variety. To *stabilize a variety* is to do breeding work with it until it becomes a stable variety.

stamen   *See* male parts of the flower.

standard variety or method   A well-known variety or method that is included in a trial or experiment as a standard for comparison, that is, a control.

standard, wings, and keel   The three distinctive kinds of petals of a legume flower. The *standard* is the outermost petal that spreads out to either side of the open flower. The *wings* are the middle layer of petals between the standard and the keel. The *keel* is the inner petal that encloses the pistil and stamens.

stigma   *See* female parts of the flower.

style   *See* female parts of the flower.

subspecies   Any subgroup within a species that seems distinctive and meaningful to whoever is talking about it. This word has no official biological meaning but is too useful to do without.

terminator technology   A combination of genes that may be engineered into a plant variety for the purpose of physiologically preventing their reproduction by seed.

tetraploidy   *See* ploidy.

translocation   A chromosomal mutation in which blocks of genes from two different chromosomes have exchanged positions.

triploid   *See* ploidy.

USDA-ARS   U.S. Department of Agriculture–Agricultural Research Service.

variety   A population of plants that are predictable and defined with respect to characteristics that matter to the grower. Open-pollinated varieties breed true; hybrid varieties do not.

vector   A pollen vector is the means by which pollen of a given species is normally dispersed. Examples: wind, water, insects, hummingbirds, gravity.

vernalization   An exposure to cold required by some plants before they will flower.

wide cross   Ordinary crosses are between individuals of the same species. Wide crosses are between more distant relatives — such as members of different species or even different genera.

wings   *See* standard, wings, and keel.

zygote   A diploid cell created when the pollen and egg fuse. It divides and grows to give rise to the embryo, which (after seed germination) develops into a whole new plant.

Annotated Bibliography

Some of the material listed here is for a lay audience. Much of it, however, is intended for professional plant breeders or plant breeding students. If you want to delve into the more advanced literature, the trick is not to let yourself be intimidated. It helps to skim as many different books as possible initially. Some authors will explain certain words and concepts, and others will explain others.

Many of the more technical books are out of print and/or very expensive, but in the United States they can be obtained through interlibrary loan.

Allan, Ken. *The Vegetable Garden Research Exchange.* (See Appendix D.)

Anderson, Luke. *Genetic Engineering, Food, and Our Environment.* White River Junction, Vermont: Chelsea Green Publishing Co., 1999. Very readable short book that provides thorough coverage of all aspects of the genetic engineering issue.

Ashworth, Suzanne. *Seed to Seed: Saving Techniques for the Vegetable Gardener.* Decorah, Iowa: Seed Saver Publications, 1991. Most complete single source of information on seed saving for all common and many rare vegetables.

Bassett, Mark J., ed. *Breeding Vegetable Crops.* Westport, Connecticut: Avi, 1986. Fourteen separate chapters covering fourteen vegetables: asparagus, beans, cabbage, carrots, corn (sweet), cucumbers, lettuce, onions, peas, peppers, squash, sweet potatoes, tomatoes, and watermelons. Each chapter includes a section on the general botany of the vegetable, floral botany and controlled pollination (doing or avoiding crosses), major breeding achievements of the past, an overview of current breeding goals, genetics, gene lists, and an extensive bibliography of the current professional literature. This book assumes an extensive background in genetics and plant breeding, but much of the book is useful to amateurs, too.

Bates, David M., Richard W. Robinson, and Charles Jeffrey, eds. *Biology and Utilization of the Cucurbitaceae.* Ithaca, New York: Cornell University Press, 1990.

Blackmon, W. J., and B. D. Reynolds. "The Crop Potential of *Apios americana* — Preliminary Evaluations." *HortScience* 21, no. 6 (1986): 1334–1336.

Burton, W. G. *The Potato*. 3d ed. New York: Halsted, 1989. Professional. History, growing, breeding, diseases, storing.

Darlington, C. D., and A. P. Wylie. *Chromosome Atlas of Flowering Plants*. London: Allen & Unwin, 1955.

Deppe, Carol. "Parching Corn." *National Gardening Magazine* May/June 1997. Describes my work on rediscovering and using true parching corn.

———. *See* my web site *www.plantbreeding.net* for a current list of my publications.

Duke, J. A., S. J. Hurst, and E. E. Terrell. *Economic Plants and Their Ecological Distribution*. 1975. A 16-page table of one thousand important plants and their chromosome numbers, life-styles, ecological characteristics, and requirements.

Duke, James A. *Handbook of Legumes of World Economic Importance*. New York: Plenum, 1981. Invaluable for anyone interested in working with established legume crops, in developing new ones, or in preserving legume biodiversity.

Facciola, Stephen. *Cornucopia II: A Source Book of Edible Plants*. Vista, California: Kampong Publications, 1998. An amazing compendium of edible plants, varieties, and sources. Encyclopedic — 700 pages of small print, large format.

Fehr, Walter R., and Henry H. Hadley, eds. *Hybridization of Crop Plants*. Madison Wisconsin: American Society of Agronomy and Crop Science Society of America, 1980. Revised in 1982. Instruction on exactly how to make crosses. Many pictures and diagrams. No background in genetics or plant breeding is required. Crops covered are bean, broad bean, chickpea, cowpea, crambe, lentil, maize/corn, pea, peanut, potato, rapeseed, mustard, soybean, sugar beet, sunflower, sweet potato, and thirty or so field and forage crops. Doesn't cover genetics or general plant breeding.

Frankel, R., and E. Galun. *Pollination Mechanisms, Reproduction, and Plant Breeding. Monographs on Theoretical and Applied Genetics 2*. New York: Springer-Verlag, 1977.

George, Raymond A. T. *Vegetable Seed Production*. 2nd ed. New York: CABI Publishing, CAB International, 1999. For the professional seed producer, but not too technical for amateurs. Researchers frequently need to deal with small lots of experimental seed, so it includes small-scale as well as large-scale methods. Covers most major and many minor crops. Includes information on pollination mechanisms and isolation requirements of the plants and on harvesting, threshing, and storing the seed.

Hawkes, J. G. *The Potato: Evolution, Biodiversity and Genetic Resources*. Washington, D.C.: Smithsonian Institute Press, 1990.

Janick, Jules, and James E. Simon, eds. *Advances in New Crops*. (Proceedings of the First National Symposium on New Crops: Research, Development, Economics, held in Indianapolis, 1988.) Portland, Oregon: Timber, 1990.

Jeavons, John. *How to Grow More Vegetables Than You Ever Thought Possible on Less Land Than You Can Imagine*. 4th ed. Berkeley, California: Ten Speed Press, 1982. Revised frequently. The classic on biodynamic gardening. Raised-bed, biodynamic growing methods for fifty-nine vegetables, twenty-seven grains or field crops, thirty cover or fodder crops, and miscellaneous others. Gives planting distances, plant requirements, vegetable/grain/fodder yields, and seed yields. The only published source on seed yields for some crops, especially on seed yields in the context of backyard rather than commercial methods.

Jones, H. A., and L. K. Mann. *Onions and Their Allies*. London: Leonard Hill, 1963. For both professionals and growers. Still the classic on the alliums and one of the few sources of breeding information for species other than *Allium cepa*.

Kalloo, Dr. *Vegetable Breeding,* Vol. I. Boca Raton, Florida: CRC Press, 1988. First of a series of three volumes by Dr. Kalloo, professor and head of the Department of Vegetable Crops, Haryana Agricultural University, Hisar, India. This first volume covers the basic techniques of vegetable breeding, including methods for hybridizing.

———. *Vegetable Breeding,* Vol. II. Boca Raton, Florida: CRC Press, 1988. This volume covers genetics of and breeding for resistance to disease, insects, and nematodes, and tolerance to environmental stress. Chapter 2, pages 95–131, "Insect Resistance in Vegetable Crops," should be of special interest to anyone wanting to develop insect-resistant varieties.

———. *Vegetable Breeding,* Vol. III. Boca Raton,

Florida: CRC Press, 1988. Chapter 2 of this volume (pages 41–59) is "Breeding for Quality and Processing Attributes in Vegetable Crops."

Kloppenburg, Jack R., Jr. *Seeds and Sovereignty: The Use and Control of Plant Genetic Resources.* Durham, North Carolina: Duke University Press in cooperation with the American Association for the Advancement of Science, 1988. Separate chapters by various authors representing all possible points of view.

Lovejoy, Ann. "Importing Plants." *Horticulture,* April 1992, pp. 16–20.

National Research Council. *Lost Crops of the Incas: Little-Known Plants of the Andes with Promise for Worldwide Cultivation.* Washington, D.C.: National Academy Press, 1989.

National Research Council. *Lost Crops of Africa: Volume I; Grains.* Washington, D.C.: National Academy Press, 1996.

Nieuwhof, M. *Cole Crops: Botany, Cultivation, and Utilization.* London: Macmillan, 1979. An introduction to genetics for beginning horticultural students.

North, C. *Plant Breeding and Genetics in Horticulture.* London: Macmillan, 1979. An introduction to genetics for beginning horticulture students.

Robinson, Raoul A. *Return to Resistance: Breeding Crops to Reduce Pesticide Dependence.* Davis, California: agAccess, 1996. Documents the efficacy of breeding for disease resistance via mass selection. Encourages farm-based plant breeding.

Ryder, Edward J. *Leafy Salad Vegetables.* Westport, Connecticut: Avi, 1979.

Shapiro, Howard-Yana, and John Harrisson. *Gardening for the Future of the Earth.* New York: Bantam Books, 2000. Sustainable agriculture one garden at a time. A tale told via visits with Deppe, Jeavons, Kapuler, Mollison, and others. Wonderful color photography.

Shinohara, Suteki. *Vegetable Seed Production Technology of Japan Elucidated with Respective Variety Development Histories, Particulars.* Vol. 1. 4-7-7 Nishiooi, Shinagawa-ku, Tokyo: Shinohara's Authorized Agricultural Consulting Engineering Office, 1984. Translation of a 1978 book by Seibundo Shinkosha Ltd. Originally intended for Japanese seed growers. The only source of breeding information for certain vegetables.

Simmonds, N. W., ed. *Evolution of Crop Plants.* New York: Longman, 1976.

Sprague, G. F., and J. W. Dudley, eds. *Corn and Corn Improvement.* 2d ed. Madison, Wisconsin: American Society of Agronomy, Crop Science Society of America, and Soil Science Society of America, 1988.

Tsunoda, S., K. Hinata, and C. Gómez-Campo. *Brassica Crops and Wild Allies: Biology and Breeding.* Tokyo: Scientific Societies Press, 1980.

United States Department of Agriculture. *1992 Yearbook of Agriculture: New Crops, New Uses, New Markets; Industrial and Commercial Products from U.S. Agriculture.* Office of Publishing and Visual Communication, USDA.

Wagoner, Peggy. "Perennial Grain Development: Past Efforts and Potential for the Future." In *Critical Reviews in Plant Sciences,* Boca Raton, Florida: CRC Press, 1990.

Whealy, Kent, ed. *The Garden Seed Inventory,* 5th ed. Decorah, Iowa: Seed Saver Publications, 1999. Revised frequently. A compilation of names, descriptions, and sources for all nonhybrid vegetable varieties available commercially in the United States and Canada. As much fun as two hundred seed catalogs, which is what it represents.

————. *Winter Yearbook.* For a lay audience. Published annually by the Seed Savers Exchange, comes with membership. Not available through bookstores or libraries. Gives you access to thousands of vegetable varieties maintained by members that are not available commercially.

Wilson, Gilbert. *Buffalo Bird Woman's Garden.* St. Paul, Minnesota: Minnesota Historical Society Press, 1987.

Wittwer, Sylvan, Yu Youtai, Sun Han, and Wang Lianzheng. *Feeding a Billion: Frontiers of Chinese Agriculture.* East Lansing: Michigan State University Press, 1987. Mind expanding. A tremendous source of ideas and perspective.

Yamaguchi, Mas. *World Vegetables: Principles, Production and Nutritive Values.* Westport, Connecticut: Avi, 1983.

Zeven, A. C., and P. M. Zhukovsky. *Dictionary of Cultivated Plants and Their Centres of Diversity: Excluding Ornamentals, Forest Trees and Lower Plants.* 2d ed. Wageningen, Netherlands: Centre for Agricultural Publishing and Documentation, 1982. Gives scientific names, chromosome numbers, centers of diversity, and a few words about use for thousands of species.

genetics and breeding of, 20, 94,
316–317
hand-pollination and crossing of,
215, 229, 315–316
harvesting, processing, and storing
seeds of, 234, 248, 256, 258, 317
hybrids of, 122, 123, 125,
272–290, 316
isolation, 315
Sandwich-Slice summer squash,
development of, 272–290
culling or roguing, 13, 60, 83,
129–130, 225, 230, 237–238, 240
plant-breeding stories of, 141, 142,
143, 145–146
cumaru tonka. *See Dipteryx odorata*
cumin, black. *See Nigella sativa*
cupuacu. *See Theobroma grandiflorum*
Curcuma:
C. *domestica,* 189
C. *zedoaria,* 189
currant:
black. *See Ribes nigrum*
red. *See Ribes rubrum*
curryleaf tree. *See Murraya koenigii*
cushaw. *See Cucurbita mixta*
cuttings, 159
Cyamopsis:
C. *pedata,* 189
C. *tetragonoloba* (= C. *psoralioides*),
189
Cyclanthera pedata, 189
Cydonia oblonga, 189
Cymbopogon citratus, 189
Cynara:
C. *cardunculus,* 297
C. *scolymus,* 297
Cyperus:
C. *esculentus,* 189
C. *rotundus,* 189
Cyphomandra betacea, 291
cypress, summer. *See Kochia scoparia*
Cyrtosperma chamissonis (= C. *edule* = C.
merkusii), 189
cytoplasm, 147, 282, 309
cytoplasmic inheritance, 104–105

Dahlia spp., 297
D. *rosea,* 189
daisy. *See Bellis perennis*
dandelion, *See Taraxacum officinale*
Russian. *See Taraxacum koksaghyz*

date (plant):
Chinese. *See Ziziphus jujuba*
desert. *See Balanites aegyptiaca*
date palm. *See Phoenix dactylifera*
Daucus carota, 12, 189–190, 216, 226
daylily. *See Hemerocallis* spp.
dehiscence, 326–327
Dendrocalamus spp., 190
desert gold. *See Lomatium utriculatum*
Desmanthus illinoensis, 190
detrimentals, 101–102
dewberry. *See Rubus* spp.
Dickson, Michael H., 314
Dictamnus albus, 190
dill. *See Anethum graveolens*
dioecious plants, 80, 118
Dioscorea spp., 27, 80, 81, 190, 193
D. *alata,* 190
D. *batatas. See D. opposita*
D. *bulbifera,* 190
D. *composita,* 190
D. *convolvulacea,* 190
D. *floribunda,* 190
D. *japonica,* 190
D. *macrostachya,* 190
D. *opposita* (= D. *batatas*), 190
D. *rotundata,* 190
Dioscoreophyllum cumminsii, 190
Diospyros spp, 80, 190
D. *digyna,* 190
D. *kaki,* 190
D. *virginiana,* 190
diploids and diploidy, 96, 97, 148,
149, 154
Dipteryx odorata, 190
disease resistance, 11, 12, 41, 75, 76,
113, 239, 281, 282, 290
breeding for, 19, 29, 30, 135–141,
240
genetically engineered resistance, 270
information about, 43
organic gardening and, 40
trials of, 58
variable expressivity and, 103, 104
diseases, 57, 64, 121, 159, 172–173
seed-borne, 217
Distichlis palmeri, 190
DNA, 93, 94, 95, 96, 104–105. *See
also* chromosomes; genes; genetics
Dolichos:
D. *biflorus. See Macrotyloma uniflorum*

D. *lablab. See Lablab purpureus*
D. *uniflorus. See Macrotyloma uniflorum*
domesticating wild plants, 163–171
dominant genes and traits, 90, 92, 100,
108, 139
codominance and, 100–101, 107,
108, 109
fixing of, 141
dragon's head. *See Lallemantia iberica*
dropseed, sand. *See Sporobolus cryptandrus*
dropwort, water. *See Oenanthe javanica*
Drowns, Glenn, 4–7, 9, 12, 15, 17, 19,
21, 27, 41, 80, 125, 236
drying seed, 248–252, 255–256
Duke, James A., 34, 35

Echinochloa:
E. *crusgalli var. frumentacea,* 190
E. *crusqalli,* 190
E. *pyramidalis,* 190
edge effects, 61, 258
eggplant. *See Solanum melongena*
native. *See Solanum macrocarpon*
eggs, 79–80, 90–91 96, 102, 104, 153
einkorn. *See Triticum monococcum*
Elaeis:
E. *oleifera,* 190
E. *quineensis,* 190
elder, American. *See Sambucus canaden-
sis*
elderberry. *See also Sambucus glauca*
Canadian. *See Sambucus canadensis*
elecampane. *See Inula helenium*
Eleocharis dulcis (= E. *tuberosa*), 190
elephant grass. *See Pennisetum pur-
pureum*
Elettaria cardamomum, 190
Eleusine:
E. *coracana,* 190
E. *indica,* 191
Eliason, Ewald, 7–9, 12, 15, 17–18,
19, 21, 41
Elymus spp., 191
E. *canadensis,* 191
E. *cinereus,* 191
E. *condensatus,* 191
E. *glaucus,* 191
E. *junceus,* 191
emasculating flowers, 7–8, 306, 328
embryo rescue, 331

standard plant breeding compared, 264–269

sustainable agriculture and, 270–271

terminator technology, 270

Genetic Engineering, Food, and Our Environment (Anderson), 265

genetic heterogeneity, 232, 241, 274–275, 290

genetics, 16, 77–110. *See also specific vegetables*

  aberrant segregation and, 102–103

  cytoplasmic inheritance and, 104–105

  early understanding of, 85–86

  exceptions to Mendel's law of, 101–105

  incomplete penetrance and variable expressivity in, 103–104

  maternal inheritance and, 104

  Mendel's experiments in, 86–92, 95, 96, 98–99, 100

  plant parenthood and, 79–81

  pleiotropy, 261, 266–268

  predictions and actuality in, 109–110

genetic variability, 59–60, 61, 83, 112, 1113, 126–127, 147, 212, 214

  environmental variability vs., 126, 128–129

  generating of, 142–143

  hybrid vigor and, 117

  inbreeding and, 112–115

  inbreeding depression and, 10

  of outbreeders, 116, 141–142

  unwanted, 134, 290

genome, 96

  cytoplasmic, 105

genotype, 90, 91, 92, 105

  variable expressivity and, 103–104

George, Raymond A. T., 292, 296, 299, 308, 319, 320, 321

germination, 166, 239, 330

germplasm, 14, 15, 21, 38

  evaluating of, 42–43, 54–76. *See also* garden trials

  finding of, 42–53

  Germplasm Resources Information Network (GRIN), use of, 334

  inconsistent variety names and, 51–53, 55–56

  mid-project working material, 47–48

selecting for preservation of, 241

USDA-ARS germplasm collections, 332–334

gherkin, West Indian. *See Cucmis anguria*

gilo. *See Solanum gilo*

ginger. *See also Zingiber officinale*

  Japense wild or temperate. *See Zingiber mioga*

Gladiolus spp., 192

Glycine:

  G. apios. See Apios americana

  G. max, 192

  G. wightii, 192

Glycyrrhiza:

  G. glabra, 192

  G. lepidota, 192

gobo. *See Arctium lappa*

Good King Henry. *See Chenopodium bonus-henricus*

gooseberry:

  European. *See Ribes uva-crispa*

  hairy. *See Ribes hirtellum*

goosegrass. *See Eleusine indica*

gourd. *See Cucurbita* spp.

  bitter. *See Momordica charantia*

  buffalo, figleaf, or Malabar. *See Cucurbita ficifolia*

  club or snake. *See Trichosanthes cucumerina*

  edible or white-flowered. *See Lagenaria siceraria*

grafting, 159

gram:

  black. *See Vigna mungo*

  golden or green. *See Vigna radiata*

  Madras. *See Macrotyloma uniflorum*

granadilla. *See Passiflora* spp.

grape:

  Brazilian. *See Myrciaria cauliflora*

  fox. *See Vitis labrusca*

  muscadine. *See Vitis rotundifolia*

  wine. *See Vitis vinifera*

grasses. *See also* bunchgrass

  antelope. *See Echinochloa pyramidalis*

  barnyard. *See Echinochloa crusgalli*

  billion dollar. *See Echinochloa crusgalli* var. *frumentacea*

  Boer lovegrass. *See Eragrostis chloromelas*

  eastern gamagrass. *See Tripsacum dactyloides*

elephant. *See Pennisetum purpureum*

goosegrass. *See Eleusine indica*

johnsongrass. *See Sorghum halepense*

kikuyugrass. *See Pennisetum clandestinum*

rice. *See Oryzopsis hymenoides*

smilograss. *See Oryzopsis miliacea*

Sudan. *See Sorghum sudanense; S. vulgare* var. *sudanensis*

switchgrass. *See Panicum virgatum*

greens:

  African, bush, Chinese, or Spanish. *See Amaranthus cruentus; A. dubius; A. tricolor*

  creasy. *See Barbarea verna*

Grewia asiatica, 192

Gritton, Earl T., 299, 302

groundbean. *See Macrotyloma geocarpum*

groundnut. *See also Arachis hypogaea*

  American. *See Apios americana*

  bambara. *See Voandzeia subterranea*

  Kersting's. *See Macrotyloma geocarpum*

growing conditions:

  hybrids and, 122–123

  marginal, for seed production, 252

growing methods:

  selecting for, 29–31, 39

  trials of, 58

growth rate, 109

guar. *See Cyamopsis tetragonoloba*

guava. *See Psidium guajava*

gynmigit. *See Allium tuberosum*

hairiness, pest resistance and, 68–70

Handbook of Legumes of World Economic Importance, 34, 35

hand-pollination, 5, 7, 8, 17, 83, 162, 215, 229–230, 322–329. *See also specific vegetables*

  of buds, 9–10, 120

  six steps to, 326

  tools for, 327–328

Hannan, Rich, 32–33, 35, 38, 50, 164, 224

harvests, 57. *See also specific vegetables*

  extended, 239

  by machine, 25–26, 113, 122

  of seed, 243–247

hazelnut. *See Corylus* spp.

heat tolerance, 31, 95

Vaccinium. 204
valerian, red. *See Centranthus ruber*
Valerianella locusta (= *V. olitoria*), 204
variable expressivity, 103–104
varieties:
    inconsistencies in names of, 51–53,
        55–56
    maintaining of, 129–130
    preventing contamination of, 225.
        See also isolation distances
    "running out" of, 259
vegetative propagation, 7, 17, 81, 97,
    150, 159–160
velvetbean. *See Mucuna* spp.
vetch. *See also Vicia sativa*
    bard. *See Vicia monantha*
    blackpod. *See Vicia sativa* ssp. *nigra*
    hairy or winter. *See Vicia villosa*
    Hungarian. *See Vicia pannonica*
Vicia spp., 204
    V. faba, 28–29, 49, 61, 161, 204,
        227, 230
    V. monantha, 204
    V. pannonica, 204
    V. sativa, 204
    V. sativa, ssp. *nigra*, 204
    V. villosa, 204
Vigna spp.
    V. aconitifolia, 204
    V. angularis, 33–34, 35, 204
    V. mungo (= *Phaseolus mungo*), 204
    V. radiata, 204
    V. umbellata, 205
    V. unguiculata, 205
    V. unguiculata ssp. *cylindrica*, 205
    V. unguiculata ssp. *sesquipedalis*,
        205
    V. vexillata, 205
vigor, 108, 109
    of hybrids, 117, 121, 124
Vitis:
    V. labrusca, 205
    V. rotundifolia, 80, 205
    V. vinifera, 205
Voandezia:
    *V. poissonii. See Macrotyloma geo-
        carpum*
    V. subterranea, 205
volunteers. 57, 240

Wallace, D. H., 314
Wallbrunn, Henry, 263, 331

walnut:
    black. *See Juglans nigra*
    English. *See Juglans regia*
wampi. *See Clausena lansium*
Wasabia japonica, 205, 311
water chestnut. *See Eleocharis dulcis;
    Trapa natans*
watercress. *See Rorippa* spp.
water lily:
    American. *See Nymphaea odorata*
    Australian. *See Nymphaea gigantea*
    Cape. *See Nymphaea capensis*
    Egyptian. *See Nymphaea lotus*
    Gorgon. *See Euryale ferox*
    magnolia. *See Nymphaea tuberosa*
    pygmy. *See Nymphaea tetragona*
    red. *See Nymphaea rubra*
    yellow. *See Nuphar lutea*
watermelon, 80, 117, 125. *See also
    Citrullus lanatus; Passiflora* spp.
    breeding of, 4–6, 17, 19, 27
waternut. *See Eleocharis dulcis*
waxgourd. *See Benincasa hispida*
wet processing, seed, 252–255
Whealy, Kent. 15, 43, 44, 251
wheat. *See Triticum* spp.
    duck or India. *See Fagopyrum tatar-
        icum*
    Inca. *See Amaranthus caudatus*
    strand. *See Leymus arenarius*
wide crosses, 40–41, 76, 102–103,
    105, 109, 117, 151, 152–155, 329
    between species, 153–155
    within species, 152–153
wild plants, 140
    collecting of, 165–166, 167–168
    domesticating of, 41, 163–171
    isolation distances and, 226
    precautions for, 170–171
    untapped potential of, 163–164
willow. *See Salix* spp.
winter gardening, 48
    overwintering and, 28, 30–31, 34,
        38, 39, 58, 128
wintergreen. *See Gaultheria procumbens*
winter hardiness, 30–31, 142–143
Winter Yearbook, 15, 43, 157
wonderberry. *See Solanum melanocera-
    sum*
woodruff, sweet. *See Asperula odorata*
wormweed, American or Indian. *See
    Chenopodium ambrosioides*

Xanthosoma spp., 205
    X. sagittifolium, 205

yacon. *See Polymnia sonchifolia*
yam. *See Dioscorea* spp.
yam bean or yam pea. *See Pachyrhizus
    erosus; Sphenostylis stenocarpa*
yautia. *See Xanthosoma sagittifolium;
    Xanthosoma* spp.
yield, 109, 113, 124, 130
    evaluating of, 61–62, 70–71
Yucca elephantipes, 205

zanier. *See Xanthosoma* spp.
Zea:
    Z. diploperennis, 205
    Z. mays, 205
    Z. mays ssp. *mays*, 20, 80, 81,
        146–147, 159, 162, 165, 205,
        317–321
    breeding system of, 111, 119,
        127, 233, 318
    controlling pollination of, 83
    genetics and breeding of, 93–94,
        117, 320–321
    hand-pollination and crossing of,
        318–319
    harvesting, processing, and stor-
        ing seeds of, 216, 217, 234,
        235, 237–240, 255, 321
    hybrids of, 117, 121–122,
        124–125, 145–146, 319–320
    inbreeding depression, 318
    isolation, 230–318
    roguing, 237–238
    Z. mays ssp. *mexicana*, 205
    Z. perennis, 205
    spacing trials of, 70–71
zedoary. *See Curcuma zedoaria*
Zingiber:
    Z. mioga, 205
    Z. officinale, 205
Zizania:
    Z. aquatica, 205
    Z. palustris, 205
Ziziphus:
    Z. jujuba, 205
    Z. mauritiana, 205
zygotes, 90, 97, 98, 153